From Geometry to Behavior

An Introduction to Spatial Cognition

Hanspeter A. Mallot

The MIT Press
Cambridge, Massachusetts
London, England

This book was set in Times New Roman by Hanspeter Mallot. Printed and bound in the United States of America.

Library of Congress Cataloging-in-Publication Data

Names: Mallot, Hanspeter A., author. Title: From geometry to behavior : an introduction to spatial cognition / Hanspeter A. Mallot.
Description: Cambridge : The MIT Press, 2023. | Includes bibliographical references and index.
Identifiers: LCCN 2023003772 (print) | LCCN 2023003773 (ebook) | ISBN 9780262547116 (paperback) | ISBN 9780262377317 (epub) | ISBN 9780262377300 (pdf)
Subjects: LCSH: Spatial behavior. | Cognition.
Classification: LCC BF469 .M26 2023 (print) | LCC BF469 (ebook) | DDC 153.7/52–dc23/eng/20230504
LC record available at https://lccn.loc.gov/2023003772
LC ebook record available at https://lccn.loc.gov/2023003773

10 9 8 7 6 5 4 3 2 1

From Geometry to Behavior

Men go forth to marvel at the heights of mountains and the huge waves of the sea, the broad flow of the rivers, the vastness of the ocean, the orbits of the stars, and yet they neglect to marvel at themselves. Nor do they wonder how it is that, when I spoke of all these things, I was not looking at them with my eyes—and yet I could not have spoken about them had it not been that I was actually seeing within, in my memory, those mountains and waves and rivers and stars .... But when I saw them outside me, I did not take them in to me by seeing them; and the things themselves are not inside me, but only their images.
—Augustine of Hippo (354–430), *Confessions X.8.15*, translated by Albert C. Outler

"Thought," reason, intelligence, whatever word we choose to use, is existentially an adjective (or better an adverb), not a noun. It is disposition of activity, a quality of that conduct which foresees consequences of existing events, and which uses what is foreseen as a plan and method of administering affairs.
—John Dewey, *Experience and Nature*, Chicago and London 1926, p. 158f

Contents

Preface ix

1 Introduction 1

1.1 Space and Mind 1
1.2 Behavior 4
1.3 Space and Mathematics 11
1.4 Neurophysiology 18
1.5 Topics in Spatial Cognition 21

2 Egomotion 31

2.1 The Space for Motion 31
2.2 Perceiving Egomotion 34
2.3 Optic Flow 38
2.4 Neural Mechanisms 45
2.5 Performance 47
2.6 Cue Integration 52

3 Peripersonal Space 63

3.1 A Behavioral View 63
3.2 Visual Space Cues 64
3.3 The Intrinsic Geometry of Peripersonal Space 70
3.4 Mental Transformations: Predictive Coding of Space 75
3.5 Recalibration in Peripersonal Space 83

4 In the Loop 89

4.1 Directed Movement 89
4.2 Left–Right Balancing 92
4.3 Cognitive Components 100
4.4 Augmented Action–Perception Cycles 110

5 Path Integration 119
5.1 Dead Reckoning 119
5.2 The Home Vector 120
5.3 Path Integration in Humans 131
5.4 The Computational Neuroscience of Path Integration 138

6 Places and Landmarks 157
6.1 Here and There 157
6.2 Snapshot Homing 162
6.3 Including Depth Information 175
6.4 Identified Landmark Objects 184
6.5 Neurophysiology of Place Recognition 190

7 Spatial Memory 205
7.1 What Is Working Memory? 205
7.2 Working Memory Tasks 210
7.3 Models and Mechanisms for Spatial Working Memory 225
7.4 Routes 232
7.5 From Routes to Maps 241

8 Maps and Graphs 257
8.1 Spatial Problem Solving 257
8.2 Graphs: Basic Concepts 259
8.3 Metric Maps 264
8.4 Regions and Spatial Hierarchies 281

9 Epilogue: Reason Evolves 301

Index 311

Preface

The ability for dealing with space is essential for all behaving or "animate" agents, be they animals, humans, or robots. It is based on a set of mechanisms that allow the animats to discover, store, and make use of spatial knowledge. As a result, animats remember places and find back to them after excursions. They infer routes and route networks between multiple places and communicate with other animates about distant places and what to do there. They are able to plan future actions in space, either temporary ones, such as navigating a route to a distant goal, or more enduring ones, such as constructing a street or developing a residential area. Advanced planning and land surveying triggered the development of geometry as a mathematical theory, which may be considered the ultimate manifestation of spatial thinking.

Spatial cognition is the study of these abilities and their underlying mechanisms. Explanations are not sought in the laws of geometry but in the cognitive processes controlling behavior. Known places, for example, are not geometrical points but condensed memories of events that "took place" in the same location. In violation of basic rules of geometry, perceived distances between places may depend on direction, such that place A may seem close when looking from B while B seems far when looking from A. Cognitive mechanisms may even accommodate impossible geometrical relations such as "wormholes" presented in virtual environment navigation. In other cases, the cognitive mechanisms may approximate the laws of Euclidean geometry, but this is not generally the case.

Understanding how we deal with space requires input from many fields, including ethology, neuroscience, psychology, cognitive science, linguistics, geography, and spatial information theory. This text gives an overview of the basic mechanisms of spatial behavior in animals and humans and shows how they are combined to support higher-level performances. It should be of interest for advanced students and scholars in cognitive science and related fields.

The general approach of this text may be called "psychophysical": that is, it focuses on quantitative descriptions of behavioral performance and their real-world determinants and tries to make contact to theorizing in computational neuroscience, robotics, and computational geometry. Emphasis is also on a comparative view of spatial behavior in animals and humans and on the evolution of complex cognitive abilities from simpler ones. The text assumes some prior knowledge in neuroscience, as well as college mathematics, and occasionally uses technical formulations from both fields. It should be possible to skip these parts without loosing track of the overall progression of ideas. In case of problems, the reader is referred to standard textbooks or Wikipedia.

This book is not an encyclopedia of spatial cognition but a series of lectures. It grew out of a course for graduate students of neurobiology and cognitive science taught at the University of Tübingen since 2010. Each chapter focuses on one of the basic phenomena in spatial cognition and can to some extent be read independently of the others. The chapters are meant to give a balanced tour through the various types and levels of spatial behavior and may not do complete justice to some topics that I found to lie a bit out of my way. I apologize to those who feel that the resulting gaps are too large.

The organization of the material roughly follows the order of increasing memory involvement. The perception of egomotion (chapter 2) and peripersonal space (chapter 3) is largely independent of memory, at least of memories of the specific locations in which they take place. Control behavior in closed loop requires some monitoring of the state of affairs; it is discussed in chapter 4. Basic elements of spatial memory—that is, the accumulation of traveled distances and the recognition of places—are studied in chapters 5 and 6. Building on these elements, spatial working memory provides a chart of the current environment in which immediate behavior can be planned and monitored. In addition, state–action associations are stored in long-term memories for routes and provide knowledge for planning extended travels. Both types of memory are presented in chapter 7. Chapter 8 discusses further aspects of spatial long-term memories: that is, metric and hierarchical structuring, which add to the overall map-like knowledge of space. Finally, an epilogue summarizes some ideas derived from spatial cognition for the evolution of cognition and reasoning at large.

This book would not have been possible without support and contributions from many people. I am grateful for discussions with my students and colleagues at the University of Tübingen, the Max Planck Institute for Biological Cybernetics, and in the larger community. In particular, I am indebted to the late Christian Freksa and the Research Center "Raumkognition" of the Deutsche Forschungsgemeinschaft (DFG), who fueled my initial interest in the field and drew my attention to the computational and psychological sides of the topic; to Heinrich Bülthoff, who introduced me to virtual environments technology as a tool for "cognitive psychophysics"; and to Gregor Hardieß, who developed the laboratory course from which this text emerged. I am also grateful to Hansjürgen Dahmen, Ann Kathrin Mallot, and Michaela Mohr for valuable comments on the text.

Tübingen, September 2022
HAM

1 Introduction

This chapter gives an overview of how the cognitive apparatus deals with space and how cognitive space is studied in philosophy and in different fields of science. It is also an overview of the later chapters. Within cognitive science, the understanding of space is considered as one *domain* of cognition that can be distinguished from other such domains concerned with the understanding of objects, events, or social others. In relation to mathematics, spatial cognition describes a kind of folk geometry focusing not on the validity of mathematical theorems but on human and animal thinking and behavior. It can thus be considered part of the behavioral sciences to which concepts of psychology and ethology apply. In neuroscience, spatial cognition provides one of the best-studied cases of a cognitive performance where neural correlates of cognitive processes and representations have indeed been identified.

1.1 Space and Mind

Spatial cognition deals with the cognitive processing that underlies advanced spatial behavior such as directed search, wayfinding, spatial planning, spatial reasoning, building and object manipulation, or communication about space. These mechanisms are part of the larger "cognitive apparatus" that subserves also other domains of cognition such as visual and object cognition; the understanding of events, actions, and causality; or the large field of social cognition, including language. There is good reason to assume that among these, spatial cognition is the one that occurred first in the course of evolution and is now the most widespread throughout the animal kingdom. In this sense, the study of spatial cognition is also an exercise in cognitive evolution at large with respect to both the first origins of cognition and the phylogenetic tracing of individual mechanisms across the various cognitive domains. In humans, spatial concepts play a central role in all fields of thinking. One interesting example is the notion of psychological distance (Trope and Liberman 2010), in which an originally spatial concept is transferred to thinking about the future, the past, other people's perspectives, or counterfactual ideas. When we say that somebody's thinking is "close" to our own, or that an idea be "far-fetched," we use spatial metaphors in an assumed space of ideas. Another example is the mnemonic "method of loci" in which large numbers of arbitrary items can be memorized by associating them with a location and spatial ordering in an imagined "memory palace." Even in the evolution of language, which may be considered the most complex part of cognition, spatial abilities such as the sequencing of actions into routes, the recognition of places, or the ability to optimally search and forage an environment have been discussed as preadaptations[1] to more explicit thinking capacities. Hauser, Chomsky, and Fitch (2002), for example, treat spatial cognition as part of their "faculty of language in a broad sense" (FLB), which is a general cognitive basis from which language evolution is thought to have proceeded.

The prominent role of spatial intuition in the cognitive apparatus was well recognized by early modern philosophers such as Berkeley, Hume, or Kant, who considered space as something that is not so much a matter of experience but rather a framework in our mind that structures our sensations and without which experience would be impossible (see Jammer 1993). The question whether space "as such" can at all be perceived became an issue with Newton's notion of the absolute space, which is thought to exist independent of the contained objects and, indeed, independent of the observer. This idea has proven extremely useful in physics and still pervades discussions of "allocentric" versus "egocentric" representations in spatial cognition (see section 1.3.3). However, a space that is independent of the observer would be unobservable by definition. Immanuel Kant, in his "Critique of Pure Reason," therefore abandoned the idea of observer-independent space (and time) and replaced it with an inner sense of space: that is, a mode of perceiving, or a structure of the observer's cognitive apparatus. To quote from Kant (1781) in the translation of Müller (1922, 18f):

> Space is not an empirical concept which has been derived from external experience. For in order that certain sensations should be referred to something outside myself, i.e. to some thing in a different part of space from that where I am; again, in order that I may be able to represent them as side by side, that is, not only as different, but as in different places, the representation of space must already be there. Therefore the representation of space cannot be borrowed through experience from relations of external phenomena, but, on the contrary, this external experience becomes possible only by means of the representation of space.

If the sense of space is a part of our cognitive apparatus, its structure becomes an issue of psychology and can again be studied with empirical methods. Kant himself avoided this conclusion by assuming that Euclid's axioms and thus Euclidean geometry were the only possible structure that the a priori understanding of space must assume by necessity.

This idea, however, was challenged by Helmholtz (1876), who observed that a being living on a curved surface might develop geometrical intuitions of straight lines that are

1. A preadaptation is a trait developed in response to a given ecological demand but later found to be also useful for other applications in a changing environment. An example is the feather that evolved for thermal insulation in dinosaurs but, once invented, proved useful for the airborne lifestyle of birds. It is also called exaptation to stress the fact that its eventual use was not the driving force in its early evolution.

actually "geodesics": that is, the shortest connections of two points *not leaving the surface*. In a plane, geodesics are indeed straight lines in the ordinary sense, but on curved surfaces, they are generally not. For example, the shortest connection between two points on a sphere is contained in the great circle passing through these points: that is, it is a circular arc. The straight lines so defined violate one of the Euclidean axioms, and the resulting geometry is therefore called non-Euclidean. The axiom in question states that two straight lines in a plane may coincide, intersect in one point, or be parallel, in which case they have no point in common. For any point outside a straight line, one such parallel exists passing through this point. On the sphere, however, any two non-coincident great circles will intersect in two points, marking a diameter of the sphere. Parallels can therefore not exist, and the axiom of parallels is violated. (Note that circles of latitude are not parallels in this sense, since, with the exception of the equator, they are not great circles. Therefore, they are not shortest connections between contained points, but "curves" in spherical geometry.) Indeed, Helmholtz presented evidence for the non-Euclidean structure of perceptual, or peripersonal space, which will be discussed in more detail in chapter 3.

The general conclusion—that is, that the laws of spatial intuition and thinking are subject to empirical research—is also supported by more recent ideas on the multiplicity of spatial representations given by the distinctions of perceptual versus memorized space, or close (peripersonal) versus large (navigational) space. These multiple representations seem to employ different concepts of space, indicating that there is no unique set of rules that thinking would have to employ by necessity.

Indeed, Helmholtz's argument about the sense of space developed by creatures confined to curved surfaces evokes a representation of space that is different from the one implied by Kant's argument about object configurations. It puts the emphasis on egomotion rather than on the recognition of objects and scenes. As we will see in the later chapters of this book, the ability of an agent to orient and move toward some goal is simpler than the ability to simultaneously contemplate multiple targets in a visual space, which was the task in Kant's argument. Egomotion and the sense of heading logically and phylogenetically precede the distinction of objects by their location. The above quote from Kant states that an a priori concept of space is required for the latter task, but does not explain the origin of such concepts, which may indeed be derived from simpler behavioral schemata subserving orientation and directed movement.

In an evolutionary view, the ability to deal with space, which is built into ours and other animals' cognitive apparatus as a prerequisite of our various perceptions, experiences, and behaviors, is a set of evolutionary *traits*: that is, genetically transmitted adaptations that are inherited from ancestral species and refined in the process of evolution. It is therefore subject to the famous four questions of biological science formulated by Tinbergen (1963): adaptive value (or ecology), phylogeny, physiological mechanisms, and ontogenetic processes, including learning. In philosophy, the idea of cognitive traits having an adaptive value allows one to ground their rules (i.e., "pure reason") in the corroboration provided by their success or failure in an animal's life. Only if reasoning about the world is at least partially appropriate will it exert an adaptive pressure on brain evolution. This is the theme of evolutionary epistemology pioneered by Popper (1972) and Lorenz (1973).

Tinbergen's four questions cannot be answered comprehensively at this point, but many aspects will be discussed in the subsequent chapters. Some emphasis will be put on the comparative study of spatial behavior throughout the animal kingdom as a major source of information about the phylogeny of spatial cognition. Cognitive traits such as path integration and place recognition have been shown to exist in many animal groups, including insects, spiders, cephalopods, and vertebrates, and may be related to various extents. Within the mammals, hippocampal place cells have been found in rodents, bats, and primates and are thought to be homologous. By keeping an eye on these questions, we will strive for a view of spatial cognition that is informed by evolutionary thinking.

1.2 Behavior

1.2.1 Stimulus–Response Behavior

Simple behaviors, including also basic forms of spatial behavior, can be explained as associations of a stimulus and a response, either as innate reflexes or acquired stimulus–response pairs. In spatial behavior, reflexes are also called "taxes" (singular "taxis") and occur in cases such as the attraction to light or chemical substances (photo- and chemotaxis), keeping in touch to a wall (thigmotaxis), keeping the sun or another object at a fixed bearing or point of the retina (telotaxis), and so on.[2] Indeed, a large body of literature in the tradition of behaviorism has attempted to explain animal spatial behavior exclusively in terms of such taxes: see, for example, Loeb (1912), Kühn (1919), Fraenkel and Gunn (1940), Merkel (1980), and chapter 4.

In stimulus–response (SR) behavior, the "response" is determined by the stimulus and the particular SR schema activated in each case. Learning of SR schemata is called "associative" and comes in two major types: In classical or Pavlovian conditioning a new trigger stimulus is associated with an existing response type, such that the response is now elicited by the novel stimulus. This is basically the memory encoding of correlations within various stimulus dimensions or between stimuli and responses and may therefore also be called Hebbian learning. In operational conditioning or reinforcement learning, the second type of associative learning, a novel response is associated with a given stimulus. In this case, novel responses are actively explored in a "trial-and-error" behavior, and successful trials are rewarded. The learning then happens by reinforcing the SR schema that was used in the rewarded trial. This goes beyond Hebbian learning, since reinforcement depends not only on SR correlations, but also on the generated reward. Associative learning adds a fair

2. A related term is "tropism," which denotes growth movements of plants and other sessile organisms, while "taxis" is reserved for freely moving organisms.

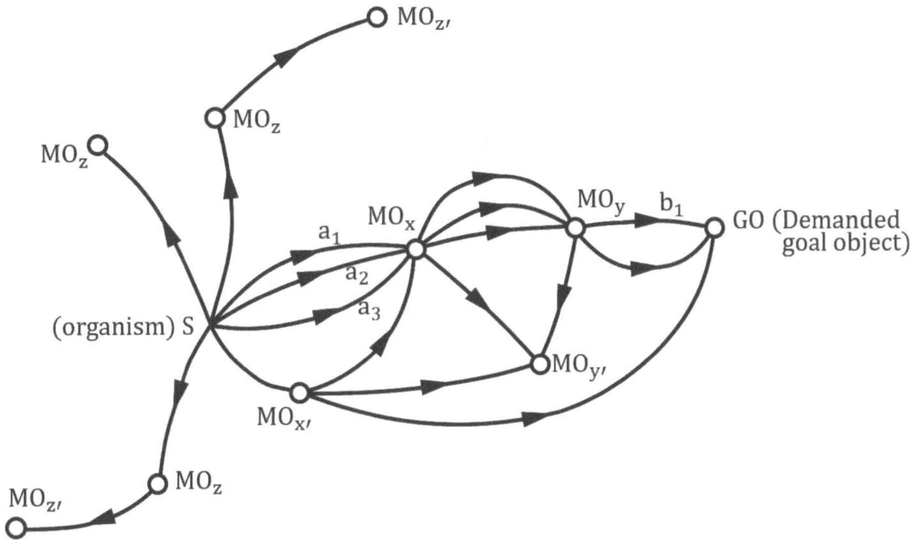

Figure 1.1
Tolman's "means-ends-field." S: start state of the organism; GO: goal object; MO: means objects; arrows: means-ends-relations. The overall action can be perfomed as a chain S–MO$_X$–MO$_Y$–GO or by the longer chain S–MO$_{X'}$–MO$_X$–MO$_Y$–GO. The means-ends-relations a_1, a_2, and a_3 are mutually "substitutable." The other MOs do not lead to the goal. Redrawn from Tolman (1932, 177).

amount of flexibility to the SR behavior and, in the behaviorist thinking of the past century, has been assumed to provide sufficient explanation of even the most complex levels of animal and human behavior.

1.2.2 Inner States

It turned out, however, that the idea of the stimulus completely determining the response severely limits the range of behaviors that the behaviorist approach can explain. Animals and humans can weight multiple inputs in complex ways, going far beyond simple winner-take-all models of multiple activated SR schemata. They can make goal-dependent decisions that in the same sensory situations may lead to different behaviors, such as turning left for water if the agent is thirsty or right for food if hunger is the prevailing state. Animals have explicit memories that can be combined with sensory inputs to find optimal actions. They can even experience sudden insights, which may cause them to abandon a currently pursued strategy for a better one, and so on. What all these examples have in common is that behavior is not determined by the stimulus alone but also by so-called inner states of the agent such as goals, desires, beliefs, representations, and percepts. In the case of spatial cognition, one important type of inner state is called the "cognitive map."

One possible way to think about inner states and the relation of stimulus–response and cognitive behavior is the "means-ends-field" illustrated in figure 1.1 published by Edward Chace Tolman in 1932. We will not follow the terminology used in this figure (but see the figure caption) but discuss the ideas in more standard terms. The general approach illustrated in the figure, however, is surprisingly modern.

The execution of a given SR schema leads to a new state of the organism which we will denote as S′. If these end states are known, they can be stored together with the start state S and the response R as triple associations SRS′. Two associations are connected if the outcome of the first is the start state of the second, which generates the state–action network or means-ends-field depicted in figure 1.1.

In spatial cognition, chains of SRS′-schemata are called routes (O'Keefe and Nadel 1978). Besides chaining, figure 1.1 also shows bifurcation, in which case multiple SRS′-schemata exist that share the same trigger state S. In this case, the organism will have to choose one schema to execute. Choice can be based on the expected outcome (i.e., the result S′ of the last step in the chain). In the case of bifurcation, the SRS′ schemata are therefore no longer triggered by the activation of state S alone. Rather, a representation of the goal (GO in the figure) must be assumed together with a planning procedure that selects the appropriate schema chain based on the current input and goal. The resulting memory structure is a graph from which action sequences connecting arbitrary start and goal states can be generated by the planning stage, as long as the respective graph path at all exists. This also allows for detour behavior via known path segments (i.e., wayfinding). Graphs like the means-ends-field provide parsimonious memory structures for basic forms of wayfinding behavior. Tolman (1948) replaced the term "means-ends-field" by the more intuitive term "cognitive map" with essentially the same meaning. The term "map" is now generally used to describe representations allowing detour behavior by a recombination of known way segments (O'Keefe and Nadel 1978).

It should be noted that graphs of the type shown in figure 1.1 can also be used to solve non-spatial problems. The goal need not be a place, but can be the satiation of any desire that the animal may have, and the action links can be taken from its entire behavioral repertoire. Wayfinding is thus but an example of problem solving in general (see chapter 8).

The notion of inner states is crucial to delineate cognitive from simpler behavior. This is akin to the role of inner states in automata theory, where the computations made possible by adding inner states to a system are studied with the methods of mathematics and computer science. Unlike the situation in technical systems, the definition of inner states in natural cognition is purely operational. They are said to be present if an observed behavior is not sufficiently explained by the agent's current stimulus, and are first of all just names for the unknown causes. Inner states are hidden variables that are needed to explain complex behaviors, and the investigation of their structure and nature is a problem of empirical research. Since inner states are operationally defined, experiments in the cognitive neurosciences always have to combine some behavioral protocol with the neurobiological measures. The interpretation of such experiments requires caution and will generally not reveal the inner states themselves but demonstrate correlations between neural activities, on the one hand, and the behavioral operationalizations, on the other.

In the sequel, we will use the terms "inner" and "mental" states largely as synonyms. The word "inner" state is more technical and avoids mysterious connotations that some people may associate with the term "mental." On the other hand, the term "mental state" stresses the fact that the mental world is subject to mechanistic explanations that, however, do require the assumption of inner states as hidden variables.

We will come back to the various types of inner states in the context of specific spatial performances. At this point, we will turn to a number of examples for the relation between stimulus–response and cognitive behavior in different domains of cognition.

1.2.3 Domains of Cognition

Visual cognition Convincing examples of mechanisms going beyond the level of stimulus–response associations can be found in all domains of cognition. In the case of visual cognition, consider bistable perceptions such as the Schiller illusion, the Necker cube reversal, or binocular rivalry (see, e.g., Goldstein and Brockmole 2016). In all these cases, the observer watches a static or alternating stimulus that is maintained or identically repeated throughout the experiment. However, subjects will perceive different things at different times, a phenomenon known as perceptual switching. In the Schiller illusion, for example, two frames are presented alternatingly, one with two dots in the upper left and lower right corners and another one with two dots in the upper right and lower left corners. The dots may then be perceived as jumping up and down or left and right, and perception will switch spontaneously between the two interpretations. Likewise, the Necker cube may appear as being looked at from above or below, and the single image perceived in binocular rivalry may switch between the two half-images of a stereogram or between two possible interpretations combining complementary parts from each eye. Changes of the stimulus cannot explain these changes in perception, which must therefore depend on an additional cause. This cause is an inner state in the above sense, in this case a "percept" or a "representation."

Objects Another important domain of cognition is the understanding of objects and the things that can be done with objects. This includes, for example, object permanence: that is, the knowledge that an object behind an occluder is still there and that it might be rewarding to feel for it and eventually grasp it. Pioneering work of Piaget and Inhelder (1948) in developmental psychology, based on object permanence tasks, has demonstrated that the representations of absent things develop in a well-defined sequence during infancy.

Objects are parts of larger problem-solving behaviors in the context of tool use. Wolfgang Köhler (1921) pioneered this field in chimpanzees. In one type of experiment, apes are presented with a banana hanging from a hook or pole high above the animal. Chimpanzees then use sticks to hit the target to make it fall down. When given boxes, they pile them up and mount the stack to get a hold of the food. Interestingly, they don't seem to understand that larger boxes should go below smaller ones, which yields rather unstable constructions. Of course, this is not a problem for an animal able to reach the top of a freely standing ladder before it has enough time to fall over. However, it tells us something about the chimpanzee's understanding of objects, causality, and physical events as has been studied intensively in "intuitive physics."

Additional cognitive abilities have been demonstrated in other performances. For example, if given tubes of different diameter, each too short to reach the goal, chimpanzees are able to construct longer sticks by sticking together shorter ones. This is an example of foresight, since the formation of the longer tool is not in itself the reward and the monkey must understand that rewards will become available only later in time.

In an example that combines object and social cognition, the apes form heaps of boxes in which some higher boxes are held in place by fellow chimpanzees. The reward is earned by the ape mounting the heap, not by the helpers stabilizing it. In systematic studies with nonhuman primates, "altruistic" helpers soon lose interest since they do not receive reward and the winners do not seem to understand that they should share their reward in order to receive further support in the future. Altruistic behavior is therefore rare in nonhuman primates but has been demonstrated to occur regularly in young human children playing cooperative games (Tomasello 2019).

Social cognition Social cognition is often considered the most important cognitive domain in the sense that it contains the highest levels of cognition such as language and theory of mind. Indeed, while there are ample examples of social cognition in animals in fields like cooperative hunting or social learning, theory of mind seems to be confined to just a few species of mammals and songbirds, while language is entirely restricted to humans.

As one example of social cognition, we consider a test for theory of mind used in young children (Baron-Cohen, Leslie, and Frith 1985). Children are presented with a little cartoon of two characters, Sally and Ann. Sally gets a ball and puts it into her basket. She then walks away while Ann is watching. When she is gone, Ann removes the ball from the basket and puts it into her box; then she also leaves. Now, Sally comes back, and the question is: where will she search for her ball? Children until about four years of age tend to answer "in the box," presumably because they know that the ball is actually there. Older children and adults, however, understand that, while the ball is indeed in the box, Sally cannot know this and will therefore think that it is in her basket. That is to say, children at an early age do not distinguish between what is the case (ball is in box) and what somebody thinks to be the case; they do not consider the representation of knowledge in Sally's mind. For this idea of making assumptions about somebody else's mental states, the term "theory of mind" is generally used, although it is not a scientific theory of how the mind works but a phenomenon that consists of assumptions ("theories") of other people's mental states.

In spatial cognition, theory of mind is related to perspective taking: that is, the ability to imagine how an environment would appear from a not currently occupied viewpoint. If this imagined viewpoint is the position of another person, perspective taking is no longer just a geometrical problem but amounts to the question of what the other person perceives and can subsequently know.

Other domains of cognition As a final example, we consider the understanding of numerosity. Being able to recognize objects, say, does not necessarily mean that in the presence of multiple objects, the number of objects can be reported. In humans, a disorder known as "simultanagnosia" is characterized by this problem: subjects attending to one object can recognize and deal with this object but cannot report the simultaneous presence of other objects (cf. Mazza 2017).

Numerosity is an example of an amodal representation, which does not rely on one specific sensory modality. Rather, items and events from the visual, acoustic, or touch modalities can all be counted, and the representation of numerosity seems to be independent from these modalities. Monkeys have been shown to possess a sense of numerosity that may be useful in the planning of riots, where the prospective success depends on the relative size of the own and the opposed groups (Brannon and Terrace 1998; Nieder and Dehaene 2009).

The list of domains given here is not exhaustive, and the delineation of domains is not always clear. Still, it is widely accepted that the cognitive apparatus may work differently in different domains and that it is based on different types of "core knowledge" (Spelke and Kinzler 2007; Gärdenfors 2020) in the spatial, object, and social domains.

1.2.4 Cognitive Agents

Within a given behavioral domain, simple and more complex mechanisms may coexist. For the case of spatial cognition, simple mechanisms like obstacle avoidance, course stabilization, path integration, or landmark guidance may rely completely or mostly on stimulus–response procedures. Such mechanisms have been investigated in depth in insects and other animals, but can be demonstrated in humans in much the same way (see chapter 4). Since they are closely interwoven with more complex mechanisms and may be their evolutionary origin, we will cover also non-cognitive levels of spatial behavior. On top of these, behaviors like wayfinding and spatial planning go beyond the stimulus–response level and constitute what may be called spatial cognition proper. In humans, a third shell is formed by the interactions of spatial and social behavior such as talking about space, giving directions, producing and reading maps, reading time tables of trains and buses, urban planning, and much more (see figure 1.2).

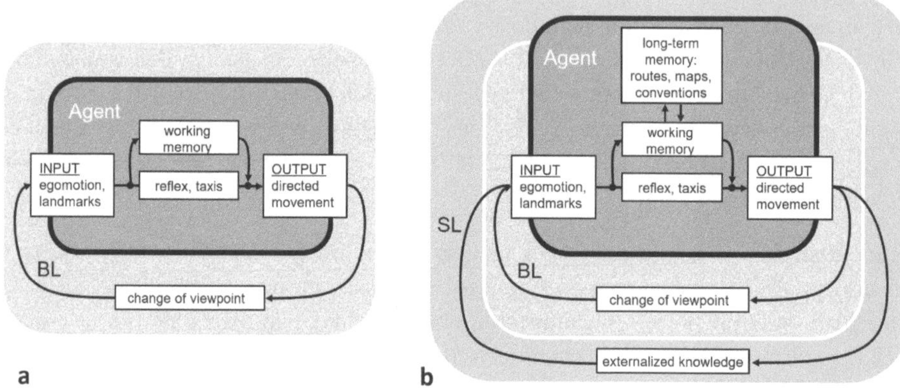

Figure 1.2
Spatial behavior in closed loop. (a) Simple spatial behavior in animals and humans happens in the "basic loop" (BL) of performed actions (bodily movements), the resulting changes of viewpoint, and the new sensory input perceived from that viewpoint. Simple stimulus–response behavior such as phototaxis occurs in all freely moving organisms. Working memory allows more complex behaviors such as path integration or systematic search. (b) The addition of spatial long-term memory supports the recognition of places, memory for routes and maps, and path planning in known (memorized) networks of places and place transitions. In humans, long-term memory also contains cultural or conventional knowledge such as the general structure of street networks, the meaning of sign posts, or how to read maps or time-tables. This memorized knowledge corresponds to "externalized knowledge": that is, the maps, time-tables, or built structures themselves. Human agents can close a second behavioral loop that interacts with other agents (i.e., map makers, airline pilots, etc.) or their productions and is therefore called the social loop (SL).

The layered structure of cognition is nicely expressed as a hierarchy of nested control loops. In the basic feedback loop (Uexküll 1926), action–perception, or stimulus–response pairs form the level of reflexes and stereotyped behavior (i.e., taxes). Working memory maintains spatial information during a maneuver, making possible behaviors such as path integration or course stabilization. Behavioral flexibility is further increased by building different types of long-term memory, including a cognitive map for wayfinding tasks. In combination with social cognition, spatial knowledge can be used for communication, cooperative behavior, or in the shaping of the environment by cooperative building (see also Mallot and Basten 2009). Action–perception cycles and their elaborations are discussed in more detail in chapter 4.

1.3 Space and Mathematics

1.3.1 Mathematical Models of Space

Mathematical constructions of space are at the origin of Western scientific thinking and form a firm basis of most of science until today. The Euclidean approach based on axioms and theorems that can be proven from these axioms is paradigmatic for the scientific method in general. For centuries, Euclid's axioms, including the ideas of line parallelism and intersection, have been considered the only possible way to think about space. In short, Euclidean geometry holds that the basic elements of space are points and straight lines, that two points define a line, and that two lines in the plane either are parallel or define a point as their intersection. Points have no extension and lines have no width.

The idea of measurement is generally ascribed to Pythagoras. It means that a stick of a given "unit" length will not change its length when moved around in space and that lengths can thus be compared and added or subtracted. Lines, or distances between points, can therefore be measured by using the unit stick as a scale. Lengths can also be triangulated, using Pythagoras' famous theorem. Angles can also be defined via the cosine in rectangular triangles. The mathematically most stringent formulation of these ideas was given by Descartes, who introduced the notion of a coordinate frame in which each point in the plane is identified with a pair of numbers: that is, its distances from the orthogonal coordinate axes. In its modern form, space is identified with the vector spaces $\mathbb{R}^2$ or $\mathbb{R}^3$ over the set of real numbers.

In section 1.1, we already discussed non-Euclidean geometries, which arise from the Euclidean case by dropping the so-called parallel postulate from the list of axioms. Non-Euclidean geometries play a role in the perception of space and will be discussed in that context in chapter 3.

Euclidean space is continuous in the sense of mathematical calculus. For example, this means that between any two points on a line, there is space for another point and in fact for a whole continuum of other points. This is related to the fact that points have diameter zero. In contrast, if we think of places in our environment, it is clear that they have a non-zero extension and that their number may be large but not infinite. Discrete sets of places and the connections between them are the topic of mathematical graph theory pioneered by Leonhard Euler. In a famous example known as the Königsberg bridge problem, he considers the map of the city of Königsberg (now Kaliningrad, Russia), which is divided into four parts by the various arms of the river Pregel (Pregolja) and their connections. In total, seven bridges existed at the time of Euler, and the question was whether it is possible to take a trip around town that passes every bridge exactly once and leads back to the starting point. Euler proved that the answer is "no" or, more specifically, that such paths exist only if every part of the city (node of the graph) would have an even number of bridges (connections to other nodes). In the Königsberg example, however, three nodes have "degree 3" (i.e., have three connections), and an Eulerian path does not exist.

Figure 1.3
Task demands for spatial representations. (a) When looking for an object (e.g., a mug) in an unfamiliar environment, verbal direction might be received (e.g., "down the hall and left") and has to be transformed into a route representation. (b) When arriving in the kitchen, a representation of peripersonal space has to be constructed, and the mug has to be localized therein. (c) When in grasping distance from the mug, a motor representation of the grasping movement will be required.

For us, it is interesting to consider not so much whether the answer in the specific case was yes or no but rather the type of the question asked. It is not about metric properties such as path length, direction, turns, or the exact location of the bridges but about connectedness and the existence of a path in a discrete set of places and connections. Networks of places (also called nodes or vertices) and connections (also called arcs or links) are studied in graph theory. Navigational space can very effectively be represented as a graph of places and place-to-place transitions, and this is of course the preferred format for path planning in practical applications as well. As opposed to metric Euclidean space, graphs are also called topological spaces, since they represent neighborhoods (i.e., the set of nodes connected to a given node) but no metric properties. Tolman's (1932) "means-ends-field" depicted in figure 1.1 is a directed graph (i.e., arcs go in one direction only) with additional labels characterizing the associated actions. We will discuss Euclidean and graph models of space in more detail in chapter 8.

1.3.2 Multiple Representations of Space

Cognitive concepts of space that underlie our everyday spatial behavior differ from the sketched mathematical concepts in a number of ways. First of all, we do not have a single, all-purpose representation of space but several such representations that subserve different tasks. This is illustrated in a little cartoon in figure 1.3. Generally, the type of spatial knowledge needed when planning an action is different from the informational needs during the various stages of the execution of the plan. As a consequence, multiple representations exist that correspond to these different needs.

One important distinction concerns the size of represented space. Peripersonal space is perceived with binocular vision and is the space in which arm and head movements take place. The visually perceived space may be much larger but is still perceived in a short sequence of views. Navigational space is not perceived at once but is inferred from many perceptions and constructed in memory. This leads to the distinction of working memories of space, which are used to control spatial maneuvers, such as keeping track of an object's position while moving in a cluttered environment, and long-term memories of space that integrate information over a large period of time and extensive exploration. Long-term memory is also needed for planning a route between two distant cities.

A similar classification of cognitive space was suggested by Montello (1993), who distinguished four scales called figural, vista, environmental, and geographical space. Figural space is smaller than the observer and includes the pictorial space of (retinal) images. Vista space covers a volume around the observer but also includes objects appearing at a distance such as an island viewed from the beach. It is close to the idea of peripersonal space introduced above. Environmental space is not seen at once but inferred from exploration and has the size of buildings or city neighborhoods. The largest category is geographical space, which contains knowledge acquired from symbolic representations of space such as maps or architectural models. The maps and models themselves, however, are part of figural space from which human navigators are able to infer properties of geographical or environmental space.

1.3.3 Coordinate Systems and Reference Frames

Coordinate systems In analytical geometry, spaces are identified with sets of coordinate values: that is, each point is characterized by an ordered list of numbers, whose length is the dimension of the considered space. For simplicity, we restrict this discussion to two-dimensional spaces. Coordinate systems have an origin, usually the point with coordinates $(0, 0)$ and two independent families of coordinate curves, along which one coordinate is fixed while the other one runs through some set of numbers (figure 1.4a). In the Cartesian system, coordinate "curves" are straight lines forming a rectangular grid. Another familiar coordinate system in the plane is the polar system, in which the coordinate curves are (i) concentric circles of equal distance from the origin and (ii) straight radial lines of equal angle. In the bipolar system for binocular vision, the curves are (i) circles of equal vergence that pass through the two eyes (Vieth–Müller circles) and (ii) the so-called Hillebrand hyperbolas of equal version, which asymptotically approach radial lines at higher distance from the center (see Mallot 2000 and section 3.2.1).

Although the question of which coordinate systems the brain and cognitive apparatus are using has received much attention, it is not entirely clear what exactly is meant by this question. Strictly speaking, we could say that the brain uses a particular coordinate system if we find neurons that have a specificity for one of the coordinate values while being indifferent to the other, and vice versa.

A case that comes close to this situation is binocular space. The original argument for the bipolar coordinates of binocular space presented by Hering is based on the dynamics of binocular eye movements, which are usually yoked and come in two types, changing either eye vergence or version. Vergence is the angle of intersection of the lines of sight of the two eyes. Pure vergence movements are antisense movements of the eyes and displace the point of fixation in depth. They belong to the slow (pursuit) type of eye movements. Version is the average direction of two eyes relative to the mid-sagittal plane of the head. Version movements turn both eyes in the same direction and occur as fast saccades or as slow pursuit movements. In combined movements, fast versions and slow vergence components can be separated by their dynamics, indicating that the oculomotor representation of binocular space indeed has a bipolar structure. Hering concluded that vergence and version are independent "inner" variables of the oculomotor system (for review, see Coubard 2013). In this case, it may therefore be fair to say that the brain uses the Hering coordinate system for the control of binocular eye movements.

The type of coordinate system used in a particular computation may also be inferred from considerations on error propagation. In path integration—that is, the continuous updating of the egocentric coordinates of a starting point from instantaneous, noisy measurements of translation and body turn—computations can be carried out in a number of different coordinate systems distinguished by the choice of the origin (ego- or geocentric) and the coordinates (Cartesian or polar). Several authors have claimed that one or another of these systems is superior to the remaining ones and is therefore likely to underlie biological path integration. However, the issues does not seem to be settled; we will discuss this problem in detail in section 5.2.

In the hippocampal place cell system (see sections 1.4 and 5.4), no coordinate system seems to be involved at all. In this case, the firing of a neuron represents a local, two-dimensional patch of space, not the grid line of a coordinate frame. As a consequence, much more than two neurons are needed to cover an extended space. Place cells and related types of neurons are therefore best described by their specificities, which are clearly spatial but do not encode space as coordinates (i.e., numbers).

Reference frames One property of coordinate systems that is also relevant in spatial cognition is the structuring they provide to large spaces. This property is also present in the more general idea of a frame of reference, which includes statements like "objects *A*, *B*, and *C* are placed on the corners of a rectangular triangle" or the "distance between objects *A* and *B* is 5 meters" but does not assume the explicit encoding of points into lists of numbers or the specification of a coordinate origin. Representations based on such descriptions of space may therefore be called coordinate free.

Frames of reference are based on recognized landmarks, places, or other spatial entities with respect to which other positions are specified. The simplest case would be to say something is close to a reference point. This idea is formalized mathematically by the "barycentric" coordinates illustrated in figure 1.4b. In the plane, distances to three non-aligned reference points uniquely define a given position. The representation of places by

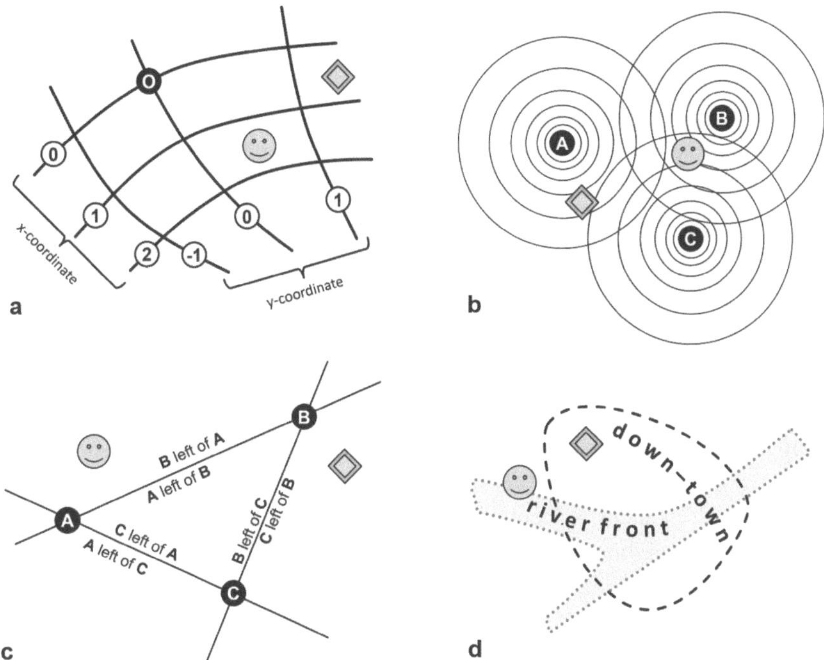

Figure 1.4
Types of reference frames used in spatial cognition. The smiley face marks the position of the agent; the diamond marks an object. (a) A two-dimensional coordinate system consists of two families of curves (coordinate lines), each parameterized with a coordinate number x or y. Each position is specified by the coordinate lines intersecting in that point. In the example, the agent and object would be at $(1.5, 0.5)$ and $(0.5, 1.3)$, respectively. The point $(0, 0)$ is the "origin" of the coordinate system. (b) Barycentric coordinates are defined by three reference points, A, B, and C. Positions are specified by the three weights placed at each reference point such that the center of gravity of the arrangement equals the represented point. (c) Tessellations of a plane can be used for qualitative place referencing. In the example, the agent is in a cell where point B appears left of points A and C while point C appears left of A. (d) Subsets or regions of space can be recognized, for example, by landmarks or other types of "local position information." In the example, the agent would be placed in region "river front," just outside the "down-town" area.

populations of hippocampal place cells is similar to this situation: the firing activity of a place cell reflects the nearness of the agent's position to the reference point (firing field center) of this place cell. It differs, however, from barycentric coordinates in that every place cell covers but a small region of space. As a consequence, many more than three cells are needed to cover an environment. Place cell firing is a population code for position: that is, the agent's position can be recovered as the mean of the place cell centers weighted by their firing activities (Wilson and McNaughton 1993).

Spatial reference points can also be used in more qualitative ways. One idea proposed by Gibson (1950) uses lines of sight connecting pairs of reference points to tessellate the plane into cells (figure 1.4c). An observer can recognize the currently occupied cell by considering the ordering of reference points as they appear from the current point of view. If a connecting line between two points is crossed, this ordering will change and may be perceived as a change of position. Different points within the same cell of the tessellation cannot be distinguished, however. Finally, reference may also be given by extended spatial entities (i.e., regions), such as the vicinity of a beacon, the valley of a river, a mountain range, or a neighborhood of a city. Places may then be specified by recognizing one or more of such overlapping regions (see figure 1.4d). This referencing system is probably best suited for specifying positions in verbal communication or qualitative spatial reasoning (e.g., Freksa 1991).

Ego- and allocentric references Coordinate systems and frames of reference can be ego- or allocentric. In the egocentric system, space is referenced with respect to the observer and their body axes. It includes concepts such as left/right, forward/backward, or above/below but also visual angles, binocular disparities, or the points in space reached by particular settings of the joints of the arm. Perceptions are egocentric by definition and so are motor commands.

Experimentally, the meaning of the term "egocentric" is not as obvious as it might seem. For example, if one asks a subject to point to a certain direction, it is not always clear whether the indicated direction is the direction from the elbow to the fingertip, from the left or right eye to the fingertip, or even from the binocular midpoint (Julesz's cyclopean point) to the fingertip (Montello et al. 1999). The problem is that "ego" is in itself a multicentered concept where spatial relations can be given with respect to the retina (i.e., they will change with eye movements), the head (i.e., they will change with head turns but not with eye movements), or the shoulders and upper body (i.e., they will change with body turns but not with movements of the head and eyes). Achieving representations in such "higher" reference frames is thus largely a matter of generating a perceptual invariance or constancy. In the parietal cortex, where egocentric representations of space are thought to reside, neurons have been identified with specificities for several different egocentric reference frames (Colby and Goldberg 1999).

Egocentric representations of places will change as the observer moves, since something that is now left of me will be to my right side if I turn around. This is not true for long-term memories: the remembered distance between two cities, for example, should not depend on the observer's current position or body orientation. Memories that do not change with observer motion are often called allocentric, although the literal meaning of the term is misleading: such representations are not centered elsewhere (allos) outside the observer, but they need not have a coordinate center at all (Klatzky 1998). We will therefore prefer the term "coordinate free" for such representations. Reference frames anchored at a fixed point in the world are also called geocentric.

The distinction of ego and allocentric reference frames can become quite confusing. Consider, for example, an automotive navigation system. This system will show a map that remains centered at the current position of the car and is egocentric in this sense. However, the driver can usually choose between two modes of map orientation. In one case, the map is turning with the curves steered by the driver such that the heading direction is always up, while alternatively a fixed north-up mode will be available. In the former case, we could say that both position and orientation are represented in an egocentric way while the latter case is egocentric with respect to position but allocentric with respect to orientation. Indeed, such "allo-oriented" representations have been suggested as models of spatial working memory. We will discuss this issue in detail in section 7.3.

To sum up, the distinction between ego- and allocentric representations of space is not binary but marks a more or less gradual transition from sensory and motor representations to more multipurpose and invariant memories of space. This transition is a form of abstraction: for example of remembered distances from experienced effort of walking or of recognized places from the various views of that place experienced during repeated visits. Abstraction may even lead to linguistic memories, in which places are represented by names; this is a form of "allocentric" memory in which coordinates seem to be completely absent.[3] The transition is also one toward reduced "modality" or increased "amodality": that is, toward memories independent of the type of sensory information (modality) from which they were originally built. We may recognize a place reached from different starting points, at different seasons or times of day, with eyes closed, or even from tales told by other people. To do this, a representation is needed that can use all available sensory channels and combines them into an amodal, invariant representation of space.

1.3.4 Continuity and Nonspatial Contents

A third aspect of cognitive spaces that distinguishes them from standard Euclidean space is their lack of continuity. While Euclidean space is continuous and homogeneous, cognitive space may be patchy: between two known areas, there may be regions of which nothing is known. The brain will not reserve storage space for such regions, as a white spot on a printed map would do, but uses a representational format that can grow together with the knowledge (as a land register does). Similarly, well-known or important regions may be represented in greater detail than others. Finally, spatial memories will not only contain "purely" spatial information, but also contents: that is, objects to find or things to do at

3. Linguistic representations of space may also involve still other types of reference systems. For example, we could say that we "stand in front of a chair" meaning that we are facing the seat. If we now turn the chair around without moving ourselves, such that we now face the backrest, we would end up standing "behind" the chair. In this case, the words "in front of" and "behind" refer to a reference frame induced by the chair itself.

certain locations. Again, these aspects of cognitive space are more reminiscent of graph theory than of Euclidean space.

1.4 Neurophysiology

The neurophysiology of spatial representations is well studied and reveals a surprisingly strong role of single neurons in spatial cognition. This section and figure 1.5 give a first overview of the topic; more detailed discussions will be given in the context of the various spatial performances, in particular path integration, place recognition, and spatial working memory.

1.4.1 Neurons with Specificities for Space

When a rat is moving around in an arena, hippocampal neurons known as "place cells" will fire in correlation with the current position of the rat (O'Keefe and Dostrovsky 1971); see figure 1.5a. In an experimental protocol known as reverse correlation, the position of the rat is tracked and a dot is marked on a map of the arena at the position occupied by the rat at the time of firing. The region in which spikes occur is called the firing field of the neuron. Other neurons will fire at other locations, and the firing fields may overlap. In an arena measuring some 40 × 40 cm, firing fields may have a diameter on the order of 10 cm such that some fifty neurons would cover the entire area with some overlap. Indeed, it has been shown that the firing activity of multiple neurons forms a clear population code for space (Wilson and McNaughton 1993; see also section 6.5.1). That is to say, if one has recorded the firing fields and then observes the activity of a population of neurons, the current position of the rat can be predicted with considerable accuracy. Place cells can reorganize or "remap" their firing fields instantaneously, for example, if the rat is put into a different maze. Place cell activity continues with the same specificity if the rats runs in darkness, indicating that they are not just driven by pictorial cues but also by path integration.

Besides the hippocampal place cells, other neurons have been found that show other spatial specificities. In head direction cells, for example, firing occurs whenever the head of the rat is oriented in a given direction, relative to path integration and orienting cues provided in the lab. In figure 1.5c, the preferred direction of a head direction cell is indicated by the pattern of white arrows covering the arena. The firing probability increases when these arrows align with the current heading of the rat. Head direction cells have been found in the hippocampus but also in the thalamus (Taube 2007).

A third type of spatial specificity is exhibited by the so-called grid cells found in entorhinal cortex (Hafting et al. 2005). Using the same protocol as for place cell recordings, these cells exhibit multiple firing fields arranged on a regular hexagonal grid. Figure 1.5b shows this grid as white lines covering the arena. The firing specificity of a given cell is characterized by its grid, which differs from the grids of other cells by grid spacing (lattice constant) and a constant offset. A grid cell will fire if the rat's position coincides with any node of the cell's grid. It thus encodes the metric position relative to the nearest grid point: that is, a

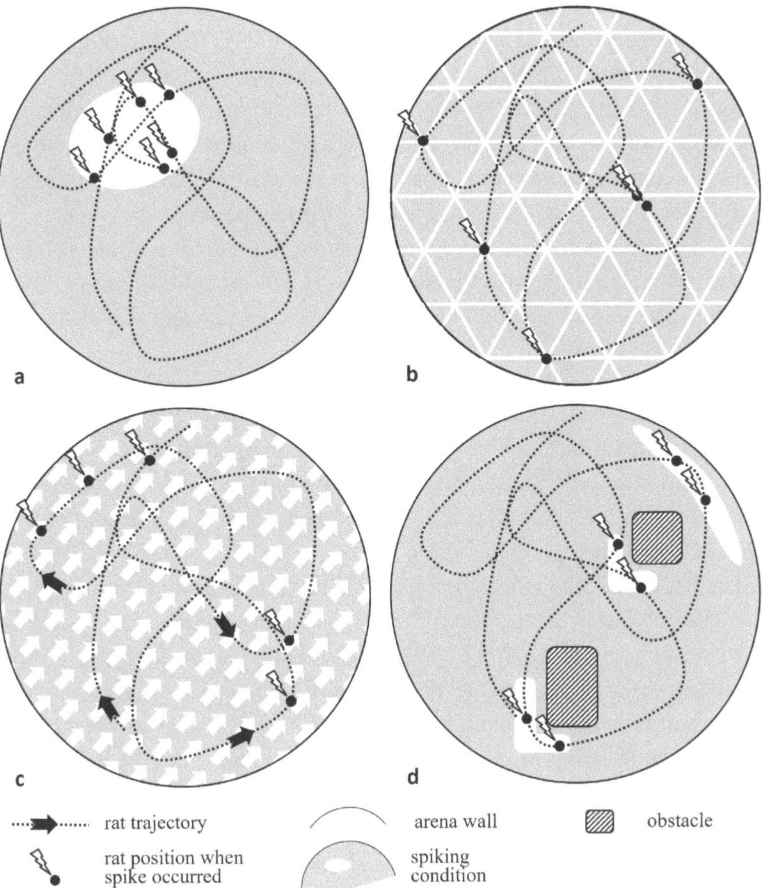

Figure 1.5
Space-related specificities of single neurons in the mammalian brain. (a) Place cells fire when the animal is at a certain location specific for each neuron (its "place field"). (b) Grid cells fire when the rat is at one of the vertices of an imaginary hexagonal grid; the grid spacing and offset are specific for each neuron while the grid orientation seems to be fixed. (c) Head direction cells fire when the rat is heading into a specific direction defined, for example, by landmark bearings. (d) Wall distance (or boundary vector) cells fire when a wall is at a specific distance and world-centered direction from the rat, irrespective of the rat's orientation or heading. For further explanations, see text.

kind of spatial "modulo" function. Grid cells with equal spacing are organized in different portions of the entorhinal cortex, within which the orientation of the grids is the same (see also section 5.4).

A final cell type found in the subiculum is called boundary vector cells and fires when the rat is at a certain distance to a wall (Solstad et al. 2008). This distance is measured in a fixed direction, such as northeast, irrespective of the body orientation of the rat: that is, an allocentric way. In figure 1.5d, regions in which the rat encounters a wall, either object or arena boundary, in the northeast direction are again marked in white. Boundary vector cells presumably rely on the head direction system, which represents the angle between the current body orientation (head direction) and an external reference direction.

The named cell types were discovered in rats and have in the meantime also been found in other rodents but also in flying animals as bats. In humans, the existence of place cells has been directly demonstrated, and indirect evidence for grid cells also exists.

1.4.2 A Model

An influential model of the functioning of the hippocampal system for place has been suggested by McNaughton et al. (2006); see also section 5.4. It takes two types of sensory inputs, local position information (pictorial cues through the visual system and allocentric wall distance through boundary vector cells) and head direction information, which is a combination of visual and egomotion cues.

The grid cell system is thought to represent the current position of the agent. Assume a rectangular array of grid cells; the current position of the rat would then be represented as a localized, reverberating peak of neuronal activity on this array. In neural network simulations, localized peaks of activity have been shown to be stable solutions or "attractors" of a simple lateral inhibition dynamics on layers of neural networks (Wilson and Cowan 1973; Amari 1977). If the rat moves, this peak receives input from the head direction cell and will move on the array in a direction determined by the head-direction cell activity (i.e., in a world-centered way). The authors now assume that when the peak reaches the margin of the array, it will move out on one side and move in again from the other side. This amounts to saying that the array is actually not rectangular but folded first into a cylinder and then into a torus such that neurons on opposite margins of the array are actually connected (Samsonovich and McNaughton 1997).

When the rat is walking for extended distances, the activity peak will move forward on the grid cell torus and every so often will return to a given cell, thus generating the multiple firing fields of the grid cells. Distance can then be measured by counting the bursts of a grid cell or, maybe more plausible, by considering simultaneous activity bursts of grid cells with different grid spacings. This will occur if the traveled distance is a common integer multiple of the grid constants of the two cells.

Grid cell activity is finally combined with the local position information from the visual system and the boundary vector cells in the place cells. The model explains important properties of place cell firing, including place cell specificity and maintained firing specificity in the absence of light. Note that the grid cell part of the model is a working memory in the sense that information is stored in the activity of neurons, not in synaptic plasticity.

1.5 Topics in Spatial Cognition

In the previous sections of this chapter, we have discussed space in the context of philosophy, evolution, behavior, psychology, mathematics, and neuroscience. *Spatial cognition*, as a field of science in its own right, is grounded on these ideas and extends them into more applied fields such as cognitive science, human factors, geography, architecture, urban planning, and sociology. Since dealing with space is a universal problem for most if not all of human behavior, the study of spatial cognition also contributes to questions of cognitive evolution and development, aging, education, neural disorders, and even the social sciences. In this text, not all of these topics will be covered or even addressed. Rather, we will take a biobehavioral viewpoint and present the basic apparatus of spatial cognition. Some guiding questions for this endeavor are summarized below.

1.5.1 Spatial Tasks and Performances

The relation between mathematical and cognitive concepts of space can be summarized by considering the purpose of spatial representations in animals and humans. As for cognition in general, this purpose lies in the organization of behavior. While there is probably no behavior completely lacking spatial components, we can define a hierarchy of spatial tasks that require more or less elaborate forms of spatial representations. In the literature, various such hierarchies have been suggested (see, e.g., Siegel and White 1975; Kuipers 2000; Wiener, Büchner, and Hölscher 2009).

Maybe the simplest spatial behavior with a clear memory component is recognizing a place. For this, a memory is needed that contains "local position information" characteristic of the place. Such local position information can be any sensory information that is reliably associated with being at that place. We will discuss place recognition in the context of landmark usage and visual snapshots.

Place recognition can be preceded by a search routine that does not require spatial long-term memory. A slightly more elaborate mechanism is "homing" to the goal location by following some sort of control mechanism. In snapshot homing, the view from the goal (the snapshot) on which recognition is based is also used to guide the approach. For example, the agent can compare the stored view with the current retinal image and then move in a direction minimizing the distance. Motion in a control loop of this type is also known as a guidance.

In a lifestyle known as "central place foraging," many animal species make excursions from a nest or hive and have to find their way back after having acquired food. This can be done with visual homing, as described above, but also by a mechanism called path integration or odometry.[4] The agent measures its instantaneous speed of translation and turn and accumulates the result. In order to return to the starting point, the agent has to

4. Odometry, from Greek "way measurement," is the technical term used in robotics. In the behavioral and neural sciences, the term "path integration" is more familiar, but basically refers to the same thing.

move in a way that reduces the accumulator's content to zero. Path integration is found in many species, including humans, and is one important source of metric information in spatial cognition.

A further step toward cognitive behavior is reached if known places together with a homing procedure between one place and the next are concatenated into chains. Such chains are known as route memory; if they bifurcate and intersect, they form a network of places and connections in which novel routes can be inferred and planned as recombinations of known route segments. This idea, originated by Tolman (1932) and O'Keefe and Nadel (1978), has already been discussed above in the context of inner states (section 1.2.2). It has also been instrumental in developing the notion of a declarative memory (Squire and Knowlton 1995): In a chain, there is only one action link or graph edge originating from each node, and memory will hold only the information needed to follow this path. If, however, multiple ways originate from a given node, memory must also hold information that allows deciding which way to pick, as in the phrase "if I walk path A, I will reach place B." For this type of declarative memory to be useful, a planning procedure is required that uses this information and the current goal to make a decision.

With the ability for spatial planning, we have reached the realm of spatial cognition proper. Further competences such as the understanding of large-scale metric layouts or the interaction of language and spatial behavior will also be touched upon in later chapters.

1.5.2 Sensory Modalities

All sensory modalities are involved in the perception of space, albeit to a different extent and with a different focus. They are used to perceive egomotion, the distance of surrounding objects, and the layout of environmental space. Places may be recognized from "local position information" such as visual landmarks but also auditory or olfactory cues. A special case is compass information, which is often not a literal compass but some rough approximation such as the visual direction of a distant landmark (Steck and Mallot 2000), the slope of an extended slanted plane, and so on. True compass senses have been found, for example, in the polarized skylight compass in insects (Rossel 1993) or the magnetic sense in migrating songbirds and pigeons (Wiltschko and Wiltschko 2005). In humans, orientation relative to the sun has also been demonstrated (Souman et al. 2009).

One systematic distinction between different types of cues can be made by the role of the observer in obtaining these cues. Optic flow, vestibular stimulation (endolymph flow in the semicircular canals), and proprioceptive information about bodily movement are generated by the observers themselves and provide information about positional change (i.e., egomotion) but not about absolute position. If the observer stands still, they go away. In contrast, landmarks can be perceived independent of observer motion and provide persistent information about position or body orientation. For these self-induced and external cues, the Greek terms *idiothetic* (self set) and *allothetic* (set by something else) are sometimes used (Mittelstaedt and Mittelstaedt 1972–1973; Merkel 1980).

1.5.3 Perception

The perception of environmental space is based on the sensory modalities mentioned above, first and foremost, however, on the depth cues contained in images, binocular image pairs, and parallactic image sequences (see, e.g., Mallot 2000; Frisby and Stone 2009). The spatial information contained in these cues is quite different. For example, the vergence angle between the two lines of sight and the accommodation state of the lens contain information about the absolute depth of the point of fixation. Disparities and motion parallax contain information about the relative depth (i.e., depth offset from the fixation point). Monocular cues such as shading or perspective hint toward surface slants while occlusion indicates depth ordering without providing quantitative information. In contrast to these cues to depth, width is much more directly perceived, via the visual angle. The construction of visual space is thus based on a multiplicity of cues that are neither commensurate with respect to their contained information nor isotropic with respect to their resolution in different spatial directions. A coherent representation of peripersonal space must be able to integrate all these cues along with cues from other modalities such as auditory direction. Space perception is thus an extreme case of cue integration and cannot be understood by considering one or a few cues in isolation. We will come back to these problems in chapters 2 and 3.

An additional problem of space perception arises from movements of the observer. Visual space is not just an egocentric view of the world but integrates information over small movements of the head and body. That is to say, the term "here" does not mean a geometric locus with diameter zero but has an extension that may coincide with the extension of a room in indoor environments, the grasp space reachable without getting up from a chair, and so on. This is similar to the famous observation by William James (1890, 609f) about the perception of the present moment in time, the "cognized present":

> In short, the practically cognized present is no knife-edge, but a saddle-back, with a certain breadth of its own on which we sit perched, and from which we look in two directions into time.

Just as the cognized present has a duration, the cognized "here" has a width. It is a construction of the cognitive apparatus rather than a "direct" perception. In spatial thinking, it can be moved around to imagined places as in the "perspective-taking" paradigm in which subjects have to report how a scene would look from a different vantage point.

1.5.4 Working Memory

In neuroscience, the term "working memory" has been adopted from the classical architecture of the von Neumann computer, where it is distinguished from input and output devices as well as from the data store and, together with the central processing unit ("executive functions" in neuroscience), runs the processes of computing. Generally, a working memory is a transient storage maintained during an ongoing process or maneuver that will be forgotten when the process is over. It supports coherent sequences of behavior over some length of time (see also figure 1.2). The neural substrate of working memory is thought to be a stable or reverberating pattern of activity, not network plasticity such as synaptic growth.

Following the work of Baddeley and Hitch (1974), human working memory is thought to be composed of two subsystems, called the phonological loop and the visuospatial sketchpad, which are sometimes also called slave systems under the mastery of the central executive. The phonological loop is used to remember words or short sentences by repeating them in silent speech. Memory for sequences would be useful also in animals, but evidence for this is scarce. The name of the visuospatial sketchpad implies that it deals with two-dimensional space in which visual information can be integrated, for example, across eye gaze saccades. More recently, Baddeley (2003) suggested a third slave system, a spatial working memory for tasks such as the Corsi block-tapping task.

In spatial cognition, there exist a large number of other working memory tasks whose relation to the Baddeley and Hitch model is not entirely clear. For example, in path integration, the minimal amount of information that has to be maintained during the task is the relative position of the starting point with respect to the agent. This knowledge is continuously updated with current egomotion data. Once the process is completed, information about the actual path traveled will be gone; if not, this memory would be of the long-term type. Updates of the positions of environmental objects that are continuously performed during path integration are known as spatial updating. Imagery, especially imagined perspective change, or the imagined ("mental") rotation of visual objects, is a related form of visuospatial working memory.

In animal research, a standard way to probe spatial working and long-term memories is the radial eight-arm maze, in which a subgroup of arms is regularly baited. If a rat is put into this maze, it will search the arms and eat the bait. In doing so, it can make two types of error: it may either visit an arm that is never baited, which would be considered a long-term memory error. If, however, it revisits an arm that was previously baited but has already been depleted in the current session, a working memory error would be reported. A related working memory task is object permanence, discussed earlier as an example of object cognition. Spatial working memory will be discussed in chapter 7.

1.5.5 Spatial Long-Term Memory

The simplest form of a spatial long-term memory is the memory of a place: that is, of local position information needed to recognize this place. This local position information may consist of landmarks (i.e., identified objects), such as a peculiar building or a conspicuous rock formation. It may, however, also be formed of simple images such as intensity edges or blobs. For this situation, the term "cue" is often used, indicating that the recognized pattern carries spatial information but is not in itself an "object."

Spatial long-term memory will also contain instructions for actions associated with the landmark knowledge as stimulus–response pairs. In the animal literature, these are also known as "recognition-triggered response" (Trullier et al. 1997). In wayfinding, they may be concatenated to chains or routes while networks of stimulus–response associations allow the inference of novel routes from known segments. As pointed out before, this is the basic idea of the cognitive map. Ongoing discussions concern the role of additional metric knowledge that may be included in this "map." We will come back to these controversies below.

Spatial memory is sometimes subsumed into the larger category of episodic memory: that is, the memory for events (Eichenbaum 2017a, 2017b). This is reasonable since taking a path decision during travel can be seen as an event in the same sense as other "autobiographic" events, such as having had breakfast or having met a friend. However, truly spatial concepts can be defined as a common property of a larger set of events, all of which took place in the same location. The concept of place would thus be constructed from a "lifeline" of events by discovering spatial features that these events have in common.

Alternatively, wayfinding may be considered just another example of problem solving as it would also occur in tool use or social interaction. Spatial memory would then be subsumed into general problem knowledge. This type of question, which asks for the delineation of different types of memory, is not just a matter of terminology. Experimentally, we can ask about neuropsychological dissociations between these different types of memory and their neural substrates: do spatial, episodic, or general problem-solving tasks recruit the same brain areas or are they "dissociated" into different neural systems?

One further argument in favor of a separate memory system for space is the representation of specifically spatial properties and relations. Metric distance, triangulation, and dimensionality are such concepts that exist in space but not in the same sense in timelines of events or problem spaces. We can go backward in space but not in time. The concept of dimensionality is probably derived from visual space, in which the third dimension is laboriously constructed. It has no equivalent in general problem spaces but is part of our spatial memory.

1.5.6 The Cognitive Map

In this text, we use the term "cognitive map" to denote a portion of spatial long-term memory that supports wayfinding and the inference of novel paths from known route sections. Alternatively, one could say that the cognitive map is the declarative part of spatial memory. We will present evidence for graph or network structure of this map. Minimally, it consists of recognized spatial entities, the places, and action links allowing to move from one node to another. Path planning is then basically a graph search. On top of this basic structure, additional types of information have been discussed that may or may not be represented. These are (i) local metric (i.e., distance information along links, which may also be represented implicitly, for example as the effort it takes to travel a link), (ii) global metric embedding of the places with coordinates or in a coordinate-free but globally consistent system based on distance geometry, and (iii) hierarchical groupings of places into regions.

The term "cognitive map" was originally introduced by Tolman (1948) and was, at the time, the first usage of the term "cognitive" in its modern sense.[5] The term replaces the earlier term "means-end-field" illustrated in figure 1.1 and is basically in agreement with the definitions given above. O'Keefe and Nadel (1978) applied the term to the rodent hippocampus and the place cell system. In addition to the network or graph aspects of the cognitive maps—O'Keefe and Nadel (1978) introduce the term "taxon" system for this part—they assume a second system for metric information. including path integration, for which they use the term "locale" system. This concept leaves open the possibility of partial metric knowledge and the organization of spatial knowledge as a labeled graph, as will be described in section 8.3.2.

Other interpretations of the concepts have led to heated debates of whether or not humans or animals have cognitive maps. For example, Wehner and Menzel (1990) claim that insects do not have cognitive maps, although they clearly have multisegment route behavior and maybe even graph-like memories. For the other extreme, Gallistel (1990) has suggested that rodents and humans have completely embedded metrical maps. Besides the modeling of partial metric knowledge, another solution to this problem may lie in the observation that spatial working memories, with performances such as path integration and spatial updating, contain more geometric information, while long-term memory is more based on routes and networks. And of course, there is no need to assume that insects and vertebrates share identical types of spatial memories.

1.5.7 Summary: The Cognitive Construction of Space

To sum up, the representation of space in the cognitive apparatus is a construction from various sensory inputs and is organized in the nervous system in a way to support behavior. Four main questions for spatial cognition can be formulated:

How is space inferred from sensory inputs? Since the most important sensory modality for space perception is vision, many of these questions are also addressed in computational vision. The idea that the senses provide information from which knowledge about the outside world is inferred has been clearly formulated by Helmholtz (1909–1911) and Mach (1922) and is the topic of computational vision (Marr 1982; Mallot 2000).

What are the "informational needs" of an agent behaving in a three-dimensional world? This is the question of the "Umwelt" of an animal in the sense of Uexküll (1926) and of ecological psychology (Gibson 1950). It may have different answers for animals with different lifestyles.

What is the logical and psychological structure of mental representations of space? This question was pioneered by Tolman (1932, 1948), who coined the term "cognitive map" and thus advocated the existence of mental representations at a time when behaviorism was the

5. However, the word may also be traced back to Decartes's term *res cogitans* for the subjective world as opposed to the physical *res extensa*.

ruling doctrine. In child psychology, the stepwise development of such representations was demonstrated by Piaget and Inhelder (1948).

Finally, how are spatial representations realized in the brain? The discovery of place cells by O'Keefe and Dostrovsky (1971) in the hippocampus of the rat was one of the first examples of a neural correlate of a mental state, namely the sense of space. With the advent of functional magnetic resonance imaging, larger cortical networks for space have been identified also in humans.

Key Points of Chapter 1

- Cognitive space is not simply an internal reflection of physical space but is actively constructed by our cognitive apparatus. The rules and mechanisms of this construction are a product of cognitive evolution.
- Cognition can be defined by the level of behavioral complexity made possible by the existence of "inner states" such as goals and purposes, expectations about the outcome of various actions in different situations, or a monitor of the current state of affairs.
- The inner states also prescribe the formats in which spatial information has to be represented in the brain. Spatial cognition is not so much about points, straight lines, and coordinates but about taxis, egomotion, place recognition, and wayfinding.
- Important mathematical approaches to the modeling of cognitive space include the continuous and metric Euclidean space, continuous spaces with non-Euclidean distance functions, and discrete graphs (topological spaces). Humans possess multiple representations of space, which may be best modeled by one or another of these approaches.
- Neural correlates of spatial representation are found in single-neuron recordings from various parts of the medial temporal lobe and adjacent regions. Different types of neurons encode heading direction (head direction cells), location within a maze (place cells), distance and bearing of walls and objects (boundary and object vector cells), and small-scale positional offsets relative to a hexagonal grid (grid cells).
- Different types of spatial performances can be distinguished by the amount of memory involvement. The perception of egomotion and visual space proceeds largely without contributions from memory. Course control, path integration, and mental transformations such as spatial updating require forms of working memory. Spatial long-term memory allows place recognition and simple route following. Finally, an interplay of spatial working and long-term memories is required for path planning and the inference of novel routes from known route segments (i.e., wayfinding).

References

Amari, S.-I. 1977. "Dynamics of pattern formation in lateral-inhibition type neural fields." *Biological Cybernetics* 27:77–87.

Baddeley, A. 2003. "Working memory: Looking back and looking forward." *Nature Reviews Neuroscience* 4:829–839.

Baddeley, A. D., and G. J. Hitch. 1974. "Working memory." In *The psychology of learning and motivation: Advances in research and theory,* edited by G. A. Bower, 47–89. New York: Academic Press.

Baron-Cohen, S., A. M. Leslie, and U. Frith. 1985. "Does the autistic child have a 'theory of mind'?" *Cognition* 21:37–46.

Brannon, E. M., and H. S. Terrace. 1998. "Ordering of the numerosities 1 to 9 by monkeys." *Science* 282:746–749.

Colby, C. L., and M. E. Goldberg. 1999. "Space and attention in parietal cortex." *Annual Review of Neuroscience* 22:319–349.

Coubard, O. A. 2013. "Saccade and vergence eye movements: A review of motor and premotor commands." *European Journal of Neuroscience* 38:3384–3397.

Eichenbaum, H. 2017a. "Prefrontal-hippocampal interactions in episodic memory." *Nature Reviews Neuroscience* 18:547–558.

Eichenbaum, H. 2017b. "The role of the hippocampus in navigation is memory." *Journal of Neurophysiology* 117:1785–1796.

Fraenkel, G. S., and D. L. Gunn. 1940. *The orientation of animals. Kineses, taxes, and compass reactions.* Oxford: Clarendon.

Freksa, C. 1991. "Qualitative spatial reasoning." In *Cognitive and linguistic aspects of geographic space,* edited by D. M. Mark and A. U. Frank, 361–372. Dordrecht: Kluwer.

Frisby, J. P., and J. V. Stone. 2009. *Seeing. The computational approach to biological vision.* Cambridge, MA: MIT Press.

Gallistel, C. R. 1990. *The organization of learning.* Cambridge, MA: MIT Press.

Gärdenfors, P. 2020. "Primary cognitive categories are determined by their invariances." *Frontiers in Psychology* 11:584017.

Gibson, J. J. 1950. *The perception of the visual world.* Boston: Houghton Mifflin.

Goldstein, E. B., and J. Brockmole. 2016. *Sensation and perception.* 10th ed. Boston: Cengage Learning.

Hafting, T., M. Fyhn, S. Molden, M.-B. Moser, and E. I. Moser. 2005. "Microstructure of a spatial map in the entorhinal cortex." *Nature* 436:801–806.

Hauser, M. D., N. Chomsky, and W. T. Fitch. 2002. "The faculty of language: What is it, who has it, and how did it evolve?" *Science* 298:1569–1579.

Helmholtz, H. 1876. "The origin and meaning of geometrical axioms." *Mind* 1:301–321. http://www.jstor.org/stable/2246591.

Helmholtz, H. von. 1909–1911. *Handbuch der physiologischen Optik.* 3rd ed. Hamburg: Voss.

James, William. 1890. *The principles of psychology.* Vol. 1. New York: Hold/Macmillan.

Jammer, M. 1993. *Concepts of space: The history of theories of space in physics.* New York: Dover.

Kant, I. 1781. *Critik der reinen Vernunft.* Riga: Johann Friedrich Hartknoch.

Klatzky, R. L. 1998. "Allocentric and egocentric spatial representations: Definitions, distinctions, and interconnections." *Lecture Notes in Artificial Intelligence* 1404:1–17.

Köhler, W. 1921. *Intelligenzprüfungen an Menschenaffen.* 2nd ed. Berlin: Julius Springer.

Kühn, A. 1919. *Die Orientierung der Tiere im Raum.* Jena: Gustav Fischer Verlag.

Kuipers, B. 2000. "The spatial semantic hierarchy." *Artificial Intelligence* 119:191–233.

Loeb, J. 1912. *The mechanistic conception of life.* Chicago: University of Chicago Press.

Lorenz, K. 1973. *Die Rückseite des Spiegels. Versuch einer Naturgeschichte menschlichen Erkennens.* München: Piper.

Mach, E. 1922. *Analyse der Empfindungen.* 9th ed. Jena: Gustav Fischer Verlag.

Mallot, H. A. 2000. *Computational vision: Information processing in perception and visual behavior.* Cambridge, MA: MIT Press.

Mallot, H. A., and K. Basten. 2009. "Embodied spatial cognition: Biological and artificial systems." *Image and Vision Computing* 27:1658–1670.

Marr, D. 1982. *Vision.* San Francisco: W. H. Freeman.

Mazza, V. 2017. "Simultanagnosia and object individuation." *Cognitive Neuropsychology* 34:430–439.

McNaughton, B. L., F. P. Battaglia, O. Jensen, E. I. Moser, and M.-B. Moser. 2006. "Path integration and the neural basis of the 'cognitive map'." *Nature Reviews Neuroscience* 7:663–678.

Merkel, F. W. 1980. *Orientierung im Tierreich.* Stuttgart: Gustav Fischer Verlag.

Mittelstaedt, H., and M.-L. Mittelstaedt. 1972–1973. "Mechanismen der Orientierung ohne richtende Außenreize." *Fortschritte der Zoologie* 21:46–58.

Montello, D. R. 1993. "Scale and multiple psychologies of space." *Lecture Notes in Computer Science* 716:312–321.

Montello, D. R., A. E. Richardson, M. Hegarty, and M. Provenza. 1999. "A comparison of methods for estimating directions in egocentric space." *Perception* 28:981–1000.

Müller, F. M. 1922. *Immanuel Kant's critique of pure reason.* 2nd ed. New York: Macmillan.

Nieder, A., and S. Dehaene. 2009. "Representation of number in the brain." *Annual Review of Neuroscience* 32:185–208.

O'Keefe, J., and J. Dostrovsky. 1971. "The hippocampus as a spatial map. Preliminary evidence from unit activity in the freely-moving rat." *Brain Research* 34:171–175.

O'Keefe, J., and L. Nadel. 1978. *The hippocampus as a cognitive map.* Oxford: Clarendon.

Piaget, J., and B. Inhelder. 1948. *La représentation de l'espace chez l'enfant.* Paris: Presses Universitaires de France.

Popper, K. 1972. *Objective knowledge: An evolutionary approach.* Oxford: Clarendon.

Rossel, S. 1993. "Navigation by bees using polarized skylight." *Comparative Biochemistry & Physiology* 104A:695–708.

Samsonovich, A., and B. L. McNaughton. 1997. "Path integration and cognitive mapping in a continuous attractor neural network model." *Journal of Neuroscience* 17:5900–5920.

Siegel, A. W., and S. H. White. 1975. "The development of spatial representations of large-scale environments." *Advances in Child Development and Behavior* 10:9–55.

Solstad, T., C. N. Boccara, E. Kropff, M.-B. Moser, and E. I. Moser. 2008. "Representation of geometric borders in the entorhinal cortex." *Science* 322:1865–1868.

Souman, J. L., I. Frissen, M. N. Sreenivasa, and M. O. Ernst. 2009. "Walking straight into circles." *Current Biology* 19:1538–1542.

Spelke, E. S., and K. D. Kinzler. 2007. "Core knowledge." *Developmental Science* 10:89–96.

Squire, L. R., and B. J. Knowlton. 1995. "Memory, hippocampus, and brain systems." In *The cognitive neurosciences,* edited by M. S. Gazzaniga, 825–837. Cambridge, MA: MIT Press.

Steck, S. D., and H. A. Mallot. 2000. "The role of global and local landmarks in virtual environment navigation." *Presence: Teleoperators and Virtual Environments* 9:69–83.

Taube, J. S. 2007. "The head direction signal: Origins and sensory-motor integration." *Annual Reviews in Neuroscience* 30:181–207.

Tinbergen, N. 1963. "On aims and methods of ethology." *Zeitschrift für Tierpsychology* 20:410–433.

Tolman, E. C. 1932. *Purposive behavior in animals and men.* New York: The Century Company.

Tolman, E. C. 1948. "Cognitive maps in rats and man." *Psychological Review* 55:189–208.

Tomasello, M. 2019. *Becoming human: A theory of ontogeny.* Cambridge, MA: Harvard University Press.

Trope, Y., and N. Liberman. 2010. "Construal-level theory of psychological distance." *Psychological Review* 117:440–463.

Trullier, O., S. I. Wiener, A. Berthoz, and J.-A. Meyer. 1997. "Biologically based artificial navigation systems: Review and prospects." *Progress in Neurobiology* 51:483–544.

Uexküll, J. von. 1926. *Theoretical biology.* New York: Hartcourt, Brace & Co.

Wehner, R., and R. Menzel. 1990. "Do insects have cognitive maps?" *Annual Review of Neuroscience* 13:403–414.

Wiener, J. M., S. Büchner, and C. Hölscher. 2009. "Taxonomy of human wayfinding: A knowledge-based approach." *Spatial Cognition and Computation* 9:152–165.

Wilson, H. R., and J. D. Cowan. 1973. "A mathematical theory of functional dynamics of cortical and thalamic nervous tissue." *Kybernetik* 13:55–80.

Wilson, M. A., and B. L. McNaughton. 1993. "Dynamics of the hippocampal ensemble code for space." *Science* 261:1055–1058.

Wiltschko, W., and R. Wiltschko. 2005. "Magnetic orientation and magnetoreception in birds and other animals." *Journal of Comparative Physiology A* 191:675–693.

2 Egomotion

Even when we close our eyes, we are aware of the movements that we perform or are able to perform in the space around us. The perception of self- or egomotion is probably our most basic sense of space and at the same time relates perception to behavior. Spatial references such as forward, backward, left, and right are often made with respect to our abilities of bodily movements. The axes so defined structure the "egocentric" reference frames in many forms of spatial representation. In this chapter, we discuss the sensory cues to egomotion and especially elaborate on the optic flow as one central source of egomotion information. We then turn to more integrative aspects, including multisensory performances in egomotion and their underlying neural mechanisms, and finish the chapter with a general discussion of cue integration or sensor fusion.

2.1 The Space for Motion

2.1.1 Origins of Spatial Knowledge

Space can be conceived of as the medium in which we perform bodily movements and in which we take distinguishable positions before and after a movement is carried out. In a constructivist logic, we could say that spatial position is the amodal perceptual or representational quality that changes upon bodily movement. Memories of visual images, sensed muscle forces, or pressures felt at the soles of the feet are "modal": that is, they keep the quality of their sensory modalities.[1] In contrast, the sense of space can be induced by many different sensory modalities, but is itself largely independent of these: that is, it does not depend on their specific visual, proprioceptive, or haptic qualities.

In any case, the perception of egomotion is a central element of the perception of space itself. It induces body-related frames of reference defining left, right, up, down, forward,

1. By the term "sensory modalities," we denote the five Aristotelian senses of vision, hearing, smell, taste, and touch plus the "kinesthetic" modalities of proprioception (sense of muscle contraction and joint settings) and the sense of pose and balance provided by the vestibulum in the inner ear. In animals, additional modalities exist, for example, electroreception or the sense of water flow provided by the lateral line system in fish. Sensory modalities are often associated with a well-defined sense organ such as the eye or ear but may also rely on distributed sensor networks as would be the case for touch.

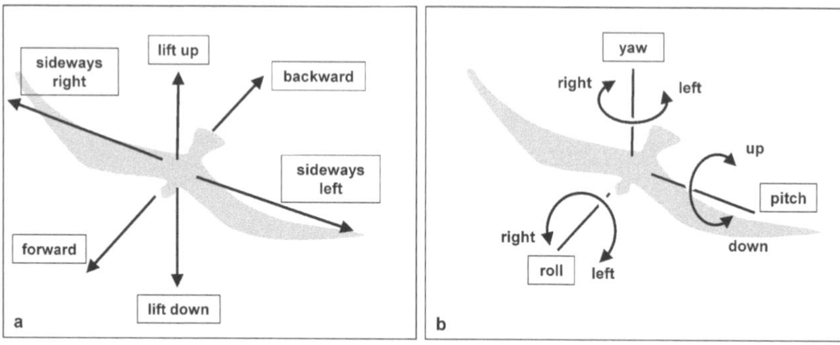

Figure 2.1
Degrees of freedom of motion. (a) Translation, (b) Rotation. These degrees of freedom describe instantaneous rigid body motion. More complex descriptions may include relative motion of body parts (such as head and eye movements or articulated walking) and maneuvers over extended periods of time (such as accelerations, braking, chasing, docking, etc.).

and backward with respect to the overall body or to smaller body parts such as the head or a hand. The term "egomotion" denotes positional or orientational changes of the entire organism; they are usually generated by movements of the limbs but may also arise from passive transport as in a vehicle or the ballistic flight phase of a jump. The perception of egomotion can therefore not rely on the monitoring of the limb movements alone but will also make use of visual and vestibular cues generated by egomotion itself.

Besides egomotion, other sources of spatial information are the perception of depth in the three-dimensional environment and the recognition of places and place changes from remembered landmark information. These can be combined with egomotion cues, for example, in motion parallax (i.e., improved depth perception from the multiple viewpoints attained during egomotion) or in haptic exploration. The combination of egomotion cues with place recognition requires more elaborate mechanisms such as path integration, landmark memories, and the representation of place adjacencies. We will come back to these topics in later chapters.

2.1.2 Degrees of Freedom of Motion

By egomotion proper, we mean bodily movement of the entire agent to a new position in space, a new body orientation, or both. Changes of position are called translations, while pure changes of orientation are of course rotations. In the terminology of classical mechanics, the agent is treated as a rigid body; instantaneously, rigid body motion can always be decomposed into a rotation and a translation, each of which is described by three "degrees of freedom" (DoFs). The total six DoFs describe the evolution of position and body orientation over time; together, their instantaneous settings are also called the "pose" of the agent.

The specific DoFs are usually expressed in a body-centered coordinate system, where each axis describes both a translation and a rotation (see figure 2.1). There are various names for the different DoFs used, for example, in animal behavior or aviation. Here we will use the terms forward/backward, sideways left/right, and lift up/down for translations, and the rotation terms yaw left/right, pitch up/down and roll left/right, for rotations about the upright (dorsoventral), left–right, and forward–backward axes respectively.

In the following, we give a formal definition of the degrees of freedom in the style of classical mechanics. It may be skipped on first reading but will be needed for a deeper understanding of optic flow as discussed below.

Rigid body motion cannot be described by a simple trajectory or space curve, because the rigid body has an extension and can therefore rotate relative to its center of mass. In contrast, the movement of a point would be completely defined by the point's change of position and has therefore just the three degrees of freedom of translation. Full rigid body motion is modeled as a moving coordinate system whose origin is the center of mass of the moving body; in addition to its movement along the trajectory, the coordinate system may also rotate about its origin. Mathematically, the situation is described as a coordinate transformation from the initial coordinate system into a new coordinate system after a movement step. Let $\boldsymbol{x} = (x_1, x_2, x_3)^\top$ denote the coordinates of a fixed point in the initial coordinate system: that is, a point not moving with the agent. The transformation can then be expressed as

$$T(\boldsymbol{x}) = \boldsymbol{R}\boldsymbol{x} + \boldsymbol{t}, \tag{2.1}$$

where $\boldsymbol{R}$ is a rotation matrix (i.e., it satisfies $\boldsymbol{R}^{-1} = \boldsymbol{R}^\top$), and $\boldsymbol{t}$ is the vector of translation. In equation 2.1, it is assumed that the transformation consists of a rotation followed by a translation or shift; if we revert the sequence of rotation and translation, a different transformation results. However, the difference gets quickly smaller for small amounts of rotation and translation and vanishes altogether in the infinitesimal case. Therefore, an arbitrary rigid motion can always be uniquely decomposed into a rotation and a translation that may, however, be constantly changing.

In order to obtain an expression for translational velocity, we write the translation vector as $\boldsymbol{t} = v\boldsymbol{u}$ with $\|\boldsymbol{u}\| = 1$. $\boldsymbol{u}$ is the current direction of movement: that is, the tangent direction of the trajectory. For the rotation, we consider the axis of rotation $\boldsymbol{r}$, again as a unit vector (i.e., $\|\boldsymbol{r}\| = 1$); $\boldsymbol{r}$ is an eigenvector of $\boldsymbol{R}$ with eigenvalue +1. We may then write

$$\frac{d\boldsymbol{x}}{dt} = \omega \boldsymbol{r} \times \boldsymbol{x} + v\boldsymbol{u}. \tag{2.2}$$

Since $\boldsymbol{r}$ and $\boldsymbol{u}$ are unit vectors, they have only two degrees of freedom each. The symbol $\times$ denotes the cross-product. ω and v are scalar quantities: that is, the rotational and translational velocities. If we express $\boldsymbol{r}$ and $\boldsymbol{u}$ in the current egocentric coordinate system, we obtain the six degrees of freedom as vu_1, vu_2, vu_3 for translation and $\omega r_1, \omega r_2, \omega r_3$ for rotation. Considered as vectors, they are the translation component (tangent or derivative of the trajectory) and the axis of rotation scaled with the rotational velocity.

For strictly earth-bound agents, rotations are only possible about the vertical axis: that is, we have $\boldsymbol{r} = (0, 0, 1)^\top$ at all times. Earth-bound translations are confined to the horizontal plane spanned by the forward/backward and the sideways direction (i.e., $u_3 = 0$). In this case, equation 2.2 reduces to

$$\frac{d\boldsymbol{x}}{dt} = \begin{pmatrix} -\omega x_2 + v u_1 \\ \omega x_1 + v u_2 \\ 0 \end{pmatrix}. \tag{2.3}$$

Only three degrees of freedom remain: the translations forward/backward (u_1) and left/right (u_2) and the yaw-rotation left/right (ω).

Applications of the theory of rigid body motion to the movements of animals and humans are complicated by the fact that these are not rigid bodies but perform articulated or "biological" motion. This type of motion is characterized by movements of joints between body parts, which in themselves are approximately rigid. Each joint may have one, two, or three degrees of freedom, resulting in a large number of DoFs for general articulated motion. Important cases in the context of spatial cognition are the strides of the legs or the eye and head movements performed to direct vision or hearing to a goal off the current movement directions.

2.2 Perceiving Egomotion

2.2.1 Percepts and Representations

The six degrees of freedom allow a complete mathematical description of whole-body egomotion and ego-acceleration. This does not mean, however, that the perception of egomotion leads to a uniform representation organized in the same way. Quite to the contrary, a fair number of different and partially independent percepts have been described—all related to egomotion—that can be distinguished experimentally and conceptually; for review, see Warren and Wertheim (1990) and Britton and Arshad (2019).

One such percept is *heading*: that is, the perceived direction of egomotion expressed either relative to the line of sight (as an angle or location in the retinal image) or with respect to some reference direction in the environment. In a study by Warren, Morris, and Kalish (1988), subjects are seated in front of a computer screen and watch a pattern of dots moving as if on a horizontal plane over which the observer is flying. The observer generates a sense of egomotion in a particular direction. After a while, a marker appears somewhere along the horizon and the observer is asked to judge in a two-alternative forced-choice task whether the marker will be passed left or right. The results show that the discrimination threshold for egomotion direction is on the order of one to two degrees. Heading can also be perceived in complete darkness, in which case it is based on vestibular information (Gu, DeAngelis, and Angelaki 2007).

The sense of being transported away from the current position is called *vection* (Dichgans and Brandt 1978; Riecke et al. 2006). Ernst Mach, in his *Analysis of Sensations*, describes a case of an illusion of vection in which his young daughter was looking out of the window into a calm but heavy snowfall. She then cried out that she was moving upward together with the entire house (Mach 1922, chap. VII.11). Another well-known version of this illusion may occur to an observer sitting in a train at a station and watching another waiting train through the coach window. When motion starts, it is often not clear which train is actually moving. Vection is to some extent independent of heading and can be perceived as a transportation without a clear sense of the direction in which the transportation goes. It is often measured by "scaling" (i.e., by asking the subjects to rate the strength of their feeling on an ordinal scale) and is therefore treated as a scalar quantity. A related percept is *cyclovection*, the sense of being rotated, usually about the vertical axis, which can be tested with an observer sitting inside a striped drum. In different experimental conditions, the drum will be spinning around the observer or the chair will be rotating inside a stationary striped drum. In both conditions, the percept may switch between that of a cyclovection (observer rotating) and object motion (drum rotating). This is also an example of a bistable perception of which we discussed the Schiller illusion and the Necker cube reversal in chapter 1. It shows that egomotion perception is constructed by the brain and that different perceptual judgments may arise from the same sensory input.

Perceived *egomotion speed* can be assessed with psychophysical procedures such as forced choice or scaling (Wist et al. 1975) or, more indirectly, by a number of speed-related percepts. One such percept is *time to collision*: that is, the nearness of obstacles expressed as the time remaining until the obstacle will be hit. As we will see below, this can be directly inferred from the optic flow. Optic flow also allows the perception of the three-dimensional layout of the environment up to a constant scaling factor, which depends on egomotion speed:. In *motion parallax*, the parallactic information obtained from subsequent viewpoints in the motion sequence is used to recover three-dimensional structure or depth ordering (see also figure 3.1), while in the *kinetic depth effect*, an object is moving (often rotating in itself) with or without observer motion, and three-dimensional shape is again inferred from the motion sequence.

Behavioral responses to egomotion cues occur, for example, as compensatory movements of the eyes, known as *optokinetic response* (OKR) if perceived vection or cyclovection is induced visually, or as *vestibulo-ocular reflex* (VOR) if perceived egomotion is induced by vestibular cues such as actual rotation on a rotating chair. Motor responses to translational optic flow are common for the stabilization of position or posture in dynamic environments in humans and animals. Waterstriders and zebrafish, for example, use optic flow to control an *optomotor response* (OMR) compensating for drifts caused by wind or water flow (Junger and Varjú 1990; Orger et al. 2004). *Postural sway*—that is, compensatory body movement intended to maintain stable stance (e.g., Peterka 2002)—occurs if an observer stands in front of a wall that is moving backward and forward in an oscillatory

Table 2.1
Cues to egomotion

	Idiothetic cues	Allothetic cues to	
		Position and orientation	Motion and acceleration
Vestibular	Inertial sense of linear and circular acceleration	Vector of gravity	–
Proprio-ceptive	Joint and muscle movement	Underground slope (via ankle joint)	–
Visual	–	Landmark bearings, celestial compasses	Optic flow
Auditory	–	Acoustic landmarks including infrasound	Doppler shift
Haptic	–	Mechanical guidance (handrail), underground (e.g., sand vs. gravel)	Foot sole pressure, self-generated airstreams
Chemical	–	Trace pheromones, scent marks	–
Magnetic	–	Magnetic compass (e.g., in birds)	–

pattern. Perceived change of egomotion speed (i.e., ego-acceleration) is also measured by asking subjects to keep a constant speed under changing environmental conditions (e.g., Snowden, Stimpson, and Ruddle 1998; Ott et al. 2016).

2.2.2 Sensory Cues to Egomotion

Sensory cues to egomotion may arise from all sensory modalities and a fair number of mechanisms. We will not discuss the physiology of the individual types of receptors and sense organs involved; readers are directed to the standard textbooks of neuroscience (e.g., Kandel et al. 2021). Rather, we will focus on the different types of information provided and the role each type plays in spatial cognition; for an overview, see table 2.1. One general distinction introduced by Mittelstaedt and Mittelstaedt (1972–1973) is the one between "idiothetic" or self-induced cues that are generated by the organism itself (i.e., independent of environmental stimuli) and "allothetic" cues indicating egomotion or position with respect to some external reference. Idiothetic cues include the inertial signals sensed in the semicircular canals of the vestibulum in the inner ear, which provide information about circular acceleration and deceleration about all possible axes. In flies and other dipteran insects, similar information is provided by so-called halteres, inertial sense organs that evolved from the hind-wings (Bender and Frye 2009). Self-induced cues are also provided by the "proprioceptive" signals indicating the actions of the motor apparatus (muscle spindles, Golgi tendon organs, etc.). Finally, the motor commands controlling bodily movements, or related signals known as "efference copies" or "corollary discharges," also provide idiothetic information about egomotion, although, strictly speaking, they are not sensory cues.

Allothetic cues can be further broken down into cues to static position and orientation, on the one hand, and direct cues to egomotion or ego-acceleration, on the other. In the vestibular modality, the most important allothetic cue is probably the sense of the direction of gravity, which allows to relate the egocentric coordinate scheme to stable world coordinates. Similar information for arbitrary directions is provided by distant landmarks relative to which egomotion can be judged. These may be visually perceived items such as distant objects ("beacons"), the sun azimuth in relation to the time of day, or the center of rotation of the starry sky used by night-migrating birds to keep their direction. Such cues are usually called "compasses" because they provide an approximate estimate of geo-centered direction. Magnetic compasses have also been found in a number of species and do play a role in bird migration. In the auditory modality, moving sound sources have been shown to induce vection, cyclovection, and even reflectory eye movements called audiokinetic nystagmus (Väljamäe 2009). Also, infrasound from distant sources has been suggested to play a role, for example, in pigeon homing (Hagstrum 2016).

Surface slope is another allothetic cue that defines local orientation; it may also contribute to place recognition in the sense that a city square with a slope will hardly be confused with a level square, even if the surrounding buildings look similar. Slope can be perceived in various ways, including vision, walking (or cycling) effort, or proprioception of the angle between foot and leg during stance or stride (Proffitt, Creem, and Zosh 2001; Restat et al. 2004; Laitin et al. 2019). In rodents, it has been shown that slope information is employed also in complete darkness and affects the spatial specificity of place cell firing (Moghaddam et al. 1996; Jeffery, Anand, and Anderson 2006).

Direct allothetic cues to egomotion are provided by optic flow in the visual domain, Doppler shifts in active auditory sensing (e.g., in bats) and stepping-dependent pressure patterns sensed in the soles of the feet. In humans, the most important of these is probably optic flow, which will be discussed in some detail below.

The multitude of sensory cues to egomotion and the many different percepts related to egomotion raise the question of cue integration or sensor fusion. We will give a first account of cue integration in egomotion perception in section 2.6, but issues of multisensor integration will come up in many other contexts.

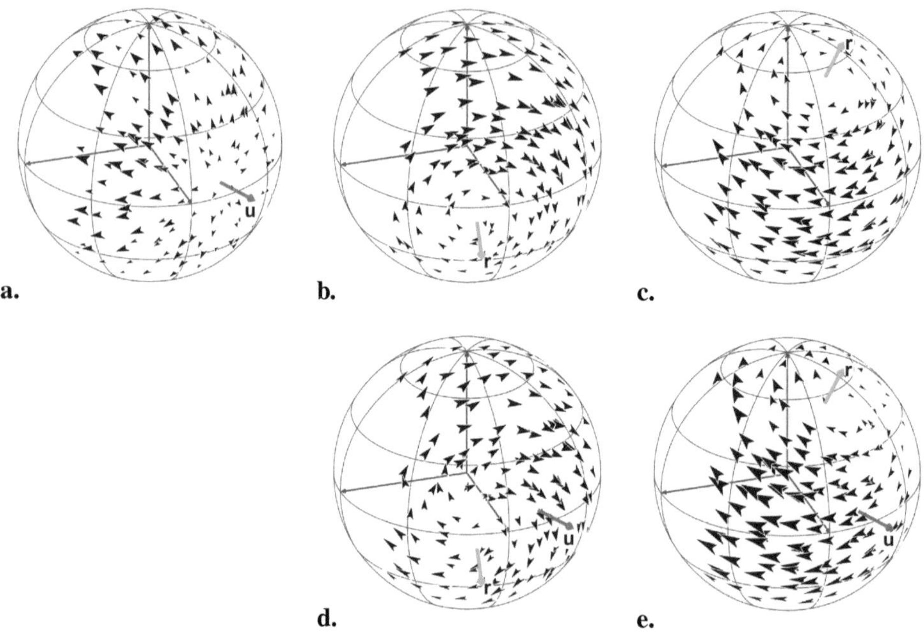

Figure 2.2
Structure of optic flow. Red and green arrows show vectors of translation and rotation (***u*** and ***r***) from equation 2.2. Each panel shows the view sphere of the observer with current axes of translation and rotation, if any. The observer is at the center of the sphere, and the axes of the egocentric coordinate frame are shown inside the sphere. The small arrows on the surface of the sphere show image motion of a random cloud of dots located in front of the observer. (a) Pure translation in direction ***u***. All flow vectors emanate from the focus of expansion, which coincides with the intersection of ***u*** and the view sphere. (b) Pure rotation about an axis close to the observer's forward direction (roll movement). The flow field is a vortex about that axis. (c) Pure rotation about an axis close to the observer's vertical axis (yaw movement). The flow field in the forward direction is laminar toward the right. (d) Combined flow with the pure movements from parts a and b. The flow field forms a spiral pattern and vanishes at a point between the axes of rotation and translation. (e) Combined flow with the pure movements from parts a and c. Note the absence of a focus of expansion.

2.3 Optic Flow

2.3.1 James J. Gibson and "Direct Perception"

Optic flow is the pattern of visual motion on the visual field generated by observer motion in a static environment but also in the presence of "independently" moving environmental objects. The importance of optic flow for the control of behavior was first discovered by J. J. Gibson (1950), who used it as a study case for his ideas in ecological psychology and direct perception. Optic flow is ubiquitous when an observer moves but is most obvious in the case of flying agents such as insects, birds, or airplane pilots. In one example, Gibson, Olum, and Rosenblatt (1955) argue that the perceptual judgments needed to land an aircraft on a runway are judgments of "surface approach" derived directly from the "motion perspective" (optic flow) without taking a detour via percepts such as egomotion, depth, and three-dimensional layout of the environment. The perception of surface approach is part of a feedback loop including also the pilot's actions and the stimulus updates resulting from each action. Aircraft landing thus proceeds in an "action–perception cycle" in which intervening perceptions of motion or depth are not needed and might indeed be absent. Perception is "ecological" (Gibson 1979) in the sense that it concerns only those aspects of the environment that are relevant for the task of landing the airplane.

Descriptions of behavioral maneuvers as feedback control of some simple perceptual parameter have been very fruitful, especially in the context of course control and simple navigation (see, for example, Srinivasan 1998; Warren 2006), and provide an approach to defining minimal or most parsimonious mechanisms of behavior. We will come back to this issue in chapter 4.

Gibson's idea of direct perception posits that perceptual judgments are obtained without the cognitive representation of, or conscious awareness about, intervening variables such as motion and depth in the above example. This does not mean, however, that such variables are also absent as intermediate stages of computation or neural processing (Ullman 1980). Indeed, neurons responding to both egomotion and object motion have been found in various areas of the brain and play an important role in the interaction of visual and vestibular cues (Smith et al. 2017). In optic flow computation, all known algorithms rely on the initial computation of local image motion, and most generate and use information about object depth (Raudies and Neumann 2012). Thus, some "intervening variables" seem to be involved as a computational necessity, even if they do not reach the level of conscious awareness in human observers.

2.3.2 Optic Flow and Perspective

The rules of optic flow are closely related to ordinary perspective, as Gibson's original term "motion perspective" clearly indicates. For a formal analysis, we start by considering the object points in three-dimensional space surrounding the observer. As the observer makes a translational movement, described by a single motion vector $v\mathbf{u}$, the object points expressed in the egocentric coordinate frame will move by $-v\mathbf{u}$. For each object point, this generates a linear motion trajectory and thus a family of straight, parallel lines in space along which the object points appear to be moving. The perspective image of this family of straight lines is a radial bundle of rays emanating from a central point. This point is the perspective vanishing point of the family of parallels and also the locus of the image of all object points on the observer's line of motion. In optic flow, we do not see the parallels themselves as image structures but as the trajectories along which the images of objects points are "flowing." The vanishing point defined by this family of parallels is known as the focus of expansion or contraction, depending on the direction in which the observer moves.

The situation is nicely illustrated with the use of a view sphere surrounding the observer and moving along with it; it is also known as Gibson's "ambient optical array" (see figure 2.2). The actual visual field is a subset of this sphere defined by the observer's visual perimeter or field of view. In simple translation, two vanishing points exist on the view sphere, one from which the flow lines are starting (the focus of expansion) and one on the opposite side of the sphere to which the flow lines converge (the focus of contraction). They are the images of the line $\lambda \boldsymbol{u}$: that is, of object points moving straight toward or away from the observer. Imaging of this central line simply amounts to taking the intersection with the view sphere, while the images of objects off this line are moving along meridians (halves of great circles of the view sphere) extending between the two foci as their poles.

Figure 2.2a shows an example of a purely translational flow field. The red arrow marks the direction of translation $\boldsymbol{u}$ with its starting point put at the focus of expansion. The local motion vectors are shown for a cloud of random dots hovering in front of the observer (not shown in the figure). All motion vectors point away from the focus of expansion. In agreement with the perspective nature of the projection, flow vectors for more distant points will be shorter than flow vectors of closer points, leading to discontinuities in the flow field if adjacent image points result from objects at different depth. Flow fields of smooth surfaces without occlusions are continuous.

Assume that we have a discrete set of object points from which local vectors of image motion can be measured. Let us denote the visual direction (i.e., the direction from the observer to the object point) of the ith object point by the unit vector $\boldsymbol{x}_i$ and its distance by d_i. Thus, $d_i\boldsymbol{x}_i$ is the three-dimensional position of the object point and $\boldsymbol{x}_i$ its image on the unit sphere surrounding the observer. As pointed out before, all object points will be moving with the vector $-v\boldsymbol{u}$ when expressed in the moving coordinate system of the observer. The local flow vector measured at $\boldsymbol{x}_i$ will then have the direction of the component of $-v\boldsymbol{u}$ orthogonal to $\boldsymbol{x}_i$: that is, tangential to the view sphere, $-v(\boldsymbol{u} - (\boldsymbol{u} \cdot \boldsymbol{x}_i)\boldsymbol{x}_i)$, where $(\cdot)$ denotes the dot product; see figure 2.3a:

$$\boldsymbol{f}_{T,i} = -\frac{v}{d_i}(\boldsymbol{u} - (\boldsymbol{u} \cdot \boldsymbol{x}_i)\boldsymbol{x}_i) \qquad (2.4)$$

(Koenderink and van Doorn 1987). In this notation, the translatory flow vector $\boldsymbol{f}_{T,i}$ has three dimensions, but since it is always orthogonal to $\boldsymbol{x}_i$, it will be tangent to the view sphere, and this comprises only two degrees of freedom, as would be expected for an image motion. The fraction v/d_i is also called the "relative nearness" of the object point and increases as the point gets closer or egomotion speeds up. Note that v and d_i enter the equation only in this fraction and therefore cannot be measured independently.

The geometrical argument can also be made for rotation; see figure 2.3b. If the observer rotates about an axis $\boldsymbol{r}$, the trajectories of the object points in the egocentric system will be circles on planes orthogonal to $\boldsymbol{r}$. The intersection points of $\boldsymbol{r}$ with the view sphere mark

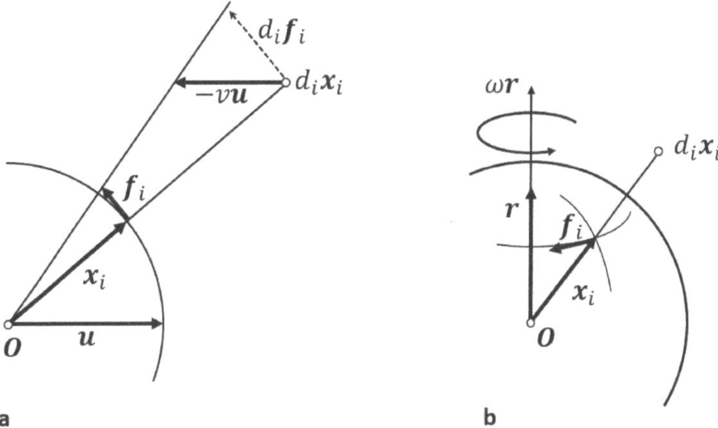

Figure 2.3
Derivation of the optic flow equation. The observer is located at $\boldsymbol{O}$ and sees an object point in direction $\boldsymbol{x}_i$, at an unknown distance d_i. The resulting flow vector is $\boldsymbol{f}_i$. (a) Translation (cf. equation 2.4). If the observer is moving with velocity v in direction $\boldsymbol{u}$, the object point at $d_i\boldsymbol{x}_i$ will move in the observer's coordinate system by $-v\boldsymbol{u}$. Of this movement, the observer can not perceive the component in direction of $\boldsymbol{x}_i$ itself, $-v(\boldsymbol{u}\cdot\boldsymbol{x}_i)\boldsymbol{x}_i$, but only the orthogonal component denoted as $d_i\boldsymbol{f}_i$. We divide by d_i and obtain Equation 2.4. (b) Rotation (cf. equation 2.5). The observer is rotating with velocity ω about an axis $\boldsymbol{r}$. In this case, the flow vector does not depend on d_i and can be obtained by the cross-product of $\omega\boldsymbol{r}$ and $\boldsymbol{x}_i$. If the observer is rotating to the left (counter-clockwise about an upright axis), the image will rotate to the right.

vortices around which the flow vectors are circulating. More general, the images of flow trajectories on the view sphere are circles of latitude, relative to the two vortices defined by the intersection of $\boldsymbol{r}$ with the sphere. Figure 2.2b,c show flow fields for two different orientations of the axis of rotation. If it is close to the forward direction (roll movement), the vortex will be inside the field of view and a curly or circulatory flow field will be seen. If, however, the axis is almost vertical (yaw movement), the equatorial region of the pattern will come into sight, resulting in a roughly constant or "laminar" field of flow vectors all going in about the same direction. If rotation is about the left–right axis (i.e., pitch) a laminar flow up or down will be seen in the forward direction. With the notations as before, we obtain the retational flow as

$$\boldsymbol{f}_{R,i} = -\omega\boldsymbol{r}\times\boldsymbol{x}_i. \tag{2.5}$$

Note that for rotation, the flow field is independent of the object distances d_i. The reason for this is that if we turn our head, the angular motion of an image point will simply equal the turning angle, irrespective of the distance of the imaged point. Alternatively, one can consider the motion of the object point expressed in egocentric world coordinates. This will increase with the distance of the point. However, this increase is exactly compensated as soon as image motion is computed by perspective projection, which involves a division by object distance.

Since rotational optic flow is independent of object distance, it does not contain information about the three-dimensional structure of the scene. Pure rotations about an axis through the center of projection (the front nodal point of the eye) do not involve any parallax and therefore do not change the lines of sight to the environmental objects. Note, however, that rotations of the head always involve a translational component at least in one eye and that even monocular eye rotations within the static orbit do not generate pure rotational fields, since the center of rotation of the eye and the front nodal point are separated by some 4 mm.

As pointed out before, egomotion can instantaneously be decomposed into a pure rotation and a pure translation. As a consequence, the flow field of a combined motion is simply the sum of the above two components, $\boldsymbol{f}_i = \boldsymbol{f}_{T,i} + \boldsymbol{f}_{R,i}$. Examples are shown in figure 2.2d, e. There is no necessary relation between the directions of translation and rotation. Depending on these directions, the flow pattern can change substantially. For example, the focus of expansion, which marks the direction of egomotion in pure translational fields, need not exist at all in combined fields or may occur at other image positions.[2] Therefore, the task of inferring egomotion from optic flow fields is often thought to amount to separating the rotational and translation components in the first place. This done, the instantaneous translation and rotation components of egomotion can be measured as the positions of the foci and vortices of the respective optic flow components.

2.3.3 Egomotion from Optic Flow: Computational Theory

Recoverable degrees of freedom: The scale–velocity ambiguity Of the six degrees of freedom of egomotion, only five are reflected in the optic flow pattern and can therefore possibly be recovered. Translation speed cannot be determined because its influence on optic flow is scaled by another unknown, the distance of the object points. Thus, to a given optic flow pattern, an identical pattern would arise if the observer would move with doubled speed in an environment of doubled size. Mathematically, this problem is apparent from equation 2.4, where translation speed v and object distance d_i enter only by their ratio. If both are multiplied with a scaling factor, this would cancel out and the flow pattern remains unchanged. In order to get an estimate of speed, an independent measurement of at least one object distance is therefore needed as a calibration. In humans, this is often assumed to be the eye height above ground (e.g., Frenz and Lappe 2005). Egomotion speed can then be determined in multiples of the calibration length per time.

2. As a consequence of the so-called hairy-ball theorem of algebraic topology, a continuous flow field on the sphere will generally have at least two foci or two vertices. This does not mean, however, that combined fields with rotational and translational components will always have a focus of expansion. Also, flow fields in cluttered environments are not continuous, due to depth discontinuities and independent object motion.

The remaining five degrees of freedom are the direction of translation $\boldsymbol{u}$ (i.e., the heading), the axis of rotation $\boldsymbol{r}$, and the angular velocity ω. In the following, we will give a brief overview of computational approaches for their recovery, with a focus on algorithms that have been considered also in the cognitive and behavioral sciences. For a more comprehensive overview covering also the machine vision literature, see Raudies and Neumann (2012) and textbooks of robot vision.

Direct solution of the optic flow equations In the presence of both translation and rotation, optic flow is determined by the sum of equations 2.4 and 2.5: that is, by the linear superposition of the translational and rotational components. For each feature point and local flow vector, the pair $(\boldsymbol{x}_i, \boldsymbol{f}_i)$ thus gives a two-dimensional constraint equation plus the relative nearness v/d_i to the feature point as an additional unknown. Therefore, the measurement of five well-placed motion vectors should in principle suffice to determine the five unknown degrees of freedom of egomotion plus the five unknown nearnesses of the used feature points. If more feature points are available, as is usually the case, better estimates can be obtained by standard numerical procedures for solving algebraic equations. This approach has been worked out for example by Koenderink and van Doorn (1987) using the spherical coordinates employed also here, or by Longuet-Higgins and Prazdny (1980) using a perspective camera model with a planar image. Heeger and Jepson (1992) have pointed out that for n local motion estimates (feature points), the $2n$-dimensional vector space of possible image flow patterns can be decomposed into two subspaces, one of which contains only the variability resulting from different translation directions. In this space, a least squares solution for translation can be found and used to recover the remaining unknowns.

The epipolar plane constraint An alternative approach pioneered by Longuet-Higgins (1981) uses a simple geometrical observation known as the epipolar plane constraint (figure 2.4a). If a given object point $\boldsymbol{p} = d_1\boldsymbol{x}_1$ is imaged in two subsequent time steps, the two image points $(\boldsymbol{x}_1, \boldsymbol{x}_2)$ or the respective view lines in the three-dimensional world are included in the epipolar plane defined by the moving center of projection and the object point. The three vectors $\boldsymbol{x}_1, \boldsymbol{x}_2$, and $\boldsymbol{u}$ are therefore coplanar, which can be expressed by saying that the triple product vanishes: $(\boldsymbol{x}_1 \cdot (\boldsymbol{x}_2 \times \boldsymbol{u})) = 0$. This equation provides a constraint for observer motion and object distance if we express $\boldsymbol{x}_2$ by means of the rigid motion transformation, equation 2.1. Longuet-Higgins (1981) suggested an algorithm based on eight points and their image correspondences, but more efficient algorithms have been suggested since. For review, see Raudies and Neumann (2012).

Exploiting field discontinuities A third idea for the decomposition of flow fields into rotatory and translational parts is based on the observation that rotational flow fields do not depend on the distance of the imaged object point (figure 2.4b). During rotation, the ray from the center of projection to the object point does not change, and image motion is given

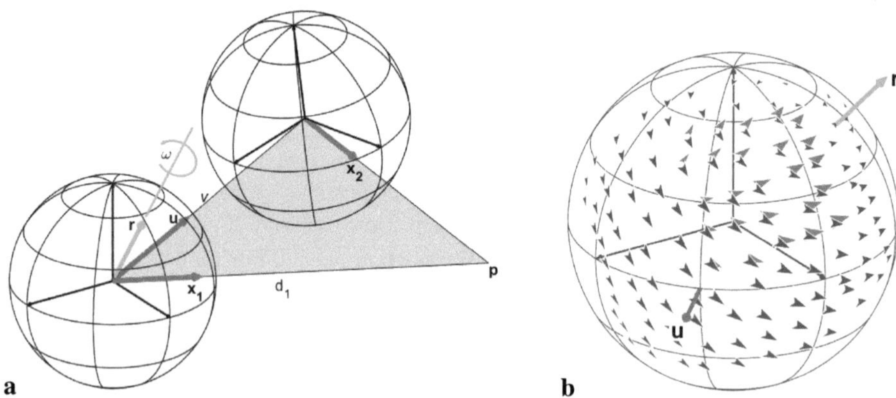

Figure 2.4
Two approaches to optic flow computation. (a) Epipolar plane constraint. The vectors $\boldsymbol{x}_1$, $\boldsymbol{x}_2$, and $\boldsymbol{u}$ are coplanar. Since the egocentric representations of $\boldsymbol{x}_1$ and $\boldsymbol{x}_2$ are related via the egomotion parameters of translation and rotation as well as by the object distance d, coplanarity provides a constraint for these parameters. (b) Flow field resulting from two sets of object points, one close to the observer (shown in blue) and one further away (shown in brown). While each of the blue and brown fields is continuous in itself, the combination is not. Discontinuities are due to the translational field component only. The local translational field can therefore be estimated as the difference of nearby flow vectors, and from this, the heading direction can be obtained.

by the length of the arc on the view sphere drawn by the ray as the view sphere rotates. Rotational fields are therefore continuous. If the motion vectors at two adjacent image points differ substantially, this must therefore result from the translational flow components, combined with a depth difference between the two object points (Longuet-Higgins and Prazdny 1980; Rieger and Lawton 1985). Since the two field components superimpose linearly and the rotational part does not depend on object distances, one can simply subtract the two motion vectors to obtain the local direction of the translational component. It will point away from the translational focus of expansion, which can thus be determined from three or more difference vectors. Once the global pattern of translational flow is known, it can again be subtracted from total flow to recover the rotational part. The algorithm predicts that egomotion perception should be better if neighboring features with strong depth differences occur (Hildreth 1992).

Template matching The above approaches to the recovery of egomotion from optic flow are motivated by the geometric and algorithmic structure of the problem and provide numerically efficient and reliable solutions. They show that solutions are possible and make predictions about expectable errors. It is not always clear, however, how operations such as the minimization of squared differences, matrix inversions, or triple vector products might be implemented in the brain. This is different in template-based approaches in which optic flow analysis is treated as a problem of pattern recognition. We first give a brief account of the idea and discuss neural mechanisms in the next section.

Detectors for characteristic types of egomotion are based on "templates": that is, typical flow patterns such as foci of extension or compression, vortices, or laminar flow areas appearing in different parts of the visual field. Global templates or matched filters that apply for just one particular egomotion can be defined for the entire view sphere and can be used to probe the presence of this very egomotion (Franz, Chahl, and Krapp 2004). Smaller templates depicting local patches of optic flow patterns can be combined to allow for more flexibility in the estimation of egomotion parameters with a smaller number of templates (Perrone and Stone 1994). Simple template-matching algorithms do not estimate relative nearnesses and will therefore confuse the narrowing of a street or corridor with increasing observer speed. This behavior has indeed been demonstrated in human observers by Festl et al. (2012).

2.4 Neural Mechanisms

Single neuron recordings have been performed in various animal species and seem to support the idea of template matching or recognition of optic flow patterns characteristic of specific egomotion cases. For example, blow flies of the genus *Calliphora* possess so-called tangential neurons in a part of their brain known as the lobula plate. The receptive fields of these neurons span almost the full visual sphere and react specifically to patterns of motion as they occur in optic flow. The system consists of some fifteen neurons, each reacting preferably to rotational motion about a specific axis such that together they form a population code for body rotation. Krapp and Hengstenberg (1996) measured the response properties of such neurons for many locations in the visual field by studying the reaction to point stimuli moving in all possible directions. For each point $\boldsymbol{x}_i$ in the visual field, let $\phi(\boldsymbol{x}_i)$ denote the motion direction eliciting the strongest response: that is. the neuron's locally preferred direction of motion. For each visual field direction $\boldsymbol{x}_i$, $\phi(\boldsymbol{x}_i)$ is a two-dimensional vector tangential to the view sphere.

Together, the locally preferred directions form a vector field that is similar to the optic flow field generated when the fly rotates about the neuron's preferred axis. If the fly does move, an optic flow will be generated, as described by the vectors $\boldsymbol{f}_i$ from equation 2.5, and the tangential neuron will compute the similarity between the actual flow pattern and its preferred directions as $\frac{1}{n}\sum_i^n(\boldsymbol{f}_i \cdot \phi(\boldsymbol{x}_i))$, where n is the number of visual field locations considered and $(\cdot)$ denotes the dot product (Franz, Chahl, and Krapp 2004). The neuron would give a strong response if the encountered flow field $(\boldsymbol{f}_i)_i$ matches its template and will thus signal that the fly is currently rotating about that neuron's preferred axis.

In addition to these matched filters for rotational flow fields, neurons with similar specificities for translational movements have also been reported. While translational flow components will be small for insects flying in open space where objects are far away, translational flow fields do play a role, for example, in landing maneuvers. For a recent review, see Mauss and Borst (2020).

Neurons reacting specifically to patterns of optic flow have also been found in many vertebrates, including bony fish (Kubo et al. 2014) and birds (Frost 2010). In primates, flow-specific neurons have been reported in the medial superior temporal (MST) area of the visual cortex which receives its input mostly from the medial temporal (MT) cortex and builds its specificity on the processing of local image motion carried out there. MST is part of the "dorsal" (parietal) stream of visual processing concerned with spatial and procedural information ("where," "how"), which can be contrasted to the "what" type of processing in the "ventral" or temporal areas.[3]

Duffy and Wurtz (1991) used three types of optic flow stimuli, called circular (showing a rotational vortex), radial (showing a focus of expansion or compression), and planar (showing parallel flow vectors) to characterize MST neurons, and found a large variety of mixed specificities (for review, see also Orban 2008). Receptive fields are much larger than in earlier visual areas, spanning on average some 60 degrees of visual angle. Note, however, that this is still substantially smaller than the receptive fields covered by tangential neurons of the fly.

The mixed specificities for a wide variety of motion components may be related to the complex structure of optical flow in cluttered environments where the self-generated background pattern of optic flow is interrupted and obscured by independently moving objects. In this case, large templates extending over the entire view sphere would be of little help, since evidence for egomotion and independent object motion is scattered and has to be collected from various image patches. Neural networks for estimating both egomotion and independent motion with MST-like neurons have been presented by Perrone and Stone (1994) and Layton and Fajen (2016).

MST neurons are not exclusively visual but can also be driven by vestibular input and do indeed integrate egomotion information from the two sensory modalities (Gu et al. 2006). For experimental assessment, physical egomotion is generated using a motion platform with six hydraulic legs. The geometry of this platform is that of an octahedron with two nonadjacent faces forming the ground plane and top of a table while the six hydraulic legs mark the edges between these two faces. By adjusting the lengths of the telescopic legs, the platform can be moved in all six degrees of freedom, albeit for small amounts (see

3. In primates, visual information is processed in a sequence of areas in the visual cortex and subsequently in two parallel processing "streams." One of these leads into parietal cortex (i.e., upward) and is therefore called the "dorsal stream"; it is concerned with tasks of object localization ("where?") and handling ("how?"). The other one proceeds into the temporal lobe (i.e., more ventral) and is concerned with object recognition tasks ("what?"); see Kravitz et al. (2011) and the textbooks of neuroscience.

figure 2.7a). While seated on this platform, a monkey watches a projection screen fixed to the platform on which optic flow patterns can be displayed. Thus, visual and vestibular cues can be presented either in isolation—that is, by presenting motion in darkness or visual stimulation at rest—or in combination. Experiments show that neurons in the dorsal part of area MST are generally tuned to visual egomotion cues and that a subpopulation is also tuned to vestibular cues. The tuning curves to both cues may or may not be overlapping. Neurons integrating visual and vestibular information are also found in the ventral intraparietal area (VIP) further downstream the dorsal path of visual processing (Chen, DeAngelis, and Angelaki 2011).

2.5 Performance

The discussed algorithms for the extraction of egomotion from optic flow make slightly different predictions about the performance in different experimental situations. Here we summarize evidence for the involvement of optic flow in egomotion perception. Psychophysical evidence favoring one or another of the sketched computational approaches, however, remains largely elusive.

2.5.1 Heading

In translational optic flow, the location of the focus of expansion marks the direction of heading. This cue to heading can easily be used if the subject is fixating a point in a stable visual direction: that is, a point not moving with the optic flow. The accuracy of heading judgments is on the order of 1 to 2 degrees if the optic flow pattern is radial (i.e., if the focus of expansion is contained in the field of view) and decreases quickly if the flow field gets more laminar (i.e., if the viewing direction is sideways to the direction of heading); see Warren, Morris, and Kalish (1988) and Royden, Banks, and Crowell (1992). Heading perception works for both foveal and peripheral presentation of the stimulus without strong differences in accuracy.

In natural viewing (for example, when driving in a car) the viewing direction will generally not be stabilized to the forward or any other direction but will follow some environmental feature along with the optical flow. The resulting eye movement is a smooth pursuit: that is, the fixation of a moving object. In this case, the optic flow field has a rotational component caused by the turning of the eyes, while the flow vector at the fovea will be zero. The flow field surrounding the fovea may look radial, mimicking a focus of expansion but in this case this "focus" is not indicative of the heading direction.[4] Still, subjects are usually able to judge perceived heading.

Heading perception in the presence of eye movements could in principle be achieved by decomposing the optic flow field into its translational and rotational parts as described

4. Computationally, it differs from a true focus of expansion in that the flow lines are constantly changing. That is to say, the flow field of a moving observer fixating a stable target is non-stationary.

above. Alternatively, it could be based on additional, "extra-retinal" motion information such as proprioceptors in the eye muscles or the motor signals causing the eye movements in the first place: that is, efference copies holding the information about the self-generated rotation. Royden, Banks, and Crowell (1992) simulated the flow patterns generated from combined translational and eye movements and presented them to observers whose eyes remained unmoved. In this case, the visual stimulus was identical to the pursuit situation, and the retinal-only hypothesis predicts that performance should be the same as before. However, results show a clear change in performance indicating that extra-retinal information is indeed needed to correctly interpret retinal flow patterns.

Heading perception from optical flow may also be affected by independently moving objects. In this case, the simple structure of the flow pattern with a focus of expansion in the heading direction is complicated by the moving objects, which may themselves generate a focus or occlude other foci generated by the background stimuli. Psychophysical measurements indicate that the effect of independent motion of small objects on perceived heading is weak (Royden and Hildreth 1996). Recent evidence further suggests that robustness against independent motion is not achieved by segmenting and ignoring the object-related parts of the field but simply by averaging over the entire field of view (Li et al. 2018). In addition, the influence of object motion seems to be modulated by additional vestibular egomotion information (Dokka et al. 2019).

Direct evidence for the usage of heading information in a navigation task has been presented in an experiment where human subjects were asked to walk toward a visual target (a door), while, in a virtual environment, they were constantly displaced sideways (Warren et al. 2001). The situation is reminiscent of crossing a river with a boat and trying to hit a landing place on the opposite side (figure 2.5). If the rowers keep heading toward the landing place, the boat will drift downstream, causing the rowers to compensate the resulting heading error by continuously turning upstream; eventually, they will cross the river in an arc (figure 2.5b). If, however, the rowers measure the actual motion of the boat, which in this case is also their egomotion, they will direct the boat somewhat upstream from the beginning and thus be able to cross the river in a straight line (figure 2.5a). The optimal turning angle for the boat can be determined by aligning the current focus of expansion with the target. In the experiment of Warren et al. (2001), the availability of optic flow information was varied by providing more or less features from which local motion vectors could be obtained. In a mostly empty environment, subjects navigated in arcs as is to be expected if they kept heading toward the target without a lead that would compensate for drift. With more flow information available, however, they produced straight paths, indicating that they did make use of the optic flow in path selection.

2.5.2 Time to Collision

As a consequence of the scale–velocity ambiguity discussed above, egomotion speed cannot be directly estimated from optic flow patterns. In the logic of Gibson's notion of direct

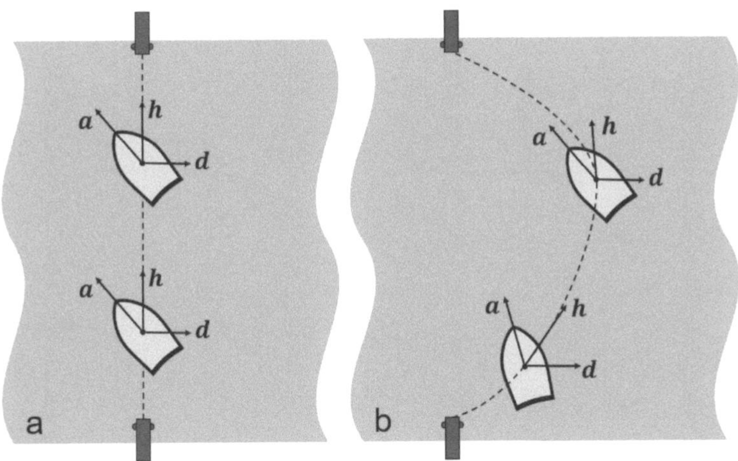

Figure 2.5
Approach under drift. (a) In order to row a boat from a start point (bottom) to a landing point (top) across a river drifting in direction ***d***, the boat has to be advanced in a direction upstream, ***a***. If correctly chosen, heading ***h*** will be in the direction of the landing place and the approach will be straight. (b) If the boat is advanced toward the landing site, it will undergo drift and advance has to be adjusted continuously. The trajectory will be curved, and heading will not align with the landing site. If optic flow cues to actual heading are available, human subjects are able to navigate the straight line.

perception, one may therefore look for other, recoverable parameters that are still useful for egomotion control. One such quantity is the time remaining until an obstacle will be hit, also known as the "time to collision", or τ, which was introduced by Lee and Kalmus (1980). In this quantity, the scale–velocity ambiguity is absent, since the unknown scaling factor cancels out. Consider an obstacle with size S at a distance x_1 from the agent (figure 2.6). Both the obstacle size and its distance will be unknown. From the triangle proportionality theorem, the size of the image of the obstacle is $s = f\,S/x$, where f is the focal length of the imaging system: that is, the distance between the camera nodal point and the image plane. Assume that the agent proceeds with a constant speed v, which again is unknown. The time to collision is then $\tau = x/v$. After some time interval Δt, the image size will be

$$s(\Delta t) = f\frac{S}{x - v\Delta t} = f\frac{S}{v(\tau - \Delta t)} \tag{2.6}$$

and we may calculate the derivative of image size with respect to time, $\dot{s}$, as

$$\dot{s}(\Delta t) = f\frac{S}{v(\tau - \Delta t)^2}. \tag{2.7}$$

From this, we see that

$$\frac{\dot{s}(0)}{s(0)} = \frac{1}{\tau}, \tag{2.8}$$

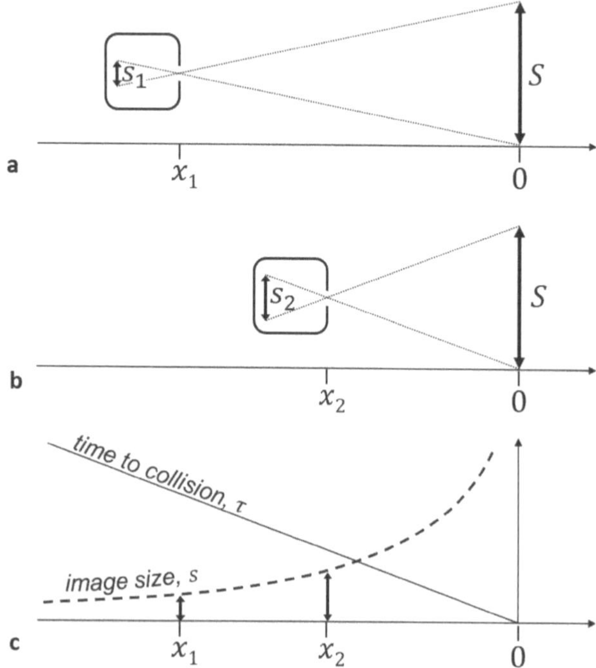

Figure 2.6
Time to collision. (a) Imaging geometry for an agent at position x_1 (camera nodal point) looking at an obstacle of size S at a distance x_1. The image size is s_1. (b) The agent has proceeded to a position x_2 closer to the obstacle. As a consequence, the image size is increased to s_2. (c) Image size and time to collision plotted as a function of agent position. The image size increases as a hyperbola with a pole at $x = 0$. If the agent is moving with a constant speed v, the time to collision decreases to zero according to $\tau = x/v$. As is shown in the text, τ can be inferred continuously from the instantaneous increase of s.

which means that time to collision can be calculated by dividing the current image size by the current change of image size and all unknown quantities cancel out. The quantity $\dot{s}/s$ is known as the expansion rate and can also be written as $d(\ln s)/dt$. Looking at figure 2.6c, the argument can be summarized by saying that when the obstacle is reached, its image size will become infinite. The time remaining until this happens is the time to collision and can be judged from the instantaneous rate of growth of the image size.

The argument can also be based on the flow field itself rather than on the size of the image. In this case, time to collision is related to the inverse of the divergence[5] of the

5. In the calculus of vector fields, the divergence is defined as $\mathrm{div}\,\boldsymbol{f} = (\nabla \cdot \boldsymbol{f}) = \partial f_1/\partial x + \partial f_2/\partial y$. It describes the infinitesimal area growth of a small patch whose outline moves with the vector field.

optic flow field, $\tau = 1/(\nabla \cdot \boldsymbol{f})$. As a less technical term for expansion rate or vector field divergence, the term "looming" is also found in the behavioral literature. Indeed, looming detection in insects has been shown to depend on local motion detection as an initial step (Schilling and Borst 2015).

One possible application of expansion rate τ in navigational behavior is braking or docking: that is, the control of speed to make smooth contacts. Lee (1980) shows that secure braking can be achieved by considering the derivative of τ with respect to time, also denoted as $\dot{\tau}$ (read "tau dot"). In unbraked collision as shown in figure 2.6, $\dot{\tau}$ would be –1, while smooth braking is achieved by keeping $\dot{\tau}$ above or equal to –0.5. This seems indeed to be the case in human braking behavior, but explanations of these findings not relying on expansion rate are possible (Yilmaz and Warren 1995; Rock, Harris, and Yates 2006). In real-world traffic situations, $\dot{\tau}$ is thought to play a role in the headway maintained by drivers from the car in front and the relation of this headway to current driving velocity (see Li and Chen 2017).

2.5.3 Speed and Acceleration

Egomotion can be perceived from vestibular cues alone (see Britton and Arshad 2019). In darkness, the perceptual threshold for the angular velocity in oscillatory yaw rotations lies on the order of 1 degree/s but is higher for roll and pitch rotations. Linear movements have thresholds on the order of 10 to 14 cm/s which are lowest for left–right movements and largest for the forward–backward direction.

In the visual modality, measurements of speed require an independent gauge with which the scale–velocity ambiguity can be resolved. In principle, this gauge can be provided in various ways, including a stereoscopic measurement of depth, the known size of the body or of body parts, or the distance walked in one step. Frenz and Lappe (2005) show that judgments of absolute speed from egomotion depend on the subjects' body size, indicating that eye elevation above ground is used.

In extended optic flow episodes, just one gauging quantity is sufficient to resolve the scale–velocity ambiguity for an entire optic flow sequence. This gauge does not change over time but can be used throughout. It is therefore well possible to detect changes of velocity (i.e., acceleration) even though velocity itself remains unknown. Optic flow can therefore be used to keep a constant speed as has been shown with simulated driving (Snowden, Stimpson, and Ruddle 1998). Interestingly, drivers increased speed when visibility dropped: that is, when simulated fog was added to the scene. This may mean that drivers underestimate velocity in low visibility and consequently speed up to keep perceived velocity constant. Perceived egomotion change in simulated driving also depends on the object distance: drivers in a narrowing tunnel report perceived acceleration even if they are actually moving with constant speed (Festl et al. 2012). Both findings are consistent with template-matching approaches to optic flow analysis, which rely on total optic

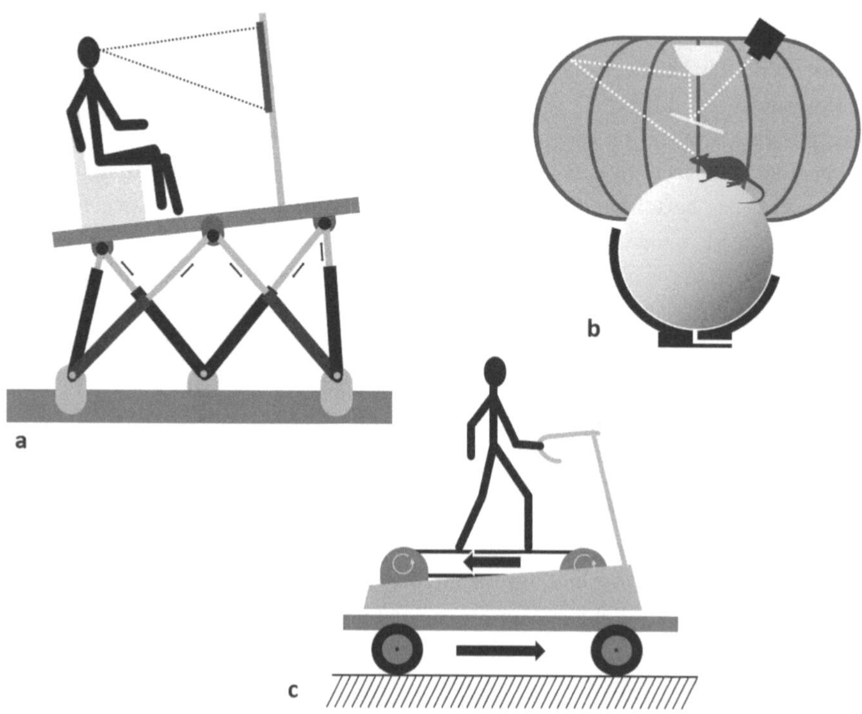

Figure 2.7
Experimental setups for the study of cue combination in egomotion perception. (a) Motion platform with six telescopic legs allowing small amplitudes of all six degrees of freedom. A subject is seated on the platform and watches a screen on which visual egomotion cues can be presented (e.g., Gu et al. 2006). (b) Air-cushioned ball with a rat running stationarily on top. Visual stimulation is provided by a panoramic projection screen (Hölscher et al. 2005). (c) Subject walking on a treadmill mounted on a trailer. This device separates consistent vision and vestibular cues from proprioception (Rieser et al. 1995).

flow rather than recovering the degrees of freedom of egomotion together with the object nearnesses.

2.6 Cue Integration

2.6.1 Types of Interaction

In table 2.1, we have listed a fair number of sensory cues that may play a role in the perception of egomotion. Some of these cues measure the same physical quantity as in the case of optic flow and Doppler shift of an acoustic signal, both of which measure egomotion speed, at least if the necessary scaling factors are assumed known. Such cues are called "commensurate" and can be accumulated by some sort of averaging. Other cues, such as the inertial sensors in the vestibulum, measure acceleration, which can be made commensurate with speed measurements by temporal integration. Indeed, integrator neurons subserving this task have been shown to exist in the vestibular pathway (Skavenski and Robinson 1973). Similarly, compass directions perceived, for example, as the bearing of a distant landmark can be made commensurate to body rotational speed by differentiation over time. In these cases, accumulation schemes may also be appropriate.

Another type of cue combination, called "cooperation" by Bülthoff and Mallot (1988), is necessary if the cues do not measure the same physical quantity but contribute to different and independent aspects of the same thing. An example for this in spatial cognition is the integration of landmark bearings, landmark distances, and egomotion estimates in a metric map of the environment. This can be achieved by the optimization of a composite cost function in which the true measurements are compared to predictions derived from the current state of the model. The model is then adjusted to better fit the data. This is the basic idea of the "simultaneous localization and mapping" (SLAM) algorithm in robotics where distance measurements to multiple targets are fitted to a model of target and observer location: that is, metric map of the environment (Durrant-Whyte and Bailey 2006; see also section 8.3.2). Cue integration here leads to the buildup of a map, as a joint representation of sensory information, in which the expected relations between the individual cues are included as part of a world model. From the point of view of cognitive evolution, it is interesting to note that the need to exploit also incommensurate sensory cues may have been an important driving force for the evolution of representations in the first place.

Other forms of cue combination such as disambiguation and vetoing occur in the context of landmark recognition. Consider, for example, the Statue of Liberty as a landmark. If you don't know where you are and encounter this monument, you will probably assume that you are in New York City. However, if you know that you have been walking just a short distance from the Eiffel Tower, you will recognize the place as the Île aux Cygnes in Paris. The problem of landmark confusion is called "aliasing" in robotics and can often be resolved ("disambiguated") by known context and distance information obtained from egomotion cues.

2.6.2 Separating the Cues

For experimental studies into the role of individual cues in an integrated percept, it is important to be able to vary individual cues independently such that they provide distinguishable information. In such setups, cues may become contradictory, which is often not the best experimental approach since subjects may notice the conflict and react in ways that differ from normal cue integration. Even silencing of one cue can cause conflict since it is generally not just a lack of sensory input but a cue to egomotion zero. A more subtle way to

Table 2.2
Sample experimental conditions for the study of three-way cue combination in egomotion perception. HMD: head mounted display as used in immersive virtual environments.

	Visual: $\hat{S}_v$	Vestibular (inertial): $\hat{S}_i$	Proprioceptive: $\hat{S}_p$
Self-controlled motion	$\hat{S}_v = \hat{S}_i = \hat{S}_p$		
Passive transportation	Transporter: $\hat{S}_v = \hat{S}_i$		$\hat{S}_p = 0$
Open-loop action plus HMD	HMD: $\hat{S}_v$	$\hat{S}_i = 0$	Motor action: $\hat{S}_p$
Walking without vision	$\hat{S}_v = 0$	Walking: $\hat{S}_i = \hat{S}_p$	
Treadmill on trailer	Trailer: $\hat{S}_v = \hat{S}_i$		Treadmill: $\hat{S}_p$

study cue integration is to vary the reliability of individual channels by adding some sort of noise, a situation that the sensory system will have to deal with also outside the laboratory.

Table 2.2 and figure 2.7 show typical experimental setups used in the study of cue integration in egomotion perception. The three channels considered are vision (i.e., optic flow or the changing distance to a landmark), vestibular or inertial signals providing "idiothetic" cues about the agent's acceleration and velocity, and proprioception (i.e., the motor commands sent to the locomotor apparatus and the feedback received from there). In natural walking situations, these three channels provide consistent information: that is, optic flow will be correlated with stimulation of the inner ear and sensations from the hip and knee joints.

If a subject is moved passively (e.g., sitting in the back-seat of a car, or being pushed in a wheelchair), proprioception is absent or at least unrelated to egomotion, while vision and the vestibulum provide consistent information about the true egomotion. The latter two can be separated by replacing veridical vision with a virtual reality presentation: for example, if a subject pushed in a wheelchair is also wearing a head-mounted display (HMD). If vestibular stimulation is to be provided for all degrees of freedom, the wheelchair can be replaced by a motion platform (figure 2.7a), as was already discussed in the context of neurons tuned to egomotion.

In open-loop action, subjects are performing a locomotor action, but this does not lead to an actual displacement. Vestibular input is therefore zero or at least unrelated to the egomotion normally generated by the motor action. This can, for example, be achieved by walking on treadmills, either one-dimensionally using an ordinary exercise device or in two dimensions using an omnidirectional treadmill (Souman et al. 2011). Visual information would then be provided via an HMD. Stationary bicycles, again from exercise equipment, have been used, for example, by Sun et al. (2003) or Restat et al. (2004). In rodents and other small animals, including insects, stationary walking can also be achieved by fixating the animal above an air-cushioned ball, which it can reach and turn with its legs (figure 2.7b and Hölscher et al. 2005). Intended egomotion can then be measured by tracking the ball movement while visual stimulation is provided via a panoramic projection.

Walking without vision is a simple but much-used technique in which a subject is blindfolded and asked to walk to a goal previously envisaged from the starting point. In this case, proprioception and vestibular cues are consistent and veridical, while vision is missing.

Finally, we mention a combination of passive transportation with open-loop walking achieved by putting a treadmill on a trailer that is pulled around by some motorized vehicle (figure 2.7c and Rieser et al. 1995). A subject walking on the treadmill will produce the motor action needed to stay on the treadmill; this motor action will also determine proprioceptive sensations. The vestibular and visual inputs, however, correspond to the speed of the trailer and are thus independent of motor action. We will discuss this experiment in more detail below.

With the advent of virtual reality (VR) in consumer electronics, many more devices have been developed in order to increase VR realism in all sensory channels. We do not attempt to give a full overview here. Clearly, these new devices may prove useful for further studies in cue combination and egomotion perception.

2.6.3 Examples

We now turn to a number of examples of cue integration in egomotion perception that also illustrate more general mechanisms of multisensory integration.

Probability summation Cue accumulation in detection tasks can be modeled as probability summation. For example, if a visual stimulus is viewed with two rather than with just one eye, visual acuity is normally increased, as is the probability of detecting a stimulus. If the probability of detecting a dim light with one eye is p, the probability of a miss is $1-p$. If the two eyes act independently, the probability of a miss in both eyes is $(1-p)^2$, which results in an overall detection probability of $1-(1-p)^2 = 2p-p^2 \geq p$. This simple rule seems to apply in the combination of visual and acoustic cues to cyclovection (the sense of being rotated). In a study by Keshavarz et al. (2014) subjects experienced rotating stimuli from each modality either individually or in combination. Acoustically evoked cyclovection was much weaker than the visually induced effect. Both cues combined led to a stronger perceived cyclovection as compared to the unimodal cues, indicating that the evidence from both channels was not averaged but combined in some sort of probability summation.

Cue accumulation If the value of a graded quantity is to be measured, as is the case in egomotion estimation from optic flow and proprioception, accumulation amounts to the calculation of a weighted average. Assume two measurements are made from the same stimulus S. We denote the measurements, or "sensations," as $\hat{S}_1$ and $\hat{S}_2$. The final percept will be determined as $\hat{S} = w_1\hat{S}_1 + w_2\hat{S}_2$ with suitable weights $w_1, w_2 \geq 0$ satisfying $w_1 + w_2 = 1$. We may now ask how the weights should be chosen to produce an optimal estimate. If the variances of the individual measurements are known, for example, by monitoring them over some period of time, an obvious idea would be to minimize the variance of the overall estimate. If $\hat{S}_1$ and $\hat{S}_2$ are statistically independent, the variance of $\hat{S}$ is given by $\mathrm{Var}(\hat{S}) = w_1^2\mathrm{Var}(\hat{S}_1) + w_2^2\mathrm{Var}(\hat{S}_2)$. Denoting the variances as σ^2, σ_1^2, and σ_2^2, it is easy to show that the weights minimizing σ are

$$w_1 = \frac{1/\sigma_1^2}{1/\sigma_1^2 + 1/\sigma_2^2} \quad \text{and} \quad w_2 = \frac{1/\sigma_2^2}{1/\sigma_1^2 + 1/\sigma_2^2} \tag{2.9}$$

and the variance of the combined "minimal variance estimator" is $\sigma^2 = \sigma_1^2\sigma_2^2/(\sigma_1^2 + \sigma_2^2)$, which is indeed less than each of the individual variances.

An alternative derivation of this equation can be given if the distributions of the variables $\hat{S}_1$ and $\hat{S}_2$ are assumed to be normal with the same expected value S and variances σ_1 and σ_2. If $\varphi(s) = (2\pi)^{-1/2}\exp\{-s^2/2\}$ denotes the standard Gaussian (density function of the normal distribution), the probability of measuring $\hat{S}_1$ and $\hat{S}_2$ is

$$P(\hat{S}_1, \hat{S}_2 \mid S) = \frac{1}{\sigma_1}\varphi\left(\frac{\hat{S}_1 - S}{\sigma_1}\right) \times \frac{1}{\sigma_2}\varphi\left(\frac{\hat{S}_2 - S}{\sigma_2}\right) \tag{2.10}$$

where $\times$ denotes multiplication. Treated as a function of S, this quantity is called the likelihood of S being the true value. It is the product of two Gaussians of width σ_1 and σ_2 centered at the measurements $\hat{S}_1$, $\hat{S}_2$. The S-value for which the likelihood function takes its maximum is called the maximum likelihood estimator (MLE) of S. It is easy to show that it is the weighted average with the same values derived above.

Maximum likelihood estimation leads the way to the Bayesian formulation of cue averaging, which allows one to model the accumulation of knowledge over multiple, also successive measurements from one or many channels. The idea is that some prior knowledge about the distribution of the parameter is available already before the measurement. This "prior" may be a general expectation about the outcome or the likelihood obtained from the last measurement. Mathematically, it takes the form of a distribution density $p(S)$, where S stands for the true value. If no prior knowledge exists, the prior is "flat": that is, the distribution is very wide. We now measure some data $\hat{S}$ and consider the "posterior probability distribution" given by its density

$$p(S \mid \hat{S}) = \frac{p(\hat{S} \mid S)p(S)}{p(\hat{S})}. \tag{2.11}$$

This equation, which is known as Bayes's formula, follows immediately from the definition of conditional probability, $p(S \mid \hat{S}) := p(S, \hat{S})/p(\hat{S})$. The density $p(S)$ is the density of the prior distribution, $p(\hat{S} \mid S)$ is the density of the expected distribution of the measurement for each value of S that takes into account sensor noise, and $p(\hat{S}) = \int p(\hat{S} \mid S)p(S)\mathrm{d}S$ is a normalizing factor not depending on S. After the measurement, $\hat{S}$ is fixed, and the maximum of the posterior distribution $p(S \mid \hat{S})$ is the resulting Bayes estimator. We can drop the conditional in the posterior and use it as a new prior in the next round of measurement. Bayesian combination of visual and haptic cues in the perception of object size has been presented by Ernst and Banks (2002). For applications of Bayesian theory in landmark-based place recognition, see Cheng et al. (2007), Mallot and Lancier (2018), and section 6.3.4.

In egomotion perception, the weighted average as a scheme for cue combination has been studied for the integration of optic flow and vestibular and proprioceptive information. While optic flow can easily be made independent of the other cues by presenting it through video goggles, the control of the body-based cues is sometimes difficult. Stationary cycling on an indoor training bike provides proprioceptive cues but no vestibular information consistent with the simulated egomotion (Sun et al. 2003). The same is true for walking on a treadmill. Vice versa, proprioceptive information is removed in passive driving, which is often realized using a wheelchair (e.g., Campos, Butler, and Bülthoff 2012). In both cited studies, linear averaging of the contributions of different sensory modalities is indeed confirmed.

Recalibration So far, we have assumed that the estimates from the individual channels are unbiased: that is, that their expected values are equal and coincide with the veridical value. For the combination of vision and proprioception, this means that the gauging factor relating both measurements is the optical flow generated by one step. This gauge factor is not a constant but subject to rapid or slower change, for example, if the subjects gets tired and starts to make shorter steps, or during growth in children when the length of the legs changes. The calibration of the gauge factor has been studied by Rieser et al. (1995); see figure 2.7c. Subjects are placed on a treadmill on which they walk with a speed specified by the speed of the conveyor belt. The whole treadmill is then placed on a trailer that can be driven around with independent speed. This speed will be perceived visually by the subject. Perceived egomotion speed is measured by an adjustment procedure known as "walking without vision." Subjects are placed in a corridor and view a target 8 meters away. Next they are blindfolded and asked to walk up to the goal. The produced distance is a direct measure of the visual-proprioceptive gain factor: how many steps are needed to walk the visually perceived distance.

In the experiment, subjects are exposed to artificial gauging factors by walking slowly on a fast-moving trailer (biomechanically slower) or by walking fast on a slowly moving trailer (biomechanically faster). After just eight minutes of adaptation, a strong effect in the walking-without-vision test is found: after adaptation in the biomechanically slower condition, subjects walk too far, because they adjusted their gain factor to indicate that fast walking generates only small visual distances and vice versa.

Recalibration is also found for rotatory movements in a study by Mulavara et al. (2005). In this case, subjects are required to step on a stepping board, keeping synchrony with a metronome. In the adaptation phase, subjects can control the direction of a simulated walk in virtual reality by body turns on the stepping board. If a bias is added to simulated heading such that simulated motion contains a left turn while the subjects are stepping straight, subsequent tests of straight walking without visual stimulation reveal a rightward bias of the subjects.

Key Points of Chapter 2

- Besides depth perception, the perception of egomotion is the central source of spatial information. The degrees of freedom of movement, three axes of translation and three axes of rotation, span a three-dimensional space of bodily position or position change.
- Although egomotion may be described sufficiently by instantaneous translations and rotations, a fair number of distinguishable percepts in egomotion have been described, including vection, heading, speed, cyclovection, postural sway, time to collision, and others.
- The perception of egomotion is based on a wide variety of sensory cues that can be classified into two groups: *idiothetic* cues are "self-generated": that is, they occur even in the absence of perceivable external objects. These are mostly the senses of movement and acceleration provided by the vestibular and proprioceptive systems. *Allothetic* cues arise from changes in the perception of external objects such as landmarks, smells, or floor cover and may occur in all sensory modalities.
- One important source of information is optic flow: that is, the self-generated pattern of image motion. It has been studied extensively in flying animals but is used for the judgment of heading and egomotion speed also in humans.
- Egomotion perception is also an example for multisensory integration. Important mechanisms include probability summation, Bayesian and maximum likelihood estimation, and recalibration.

References

Bender, J. A., and M. A. Frye. 2009. "Invertebrate solutions for sensing gravity." *Current Biology* 19:R186–190.

Britton, Z., and Q. Arshad. 2019. "Vestibular and multi-sensory influences upon self-motion perception and the consequences for human behavior." *Frontiers in Neurology* 10:63.

Bülthoff, H. H., and H. A. Mallot. 1988. "Integration of depth modules: Stereo and shading." *Journal of the Optical Society of America A* 5:1749–1758.

Campos, J. L., J. S. Butler, and H. H. Bülthoff. 2012. "Multisensory integration in the estimation of walked distances." *Experimental Brain Research* 218:551–565.

Chen, A., G. C. DeAngelis, and D. E. Angelaki. 2011. "Representation of vestibular and visual vues to self-motion in ventral intraparietal cortex." *The Journal of Neuroscience* 31:12036–12052.

Cheng, K., S. J. Shettleworth, J. Huttenlocher, and J. J. Rieser. 2007. "Bayesian integration of spatial information." *Psychological Bulletin* 133:625–637.

Dichgans, J., and T. Brandt. 1978. "Visual-vestibular interaction: Effects of self-motion perception and postural control." In *Perception. Vol. 8 of Handbook of sensory physiology,* edited by R. Held, H. W. Leibowitz, and H.-L. Teuber, 755–804. Berlin: Springer Verlag.

Dokka, K., H. Park, M. Jansen, G. C. DeAngelis, and D. E. Angelaki. 2019. "Causal inference accounts for heading perception in the presence of object motion." *PNAS* 116:9065.

Duffy, C. J., and R. H. Wurtz. 1991. "Sensitivity of MST neurons to optic flow stimuli. 1. A continuum of response selectivity to large-field stimuli." *Journal of Neurophysiology* 65:1329–1345.

Durrant-Whyte, H., and T. Bailey. 2006. "Simultaneous localization and mapping: Part I." *IEEE Robotic & Automation Magazine* 13:99–108.

Ernst, M. O., and M. S. Banks. 2002. "Humans integrate visual and haptic information in a statistically optimal fashion." *Nature* 415:429–433.

Festl, F., F. Recktenwald, C. Yuan, and H. A. Mallot. 2012. "Detection of linear ego-acceleration from optic flow." *Journal of Vision* 12 (7): 10.

Franz, M. O., J. S. Chahl, and H. G. Krapp. 2004. "Insect-inspired estimation of egomotion." *Neural Computation* 16:2245–2260.

Frenz, H., and M. Lappe. 2005. "Absolute travel distance from optic flow." *Vision Research* 45:1679–1692.

Frost, B. J. 2010. "A taxonomy of different forms of visual motion detection and their underlying neural mechanisms." *Brain, Behavior and Evolution* 75:218–235.

Gibson, J. J. 1950. *The perception of the visual world.* Boston: Houghton Mifflin.

Gibson, J. J. 1979. *The ecological approach to visual perception.* Boston: Houghton Mifflin.

Gibson, J. J., P. Olum, and F. Rosenblatt. 1955. "Parallax and perspective during aircraft landings." *American Journal of Psychology* 68:372–385.

Gu, Y., G. C. DeAngelis, and D. E. Angelaki. 2007. "A functional link between area MSTd and heading perception based on vestibular signals." *Nature Neuroscience* 10:1038–1047.

Gu, Y., P. V. Watkins, D. E. Angelaki, and G. C. DeAngelis. 2006. "Visual and nonvisual contributions to three-dimensional heading selectivity in the medial superior temporal area." *The Journal of Neuroscience* 26:73–85.

Hagstrum, J. T. 2016. "Atmospheric propagation modeling indicates homing pigeons use loft-specific infrasonic 'map' cues." *Journal of Experimental Biology* 216:687–699.

Heeger, D. J., and A. D. Jepson. 1992. "Subspace methods for recovering rigid motion I: Algorithm and implementation." *International Journal of Computer Vision* 7:95–117.

Hildreth, E. C. 1992. "Recovering heading for visually–guided navigation." *Vision Research* 32:1177–1192.

Hölscher, C., A. Schnee, H. Dahmen, L. Setia, and H. A. Mallot. 2005. "Rats are able to navigate in virtual environments." *Journal of Experimental Biology* 208:561–569.

Jeffery, K. J., R. L. Anand, and M. I. Anderson. 2006. "A role for terrain slope in orienting hippocampal place fields." *Experimental Brain Research* 169:218–225.

Junger, W., and D. Varjú. 1990. "Drift compensation and its sensory basis in waterstriders (*Gerris paludum* F.)." *Journal of Comparative Physiology A* 167:441–446.

Kandel, E. R., J. D. Koester, S. H. Mack, and S. A. Siegelbaum. 2021. *Principles of neural science.* 6th. McGraw-Hill.

Keshavarz, B., L. J. Hettinger, D. Vena, and J. L. Campos. 2014. "Combined effects of auditory and visual cues of vection." *Experimental Brain Research* 232:827–836.

Koenderink, J. J., and A. J. van Doorn. 1987. "Facts on optic flow." *Biological Cybernetics* 56:247–254.

Krapp, H. G., and R. Hengstenberg. 1996. "Estimation of self-motion by optic flow processing in single visual interneurons." *Nature* 384:463–466.

Kravitz, D. J., K. S. Saleem, C. I. Baker, and M. Mishkin. 2011. "A new neural framework for visuospatial processing." *Nature Reviews Neuroscience* 12:217–230.

Kubo, F., B. Hablitzel, M. Dal Maschio, W. Driever, H. Baier, and A. B. Arrenberg. 2014. "Functional architectre of an optic flow-responsive area that drives horizontal eye movements in zebrafish." *Neuron* 81:1344–1359.

Laitin, E. L., M. J. Tymoski, N. L. Tenhundfeld, and J. K. Witt. 2019. "The uphill battle for action-specific perception." *Attention, Perception & Psychophysics* 81:778–793.

Layton, O. W., and B. R. Fajen. 2016. "Competitive dynamics in MSTd: A mechanism for robust heading perception based on optic flow." *PLoS Computational Biology* 12 (6): e1004942.

Lee, D. N. 1980. "Visuo-motor coordination in space-time." In *Tutorials in motor behavior,* edited by G. E. Stelmach and J. Requin, 281–295. Amsterdam: North-Holland.

Lee, D. N., and H. Kalmus. 1980. "The optic flow field—The foundation of vision." *Philosophical Transactions of the Royal Society (London) B* 290:169–179.

Li, L., and X. Chen. 2017. "Vehicle headway modeling and its inferences in macroscopic/microscopic traffic flow theory: A survey." *Transportation Research Part C* 76.

Li, L., L. Ni, M. Lappe, D. C. Niehorster, and Q. Sun. 2018. "No special treatment of independent object motion for heading perception." *Journal of Vision* 18 (4): 19.

Longuet-Higgins, H. C. 1981. "A computer algorithm for reconstructing a scene from two projections." *Nature* 293:133–135.

Longuet-Higgins, H. C., and K. Prazdny. 1980. "The interpretation of a moving retinal image." *Proceedings of the Royal Society (London) B* 208:385–397.

Mach, E. 1922. *Analyse der Empfindungen.* 9th ed. Jena: Gustav Fischer Verlag.

Mallot, H. A., and S. Lancier. 2018. "Place recognition from distant landmarks: Human performance and maximum likelihood model." *Biological Cybernetics* 112:291–303.

Mauss, A. S., and A. Borst. 2020. "Optic flow-based course control in insects." *Current Opinion in Neurobiology* 60:21–27.

Mittelstaedt, H., and M.-L. Mittelstaedt. 1972–1973. "Mechanismen der Orientierung ohne richtende Außenreize." *Fortschritte der Zoologie* 21:46–58.

Moghaddam, M., Y. L. Kaminsky, A. Zahalka, and J. Bures. 1996. "Vestibular navigation directed by the slope of terrain." *PNAS* 93:3439–3443.

Mulavara, A. P., J. T. Richards, T. Ruttley, A. Marshburn, Y. Nomura, and J. J. Bloomberg. 2005. "Exposure to a rotating virtual environment during treadmill locomotion causes adaptation in heading direction." *Experimental Brain Research* 166:210–219.

Orban, G. A. 2008. "Higher order visual processing in macaque extrastriate cortex." *Physiological Review* 88:59–89.

Orger, M. B., E. Gahtan, A. Muto, P. Page-McCaw, M. C. Smear, and H. Baier. 2004. "Behavioral screen assays in zebrafish." In *Methods in cell biology,* 2nd ed., edited by H. W. Detrich III, L. Zon, and M. Westerfeld, 77:53–68. New York: Academic Press.

Ott, F., L. Pohl, M. Halfmann, G. Hardiess, and H. A. Mallot. 2016. "The perception of ego-motion change in environments with varying depth: Interaction of stereo and optic flow." *Journal of Vision* 16 (9): 4.

Perrone, J. A., and L. S. Stone. 1994. "A model of self-motion estimation within primate extrastriate visual cortex." *Vision Research* 34:2917–2938.

Peterka, R. J. 2002. "Sensorimotor integration in human postural control." *Journal of Neurophysiology* 88:1097–1118.

Proffitt, D. R., S. H. Creem, and W. D. Zosh. 2001. "Seeing mountains in mole hills: Geographical-slant perception." *Psychological Science* 12:418–423.

Raudies, F., and H. Neumann. 2012. "A review and evaluation of methods estimating ego-motion." *Computer Vision and Image Understanding* 116:606–633.

Restat, J., S. D. Steck, H. F. Mochnatzki, and H. A. Mallot. 2004. "Geographical slant facilitates navigation and orientation in virtual environments." *Perception* 33:667–687.

Riecke, B. E., J. Schulte-Pelkum, M. N. Avraamides, M. von der Heyde, and H. H. Bülthoff. 2006. "Cognitive factors can influence self-motion perception (vection) in virtual reality." *ACM Transaction on Applied Perception* 3:194–216.

Rieger, J. H., and D. T. Lawton. 1985. "Processing differential image motion." *Journal of the Optical Society of America A* 2:354–360.

Rieser, J. J., H. L. Pick, D. H. Ashmead, and A. E. Garing. 1995. "Calibration of human locomotion and models of perceptual-motor organization." *Journal of Experimental Psychology: Human Perception and Performance* 21:480–497.

Rock, P. B., M. G. Harris, and T. Yates. 2006. "A test of the tau-dot hypothesis of braking control in the real world." *Journal of Experimental Psychology: Human Perception and Performance* 32:1479–1484.

Royden, C. S., M. S. Banks, and J. A. Crowell. 1992. "The perception of heading during eye movements." *Nature* 360:583–585.

Royden, C. S., and E. C. Hildreth. 1996. "Human heading judgments in the presence of moving objects." *Perception & Psychophysics* 58:836–856.

Schilling, T., and A. Borst. 2015. "Local motion detectors are required for the computation of expansion flow-fields." *Biology Open* 4:1105–1108.

Skavenski, A. A., and D. A. Robinson. 1973. "Role of abducens neurons in vestibuloocular reflex." *Journal of Neurophysiology* 36:726–738.

Smith, A. T., M. W. Greenlee, G. C. DeAngelis, and D. E. Angelaki. 2017. "Distributed visual-vestibular processing in the cerebral cortex of man and macaque." *Multisensory Research* 30:91–120.

Snowden, R. J., N. Stimpson, and R. A. Ruddle. 1998. "Speed perception fogs up as visibility drops." *Nature* 392:450.

Souman, J. L., R P Giordano, M. Schwaiger, I. Frissen, T. Thümmel, H. Ulbrich, A. De Luca, H. H. Bülthoff, and M. O. Ernst. 2011. "CyberWalk: Enabling unconstrained omnidirectional walking through visual environments." *ACM Transactions of Applied Perception* 8 (4): 1–22.

Srinivasan, M. V. 1998. "Insects as Gibsonian animals." *Ecological Psychology* 10:251–270.

Sun, H.-J., A. J. Lee, J. L. Campos, G. W. Chan, and D.-H. Zhuang. 2003. "Multisensory integration in speed estimation during self-motion." *CyberPsychology & Behavior* 6:509–518.

Ullman, S. 1980. "Against direct perception." *Behavioral and Brain Sciences* 3:373–381.

Väljamäe, A. 2009. "Auditorily-induced illusory self-motion: A review." *Brain Research Reviews* 61:240–255.

Warren, R., and A. H. Wertheim, eds. 1990. *Perception and control of self-motion.* Hillsdale NJ: Lawrence Erlbaum Associates.

Warren, W. H. 2006. "The dynamics of perception and action." *Psychological Review* 113:358–389.

Warren, W. H., B. A. Kay, W. D. Zosh, A. P. Duchon, and S. Sahuc. 2001. "Optic flow is used to control human walking." *Nature Neuroscience* 4:213–216.

Warren, W. H., M. W. Morris, and M. Kalish. 1988. "Perception of translational heading from optic flow." *Journal of Experimental Psychology: Human Perception and Performance* 14:646–660.

Wist, E. R., H. C. Diener, J. Dichgans, and T. Brandt. 1975. "Perceived distance and the perceived speed of self-motion: Linear vs. angular velocity?" *Perception & Psychophysics* 17:549–554.

Yilmaz, E. H., and W. H. Warren. 1995. "Visual control of braking." *Journal of Experimental Psychology: Human Perception and Performance* 21:996–1014.

3 Peripersonal Space

In this chapter, we discuss the perception and initial understanding of the space surrounding us. This is largely a matter of perception, which will be the topic of the first part of this chapter. We will see that the sense of space is constructed from a number of cues providing various, often non-commensurate types of information. Also, perceptions of depth and width are vastly anisotropic, which is probably the reason for a surprising property of visual space: that is, its non-Euclidean metric. Imagery in the resulting representation of space is also a tool of spatial reasoning and prediction making ("mental travel"), which can be based on various forms of mental transformations. Of these, we discuss mental rotation, perspective taking, and spatial updating and give examples of experimental paradigms.

Spatial imagination is relevant also in relation to further topics such as path integration, spatial working- and long-term memories, path planning, or the perception of presence in self-consciousness. These will be postponed to later chapters.

3.1 A Behavioral View

When we say we are "here," as opposed to somewhere else, we do not mean that the center of mass of our body occupies a specific geometric point with well-defined coordinates and that we would be somewhere else as soon as we would move ever so slightly away from that point. Rather, being here means that we are "in a position" to do certain things such as responding to another person's call; taking care of arising needs in a household, work, or leisure environment, making decisions about the continuation of a route, and so on. "Here" thus denotes a part of the environment, together with the behavioral opportunities available therein. Reporting to be "here" always has a connotation of being ready to do something, be it from the social, object, problem solving, or any other domain of behavior. The originally spatial concept thus extends to a general notion of readiness in both spatial and nonspatial domains of behavior and cognition.

In chapter 1, we already introduced the term "peripersonal" space for the representation of the environment in which the current action is taking place (see also figure 1.3). As a provisional definition, we state the following points:

1. Peripersonal space includes what we can reach with the arms or a few steps of walking as well as more distant points that we can see from the current viewpoint. It also extends into the space behind us.
2. It is the result of multisensory as well as temporal integration, including information obtained from different viewpoints.
3. Egocentric reference systems such as distance from the observer and angle from the midline play an important role.
4. Peripersonal space is an active representation (i.e., a working memory) in the sense that we can imagine—or "simulate"—movements of ourselves as well as of the contained objects.

In this chapter, we will discuss aspects related to perception, geometry, and object and scene cognition in peripersonal space. For a discussion of peripersonal space in social cognition, see, for example, Hunley and Lourenco (2018).

3.2 Visual Space Cues

Visual depth (i.e., the position of objects and points in three-dimensional space) is inferred from the two-dimensional retinal images. In the process of image formation by central projection, information about the depth dimension is lost since an image occurring at a given retinal position can represent objects anywhere along the visual ray from the retinal position through the center of projection. Therefore, depth has to be inferred from a number of "depth cues," some of which are shown in figure 3.1. Most of these were already discussed by Helmholtz (1909–1911) and provide support for his ideas on "unconscious inferences": unconscious processes collect evidences of depth in images (sensations), draw conclusions, and thereby prepare for a subsequent "judgment," which may then reach the level of conscious perception. The judgment theory of perception is the basis of the modern computational theory of vision pioneered by David Marr (1982). For modern accounts of visual perception, including depth perception, see Frisby and Stone (2009) and Goldstein and Brockmole (2016). Here we give a brief overview of some important cues.

3.2.1 Stereopsis

The centers of projection of the two eyes are offset by some 6.3 cm (in humans), leading to "parallactic" differences between the two images (figure 3.1a). Two sticks stacked in depth such that they appear aligned in one eye will therefore be seen with an angular offset in the other eye. As usual, we express retinal positions as angles relative to the fovea or, equivalently, to the line of sight, which is the ray from the center of the fovea through the center of projection and into visual space. Each stick produces an image in the left and right eye, whose position is measured as an angle relative to the line of sight of the respective eye. The difference between these angles (i.e., between the left and right image positions of the same object) is called the "disparity" of the binocular image of the object.

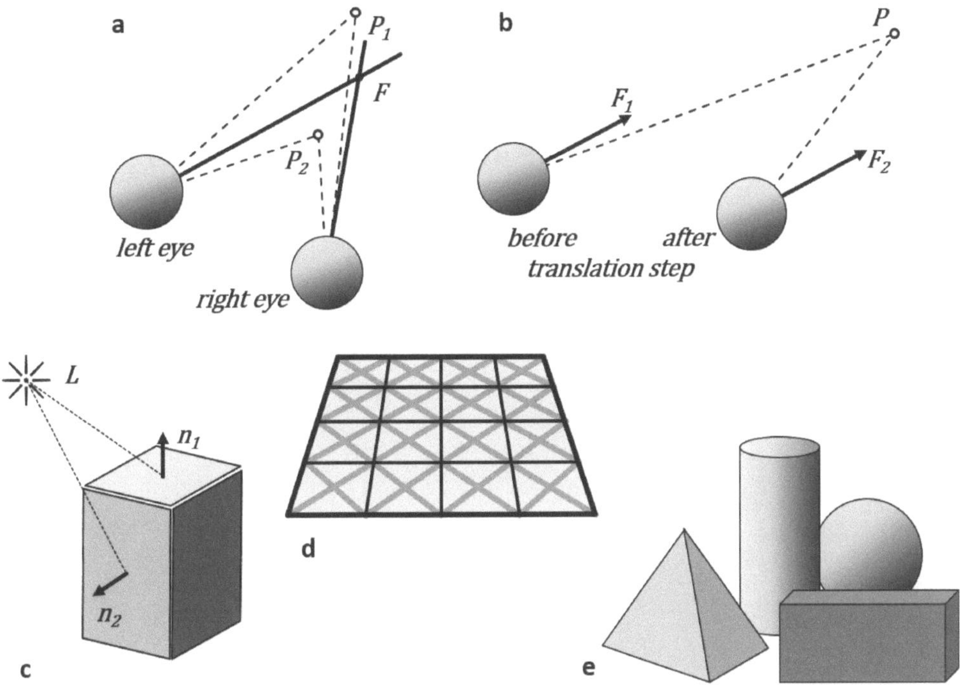

Figure 3.1
Overview of visual depth cues. (a) Stereopsis. With both eyes fixating $\boldsymbol{F}$, the object points $\boldsymbol{P}_1$, $\boldsymbol{P}_2$ will be imaged at different ("disparate") relative position to the fovea, depending on their depth. (b) Motion parallax. In an eye directed toward $\boldsymbol{F}_1$ and $\boldsymbol{F}_2$ before and after a translation step, the object point $\boldsymbol{P}$ will appear first left and then right of the fixation direction. The resulting image motion depends on the egomotion and depth. (c) Shape from shading. For surfaces of a given color (surface reflectance), the amount of light reflected from any one point depends on the direction of illumination ($\boldsymbol{L}$), the local surface normal $\boldsymbol{n}$, and for some materials also on the direction to the observer. (d) Texture gradient. The orientation of surfaces can be judged from perspective cues. (e) Occlusion. Occlusion can be detected from recognized object shapes or local properties of occluding contours. It contains ordinal depth information: that is, information about the depth ordering.

This binocular image consists of two "half-images", one in each eye, which are said to be corresponding if they show the same object from the outside world (see, e.g., Howard and Rogers 1995; Mallot 2000).

The lines of sight of the two eyes usually intersect in the so-called fixation point, which is imaged at the fovea in both eyes. Clearly, its disparity is zero by definition. As a consequence, disparity of a given object point changes if eye fixation moves from one target point to another; disparity is therefore a measure of *relative*, not absolute, depth: that is,

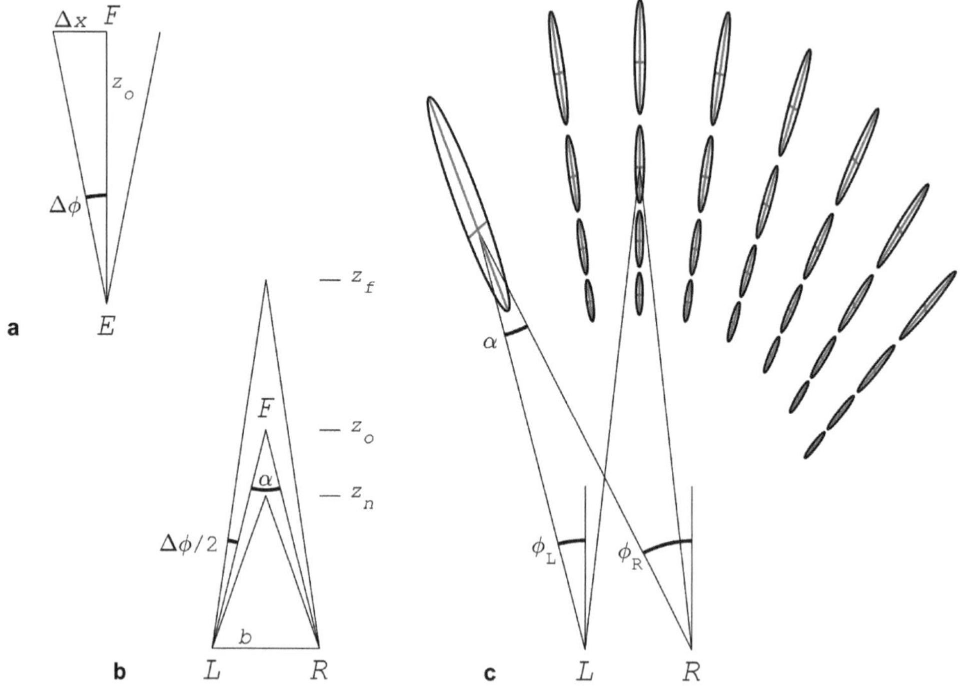

Figure 3.2
Anisotropy of binocular space. (a) An angular offset of $\Delta\phi$ from the viewing direction corresponds to a sideways displacement of $\Delta x = z_o \tan \Delta\phi$ at distance z_o. E, F mark eye and fixation point, respectively. (b) A disparity of $\pm\Delta\phi$ when fixating a point at distance z_o corresponds to depth offsets z_n (near, "crossed" disparity) and z_f (far, uncrossed disparity) as specified in equation 3.13. L, R: left and right eye, b: eye separation (stereo base), α: vergence angle. (c) Theoretical ellipses of equal distinguishability of the viewing directions of the two eyes, ϕ_L, ϕ_R. *Left part:* If the eyes are turning with the angles ϕ_L and ϕ_R relative to the forward direction, the view axes meet in a fixation point with vergence angle $\alpha = \phi_R - \phi_L$. The ellipse shows the points in the horizontal plane with viewing directions $\phi_L + \Delta\phi_L$, $\phi_R + \Delta\phi_R$ satisfying $\Delta\phi_L^2 + \Delta\phi_R^2 = \delta^2$ where δ is set to 3 minutes of arc. The red and blue lines mark the curves of equal vergence and of equal binocular viewing direction ($\eta = (\phi_L + \phi_R)/2$; not shown in the figure). They are sections of the local Vieth–Müller circles and Hillebrand hyperbolas, respectively. *Right part:* Distinguishability ellipses ($\delta = 1$ minute of arc) for a selection of points. The values for δ in both image parts were chosen to increase visibility; realistic values for human observers are on the order of 10 seconds of arc, leading to much smaller, but equally anisotropic ellipses.

of being in front of or behind the fixation point. Put differently, stereopsis may also be said to provide a cue to "equidistance" to the fixation point. In the horizontal plane, the locus of all points with disparity zero is a circle passing through the fixation point and the centers of projection of the two eyes; it is known as the Vieth–Müller circle or theoretical horopter. Perceived equidistance of two points in different visual directions can be measured directly, and the measurements agree approximately with the predictions of stereo geometry (i.e., the Vieth–Müller circle) if the compared points are close together. However, for larger separations of the points, substantial deviations have been found, indicating that the perception of equidistance is not based on stereoscopic disparity alone (see Foley 1966; Heller 1997; and figure 3.3a).

3.2.2 The Anisotropy of Binocular Space

Binocular space is strongly anisotropic in the sense that measurements of depth or distance from the observer are far less accurate than measurements of width or visual direction. The reason for this is not so much the fact that measurements of disparity require comparisons between the two eyes, while measurements of width can be carried out with data from one eye alone. Indeed, the angular thresholds for the perception of depth and width offsets are about the same (McKee et al. 1990). The reason for the anisotropy, then, is geometrical: consider a visual angle of $\Delta\phi = 0.5$ minutes of arc, which is about the diameter of a cone photoreceptor in the fovea of the human retina. At a distance of $z_o = 10$ m, say, an offset in the image of $\Delta\phi$ corresponds to a sideways displacement of $\Delta x = z_o \tan(\pm\Delta\phi) \approx \pm 1.5$ mm (figure 3.2a). To calculate the depth offset corresponding to a disparity of $\Delta\phi$ from a fixation point at 10-m distance, we have to take into account the distance between the eyes, or base length, $b = 6.3$ cm. We first calculate the vergence angle α, which is defined as the angle at which the two lines of sight intersect at the fixation point. It is acute and decreases with the distance of the fixation point. In our example, we obtain:

$$\alpha = 2\tan^{-1}\frac{b}{2z_o} \approx 21.7 \text{ minutes of arc.} \tag{3.12}$$

Next, we calculate the depth of objects generating disparities of $\pm\Delta\phi$ as

$$z_{\text{near,far}} = \frac{b}{2}\cot\frac{\alpha \pm \Delta\phi}{2} \approx \{9.80 \text{ m},\ 10.22 \text{ m}\}, \tag{3.13}$$

see figure 3.2b. That is to say, if we assume the same accuracy for measurements of retinal displacement within and between the two eyes, as is indeed the case (McKee et al. 1990), we obtain errors for width and depth measurements at a 10-m distance, which are in the order of 3 mm and 42 cm, respectively. An error ellipse for binocular localization from the retinal coordinates of a point would therefore be elongated in the depth direction with an aspect ratio of 420:3 and the anisotropy will be the larger the further the target point is away. Expected error ellipses for a number of reference points illustrating this anisotropy of binocular space are shown in figure 3.2c. The anisotropy of binocular space is not just an issue of perception but can be demonstrated also with grasping movements, which show systematic distortions when going to different locations (Campagnoli, Croom, and Domini 2017).

With the geometric approach illustrated in figure 3.2a,b, it is easy to show that in the midline of the horizontal plane, the disparity generated by a certain depth offset Δz from the fixation point scales with the distance of the fixation point from the observer, z_o, by a factor of $1/z_o^2$. Stereopsis is therefore most important in a range of a few meters from the observer while judging depth differences from disparity quickly becomes less and less reliable at larger distances.

The perception of absolute depth can be based on the vergence angle α as introduced above. In principle, vergence can be estimated from retinal cues, namely, from the distribution of the vertical components of stereoscopic disparities, or from extraretinal cues such as proprioception or efference copies of the eye muscles. Psychophysical measurements show, however, that changes in the distribution of vertical disparities in the absence of true vergence movements of the eye muscles do not affect the perceived shape of surfaces, indicating that vertical disparities are not used as a cue to vergence and absolute depth (Cumming, Johnston, and Parker 1991). In any case, depth perception from vergence is not very accurate and is subject to the same z^{-2}-scaling discussed above for relative depth from disparity. Besides vergence, accommodation may also contribute to the perception of absolute depth. Perceptual thresholds for the perception of absolute distance have been determined and, when expressed by the vergence angle, are on the order of a minute of arc (Foley 1978), which results in thresholds of about 1 m already at an overall distance of just 10 m.

3.2.3 Other Depth Cues

Information from multiple viewpoints (i.e., parallax) can also be obtained from motion of the observer, the environmental objects, or both (figure 3.1b). The case of observer motion is of course optic flow, which was discussed already in chapter 2. In the context of the perception of depth and object shape, it is often called "motion parallax," with a focus on object motion also "structure from motion" or the "kinetic depth effect" (Wallach and O'Connell 1953). If a moving observer fixates a target point while moving, as is generally the case for humans, optic flow at the fovea will be zero and the off-center flow vectors provide relative depth information of an object being nearer or farther away than the fixation point. This is completely analogous to the situation in stereopsis. An additional uncertainty of optic flow–based depth perception is egomotion speed, which cannot be judged independently from depth. This problem was discussed in chapter 2 as the scale–velocity ambiguity. In stereopsis, no analog of this ambiguity exists, since egomotion speed can be replaced by the interocular separation, which is of course fixed.

Depth cues in static monocular images include, for example, shading or attached shadows, which contain information about the orientation of a surface with respect to an illuminant or the perspective foreshortening on slanted surfaces resulting in so-called texture gradients in the depth direction (figure 3.1c,d). Local surface orientation is especially important in the perception of object shape; for example, the curvature of a ball, say, becomes apparent from the changing image intensity (lightness) between the sides exposed to and occluded from the light source. In large, also outdoor, environments, texture gradients support the perception of depth, for example, on a plain extending in front of the observer or a field of wheat where the spikes form a regular pattern of increasing density toward the horizon.

A depth cue closely related to object and scene recognition is the partial occlusion of one object by another (figure 3.1e). It requires assumptions about the object shapes, for example, the assumption that the object extends in a continuous or regular way behind the occluder. Occlusion is an ordinal variable: that is, it gives depth ordering rather than quantitative depth. Occlusion relations are transitive in the sense that if the drum is occluded by and therefore behind the pyramid, and the sphere is occluded by and therefore behind the drum, we can conclude that the sphere is also behind the pyramid, even though there is no occlusion between the latter two objects (figure 3.1e).

The types of depth information associated with each of the depth cues discussed above are quite different and not commensurate. Stereo disparity provides a distorted interval scale of distance from the fixation point. Depth from motion parallax is additionally scaled by egomotion speed. Shape from shading and texture gradients provide partial information on local surface normals, and occlusion is an ordinal measure of depth ordering. These different types of information can only be integrated if a unified representation of peripersonal space exists in which an overall consistency with the various sensory inputs can be calculated and optimized. This could be a full three-dimensional model of the visual scene, but more parsimonious suggestions are possible. One of these is David Marr's (1982) "2 1/2-D sketch": that is, an image-like array of local estimates of depth and surface orientation.

3.2.4 Size Constancy

The same depth cues that provide information about space and position are also used to infer object shape, size, and identity or to control reaching and grasping movements. When treated as geometric concepts, position, size, and shape are closely related and are indeed described by the same three-dimensional coordinates. Conceptually, however, these variables might be divided into a description of "absolute," empty space, on the one hand, and objects with three-dimensional shapes localized somewhere within that space, on the other hand.

The perception of an object's size and shape as being independent of its position in space is an example of a perceptual constancy. It is not easily achieved from the raw retinal images since these will vastly change if the observer or the object moves around in space. For example, a person's face that we look at appears to keep its size as we approach the person, although the visual angle subtended by its image increases approximately with one over distance squared. A particularly clear case of size constancy is the perceived size of retinal after-images that can be generated by looking at high-contrast patterns such as a sunlit window in an otherwise dark room. Once an after-image is established, its retinal size and position are fixed. However, if we attend to the after-image while looking at different backgrounds such as our own hand or a distant wall, it appears as a structure on that background, and its size scales roughly with the perceived distance of the background. This relation is known as Emmert's law; it requires a perception of absolute depth, which seems to rely strongly on binocular vision (Sperandio and Chouinard 2015; Millard, Sperandio, and Chouinard 2020). After distance changes, size constancy evolves with a delay of some 150 ms, indicating that high-level processes may also be involved (Chen et al. 2019).

Experimentally, perceived object size can also be assessed by the study of grasping movements, which involve a size-dependent opening of the hand. Grasping is interesting, because it is not so much associated with object recognition, but rather with the handling of an object. Milner and Goodale (1995) have therefore suggested that size perception for recognition might be dissociated from size perception for grasping as part of the dissociation[1] of the dorsal and ventral streams of visual processing. The ventral, or "what," stream leading to the temporal cortex is thought to include the processing of seen object size, while the dorsal, or "where/how," stream would control the hand opening in grasping movements. The prediction was tested with the Ebbinghaus, or Titchener, illusion in which a circle surrounded by smaller circles is perceived to be larger than the same circle surrounded by larger circles. Evidence for the dissociation, however, is mixed: Haffenden and Goodale (1998) report a dissociation, while Franz et al. (2001) suggest that both perception and grasp control rely on the same sensory representation.

3.3 The Intrinsic Geometry of Peripersonal Space

3.3.1 Geodesics, Alleys, and Exocentric Pointings

The result of the discussed processes of cue evaluation and integration is a representation of peripersonal space containing environmental objects, the observer, and some knowledge of geometry. Empirical studies show that it is not simply a Euclidean space but deviates from this in systematic ways. This can be tested by measuring "geodesic" lines, that is, perceptually shortest connections between points. In a paradigm introduced already by Helmholtz (1909–1911), subjects in a dark room are asked to direct the positioning of a dim light right in the middle of the connecting line between two other dim lights while looking from a fixed position elsewhere in the room.

This idea was elaborated in the so-called alley paradigm, in which subjects are asked to lay out a thread on a table or to position a series of markers so that they form a straight line, usually either a frontoparallel line (horopter; see figure 3.3a) or a line into depth with some

1. In neuropsychology, a dissociation between two performances is asserted if they recruit different neural systems and are thus independently affected by lesions.

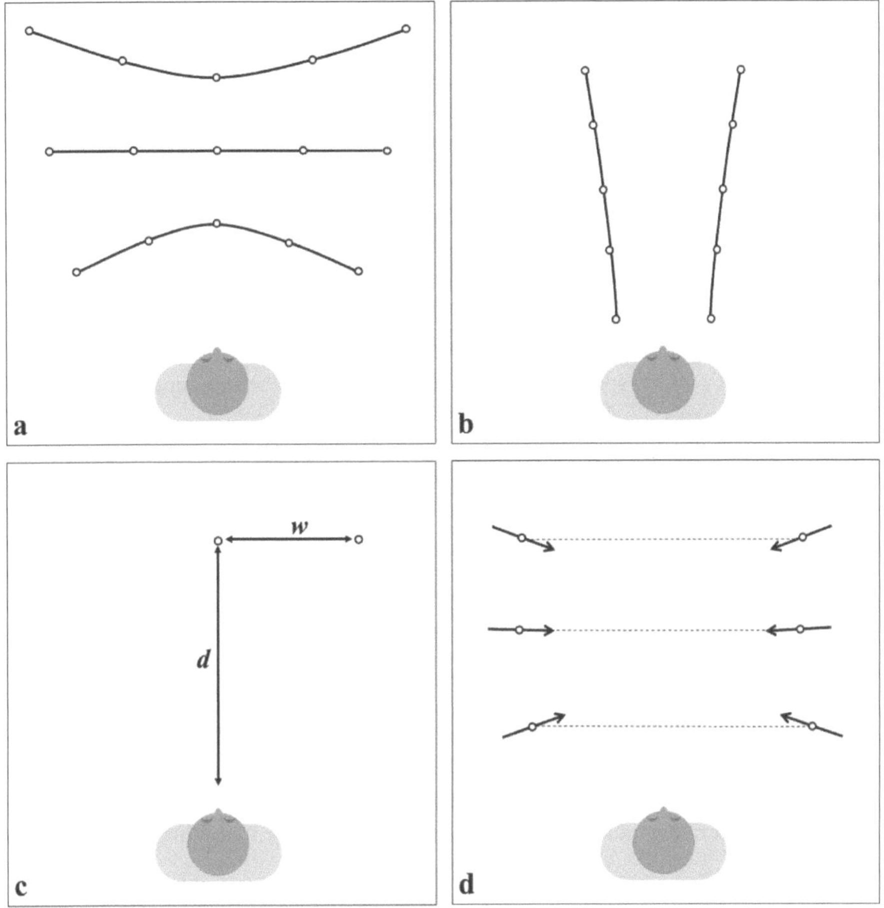

Figure 3.3
Experimental paradigms addressing the intrinsic geometry of visual space. (a) Perceived frontoparallel lines in the horizontal plane. (b) Perceived sagittal lines in the horizontal plane ("depth alleys"). (c) Perceived equality of depth and width. (d) Exocentric pointing: that is, adjustment of the perceived direction of lines connecting two distant points.

sideways offset from the observer (figure 3.3b). The latter ones are called "depth alleys" since the markers can be thought of as the trees in an alley that the observer is looking into. It is important to note that the straightness of the lines is judged from a fixed viewpoint such that at least in larger environments, the adjustments have to be performed with some apparatus or by an assistant instructed by the observer. Alley experiments of this type have been performed, for example, by Hillebrand (1901), Luneburg (1950), Zajaczkowska (1956), or Foley (1966) and reveal deviations from the straight lines expected in Euclidean geometry, as is shown schematically in figure 3.3a,b. Perceived frontoparallel lines are curved toward the observer if they are not too far away and are thus consistent with the stereoscopic horopter. However, at larger absolute distances, perceived frontoparallel lines may also be curved away from the observer, which cannot be explained by stereoscopic disparities. Depth alleys are curved away from the midline, which is consistent with the lines of constant binocular viewing direction η (see caption of figure 3.2): that is, with the hyperbolas of Hillebrand.

Another paradigm accessing the intrinsic geometry of visual space is the comparison of length estimates in different directions (figure 3.3c). For example, Foley (1972) found that the depth of an object and its width are metrically scaled as 2-to-1: that is, that distance is underestimated as compared to width by a factor of 0.5. Not surprisingly, such distortions lead to problems if views taken into different directions are to be integrated into a coherent representation of peripersonal space.

The experiments discussed so far were carried out in dark rooms with only the marker points visible as dim lights. This is done to avoid organizing effects of other objects, specifically straight walls and right-angled corners. In an outdoor setting, Koenderink, van Doorn, and Lappin (2000) used a motorized pointer and a remote control operated by subjects inspecting the scene from a position displaced from both the pointer and the target. Subjects were instructed to turn the pointer such that it appeared to point from its position directly toward the target (exocentric pointing; see figure 3.3d). The produced pointer direction is taken to be the tangent of the geodesic line at the pointer position. The authors approximate these geodesics by circular arcs and find that they are curved toward the observer (pointer adjustment behind true connecting line) for near cases, while they are curved away from the observer (pointer in front of true connecting line) for points further away. This is consistent with the curvature pattern found for the frontoparallel line adjustments discussed above.

3.3.2 The Closure of Visual Space

Perceived distance of objects that are far away from the observer is often assumed to be subject to some global limitation in the sense that the moon, the stars, and the sun are all perceived at the "sky": that is, at about the same distance. This observation is related to the idea that visual space is not open but ends at visible surfaces (ground, objects, own body) or, indeed, the sky (Indow 1999). Uexküll and Kriszat (1934) suggested that this is realized as a hard limit, which they call the "farthest plane." If an observed person or object would walk beyond this farthest plane, it would no longer be perceived as moving further away, but rather as shrinking in size. This observation is actually quite common; if looking down from a high tower, for example, cars or even houses on the ground below may appear as if they were toys: that is, shrunk, presumably because they are perceived at the distance of the farthest plane while subtending a visual angle that corresponds to a larger distance. The farthest plane would thus mark the limit of the perception of size constancy.

Gilinsky (1951) has directly measured the upper limit of perceived distance using constant distance increments. Subjects were standing in a large indoor environment and watched an experimenter pointing to the floor with a stick. The experimenter would then move further away until the observer reported that the perceived distance to the previous pointing equaled some standard length. It turns out that the increments perceived as standard lengths grow larger and larger as the experimenter moves further out. The result can be quantitatively described by the hyperbolic compression formula

$$d = \frac{DA}{D+A} \tag{3.14}$$

where d and D are perceived and true distance, respectively, and A is a constant representing the largest possibly perceived distance, or the distance of the farthest plane. It was found to be roughly constant at about 30 m within a series of measurement but depends on viewing condition (monocular vs. binocular) and also on the availability of depth cues in the environment. For a recent review of Gilinsky's and other models of perceived depth compression, see Erkelens (2017).

Gilinsky (1951) used her measurements of perceived depth also to make predictions about the perceived size of the moon. While this question may seem somewhat odd, many observers would indeed agree that the moon appears larger than an apple, say, but smaller than a wagon wheel. With the true visual angle subtended by the moon (about 0.5 degrees), it is easy to calculate that a perceived distance of some 30 m corresponds to a diameter of 26 cm, which is indeed in line with the perceived relative sizes.

Depth compression does not occur if the observer is allowed to walk to the goal. Rieser et al. (1990) and Philbeck and Loomis (1997) used the "walking without vision" paradigm in which a subject first views a target and is then blindfolded and asked to walk to the goal. Of course, this task is subject to substantial errors, but the Gilinksky-type compression is not observed. While this result implies that the perceived distance is not directly translated into walking distance, it is consistent with the idea that the representation of peripersonal space is moving with the observer and that the compression of more distant parts of the environment is undone as soon these parts of the environment are approached. In virtual environments, Kelly et al. (2018) found that distance perception can be improved by active walking in the scene but not as much by stationary looking. We will come back to the effects of walking (as opposed to just looking) in the context of spatial updating.

3.3.3 Models

The distortions of visual space found in alley experiments and the compression of distance perception imply that visual space is not Euclidean. As was already pointed out in the Introduction, Helmholtz (1876) used this observation to argue that the structure of spatial thinking is not simply identical to the mathematical theory of space, which traditionally had been thought to be Euclidean by necessity. This leads, of course, to the question of alternative geometries by which visual space can be described. Luneburg (1950) developed a theory of the non-Euclidean structure of visual space, in which he suggested an elliptical geometry: that is, geometry in which no parallels exist while the remaining axioms of Euclid still obtain. In the two-dimensional case, such geometries arise on surfaces with positive Gaussian curvature[2] such as spheres or ellipsoids (hence the name). Projective geometry is also elliptical in this sense, as can be seen by the stereographic projection[3] which relates the projective plane to the sphere. In elliptical geometries, the sum of the angles in a triangle is larger than 180 degrees, which is in line with the horopter measurements for small and medium distances shown in figure 3.3a. To see this, consider the triangle made up of the observer's viewpoint and the displayed segment of the horopter. However, horopters further out may also be curved inward such that the triangle made up by the observer position and the measured segment of a distant horopter will have a sum of angles, which is less than 180 degrees. Such geometries are called hyperbolic since they occur on saddle-shaped surfaces, one-sheet hyperboloids, or concave cones, or generally on surfaces with negative Gaussian curvature.

A triangle drawn on a sphere can be moved around on that sphere without changing its shape. If, however, the sphere would be flattened into a discus or tapered to form a pear, then moving a triangle over the surface will involve deformations. Mathematically, surfaces on which figures can be moved without deformation need to have a constant Gaussian curvature, which would clearly be the case for the sphere. Luneburg (1950) therefore argued that visual space must have constant curvature since objects can be moved around in visual space without undergoing apparent deformation (see also Indow 1999). This argument, however, can be criticized since apparent deformations do occur and are indeed obvious already in figure 3.3a–c. Luneburg's theory has therefore been generalized to allow for distance-dependent curvature (e.g., Koenderink, van Doorn, and Lappin 2000).

Geometrical theories of visual space are theories of absolute space in the sense that distortions are assumed to be completely described by point-to-point transformations or distance functions and are independent of the objects or environmental structures in the environment. However, it has long been clear that if such geometric descriptions at all exist, their measurement requires "frameless" spaces, in which, except for the studied geometrical points, not much more is visible. The reason for this is probably that objects,

2. Intuitively, Gaussian curvature of a surface point can be derived from the one-dimensional curvature of lines by considering sections through the surface where the cutting plane contains the local surface normal. Each cut is a curve whose curvature is defined as the inverse of the radius of a circle touching the curve. Curvature is signed positive if the circle is on the side that the surface normal points to and negative otherwise. We now consider the set of all touching circles obtained by rotating the cutting plane about the surface normal and take its (signed) minimum and maximum. Gaussian curvature is the product of these two. It is negative if touching circles exist on both sides of the surface, as is the case for a saddle.

3. To see this, consider a pinhole camera system with a nodal point and an image plane. We place a sphere inside the system such that the optical axis (normal to the image plane passing the nodal point) marks a diameter of the sphere. Stereographic projection maps every point of the sphere (except the nodal point itself) to its projective image.

walls, or uneven terrain do influence perceived distances. Yang and Purves (2003) presented a review of such environmental interactions and provide a statistical explanation based on the distribution of visual distances in natural environments. If this distribution of and the correlations between distances seen in adjacent visual directions are used as the priors of a Bayesian inference scheme, typical interaction effects in distance estimation can be explained.

3.4 Mental Transformations: Predictive Coding of Space

So far, we have considered peripersonal space for the case of a resting observer in a static environment. In the case of movements of the observer, of the environmental objects, or both, the representation of peripersonal space does not simply reflect the occurring changes but can be used to anticipate the course of events. For example, if the observer moves, even with eyes closed, the "image" stored in visual memory is transformed according to the laws of motion parallax and the perceived egomotion such that it predicts the actual view that will be perceived if the eyes are again opened. If the prediction fails, a problem may have occurred that requires further attention.

This is reminiscent of a general principle of cognition, known as predictive coding (e.g. Srinivasan, Laughlin, and Dubbs 1982; Teufel and Fletcher 2020). In a hierarchical processing stream, lower levels of processing generate predictions that are compared to the sensory data. If a deviation occurs: that is, if the processing carried out so far does not suffice to explain the data, this deviation is passed on to the next processing step. In the context of movement in peripersonal space, predictions are not between processing steps but between subsequent instants or events in the incoming sensory stream.

The representation of peripersonal space thus differs from the mere visual image in two respects: first, it allows one to make predictions about the effects of object and observer motion and is therefore a type of active, or working, memory. Second, it integrates visual perceptions from different viewpoints and viewing directions, as well as cues from other sensory modalities such as vestibular cues to egomotion. Loomis, Klatzky, and Giudice (2013) have therefore suggested the term "spatial image," which we will occasionally make use of.

In this section, we will discuss three types of mental transformations associated with the prediction of the sensory effects of motion. In *mental rotation*, the observer is static while objects are rotating about their own axis or while the observer is imagining such object rotations. In *perspective taking*, the observer imagines moving to a new viewpoint by translations, rotation, or both. Finally, in *spatial updating*, the observer carries out such movements by bodily displacement.

3.4.1 Mental Rotation

The rotation of an imagined object in some mental representation while the observer is stationary is called "mental rotation" (figure 3.4a). It is a central part of our understanding

Figure 3.4
Different types of mental transformations illustrated by the representation of a circular plate supporting geometrical objects. (a) Mental rotation: The observer looks at the scene and imagines how it would look after a rotation. (b, c) Perspective taking: The observer imagines attaining a different viewing direction (b) or view point (c) and reasons how the scene would look from under these conditions. (d) Spatial updating: With eyes closed, the observer takes a different viewpoint or viewing direction either by passive transportation or by active walking and imagines how the scene would appear from this novel pose.

of space that allows us to decide how an object might fit into a box, how we should shape our hand to grasp an object, or how a face seen in profile might look in a frontal view. Mental rotation and related tasks are often used in tests of general intelligence and spatial imagery.

In experiments on mental rotation, subjects are presented with images of two objects or scenes that are either identical but rotated or differ in shape. In the classical studies by Shepard and Cooper (1982), the images are line drawings of objects composed of cubes, such that their three-dimensional layout is easily perceived. In three experimental conditions, objects can be identical but rotated in the image plane, identical but rotated out of the image plane (which in the image may result in mirrorings), or different. The subjects then report whether the objects have the same shape or not. The reaction times increase linearly with the rotation angle presented: that is, for every 20 degrees of rotation, one additional second of processing time is needed. This pace is about the same irrespective of whether the rotation stays within the image plane or not.

A similar effect in the recognition of scenes or object configurations has been demonstrated by Diwadkar and McNamara (1997). Subjects learned a configuration of objects on a table similar to the situation depicted in figure 3.4a. They were then led into another room, where they were presented with pictures of the same or an altered configuration (individual object moved) taken from various novel viewpoints. The task was to judge whether or not the configuration differed from the original one. As in the object rotation tasks of Shepard and Cooper (1982), response time increased linearly with the angle between the original viewing direction and the viewing direction in which the judgment was carried out.

The proportionality between the angle of rotation and the processing time has led researchers to suggest that mental rotation is based on imagery: that is, that the representation of three-dimensional shapes and shape orientations is "analogic" in the sense that small changes in the represented shape will lead to small changes in its representation. Alternatively, a representation can be propositional: that is, based on symbols or language, as would, for example, be the case if we would explicitly represent the rotation angle as a number.

In terms of neural networks and neural activities, it is hard to say what exactly should be "rotating" in the brain. The representation of a three-dimensional object is a pattern of neural activity that may or may not be three-dimensional but certainly does not obey the rules of rigid body rotations. One plausible neural mechanism for mental rotation has been suggested by Georgopoulos et al. (1982) for the motor cortex. Pointing movements of the arm are represented in motor cortex by specialized neurons tuned to specific directions in space. During the planning of a turning movement, neurons specific for the current arm directions decrease their firing activity while, at the same time, neurons tuned for future arm directions increase firing rate. The overall represented movement direction can be defined as the weighted average of each neuron's preferred direction multiplied by its current relative activity. The weighted average is known as the "population vector" or "center-of-gravity estimator" and the whole scheme as "population coding." From the point of view of the experimenter, this leads to a kind of mental rotation: if the activity of a neuron is increased, the direction represented by the population as a whole turns toward this neuron's preferred direction. If many neurons tuned for different directions are arranged in a row, mental rotation would correspond to a peak of activity moving over the arrangement, with the speed of this movement and the sampling of direction determining the speed of mental rotation.

This type of mental rotation based on population coding also plays a role in invariant object recognition, for example, the problem of recognizing a face in frontal and side views. If object memories are stored in a standard orientation, the "canonical view," the incoming image would have to be transformed into this standard orientation before it can be compared to the stored image. This alignment is a form of mental rotation that would allow to make recognition "invariant" to object orientation or to perceive the object as a "constant" entity even when it undergoes rotation. The idea that perceptual constancies of this type require mental or neural transformations was first formulated by Pitts and McCulloch (1947).

In visual recognition of objects and faces, neurons have been found in the inferotemporal cortex that respond to specific views of a given object such as the front or profile view of a face (Logothetis et al. 1994; Tanaka 2003). Mental rotation in visual space might thus be a consequence of a population code of viewing directions, in close analogy to mental rotations in the motor system. However, this type of mental rotation would rely on knowledge of rotated views of familiar objects but would not necessarily be able to predict novel views of unfamiliar objects. This is indeed in line with a study on mental rotation of complex indoor scenes, which showed that novel views can be recognized but that recognition is slower and less reliable than the recognition of familiar views (Christou and Bülthoff 1999). For an algorithm allowing the recognition of novel views from learned ones without explicit three-dimensional representations, see Blanz and Vetter (2003).

In imaging studies reviewed by Zacks (2008), neural activity related to mental rotation has been found in the parietal cortex: that is, relatively "early" in the dorsal processing stream, but not in the inferotemporal cortex concerned with direction-invariant object recognition. In some versions of the mental rotation task, activity was also found in the motor cortex, indicating a possible role of motor simulation.

Mental rotation is also one of the best-studied examples for sex differences in cognitive psychology (Nazareth et al. 2019). The strength and reliability of sex effects in spatial cognition are often overstated, especially in the non-scientific literature. For the response times in mental rotation, Lauer, Yhang, and Lourenco (2019) conclude that a male advantage exists as a weak to medium size effect and that this effect is stronger in adults than in children. This latter finding allows hypotheses about influences of education and society.

3.4.2 Perspective Taking

Perspective taking is a kind of imagery in which subjects are asked to imagine viewpoints or viewing directions that are not their current ones and then to imagine how the scenery would look from this new position. For example, the audience in a lecture hall might imagine how the scenery looks from the front of the room and whether a given person would be seen left or right from another one when viewed from the front. In one experimental paradigm used, for example, by Farrell and Robertson (1998) or Shelton and McNamara (2001), subjects are placed in the center of a circular arrangement of easily recognized objects such as office or kitchen equipment (figure 3.4b). Subjects are asked to learn the arrangement and then, with eyes closed, report the location of objects named by the experimenter. When the acquisition of a memory of the scene is thus confirmed, the subjects are asked to imagine to be facing one object and to report the relative direction to a second object (judgment of relative direction task, JRD; see also section 7.2.4 and figure 7.3). By and large, subjects are able to do this with some interesting variations.

Mou and McNamara (2002) used this paradigm to show that memory for object configurations in rooms is organized along an environmental axis, for example, the long axis of the room, or the axis from door to window, even if learning first occurred with another viewing direction. Perspective taking is best if the imagined direction aligns with the room axis, somewhat reduced if imagined viewing is orthogonal, and poor for the oblique directions. If pointing error is plotted versus imagined view angle, a characteristic zigzagging or W-shaped curve thus results. From this result, it is clear that the acquired representation of the environment is not simply an egocentric image (i.e., organized along the view axis of acquisition) but relies on a viewer-independent, "allocentric" reference axis. The situation seems to be less clear, however, for vertical locations, in which the egocentric direction of acquisition also seems to play a role (Hinterecker et al. 2019).

Many other versions of the perspective-taking task have been published; for an overview see Hegarty and Waller (2004). For example, in the "object perspective" task, subjects are given a sheet of paper with images of objects printed at certain positions and thus forming a kind of map. The subject is then asked to imagine being located at object 1, looking at object 2. The task is then to point—in egocentric coordinates—to an object 3. Another task shows a scene with an arrangement of objects and various viewpoints from which photographs of the scene are taken. The subject is then given the photographs and has to decide which photograph was taken from which viewpoint. Finally, reading on old-fashioned street map while driving in a car can be considered a perspective-taking task if the current driving direction on the map is downward and the driver has to decide whether to turn left or right.

Of course, it is not obvious that all these tasks rely on the same cognitive mechanisms and that these are different from those underlying mental rotation. This question was addressed by Hegarty and Waller (2004), who had subjects perform a battery of both mental rotation and perspective taking tasks. If mental rotation and perspective taking depend on the same cognitive resources, subjects who perform superior in mental rotation should also be good perspective takers and vice versa. If, on the other hand, the two performances rely on different mechanisms, performance should vary independently across subjects. Results from a factor analysis show that performances in perspective-taking and in mental rotation tasks are to some extent independent: that is, that they from two data groups within

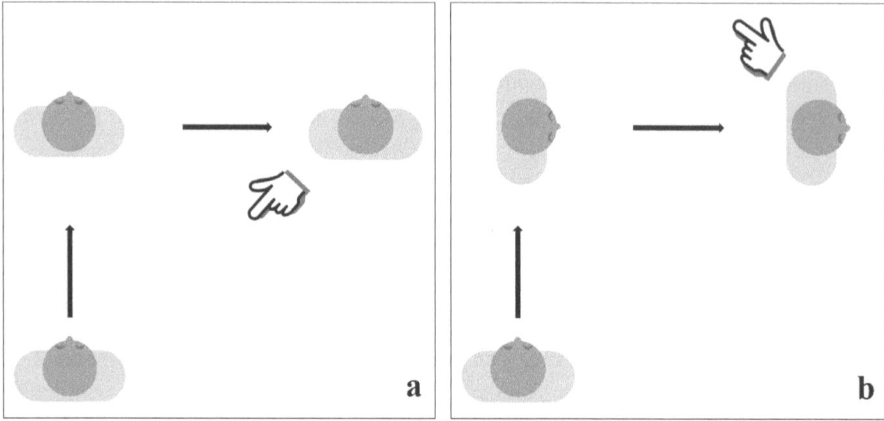

Figure 3.5
Imagined body rotations in perspective taking. (a) With your eyes closed, imagine moving move a step forward and then a step sideways to the right. Then point from your imagined end point to the start point of the movement. Most people will correctly point to the left and back. (b) Now imagine moving a step forward, thening turn to the right and moving another step forward. You will end up in the same location as before but with a different body orientation. Now the start point is behind you on your right, but most people will again point in the direction left behind. That is, pointing is egocentrically the same as in case (a), indicating that the imagined rotation has been ignored (see Klatzky et al. 1998).

which correlation is higher than between these two groups. However, overall correlation is also substantial. This suggests at least a partial independence of the two types of mental transformation.

The imagined perspective can be defined in various ways. The subjects may be seated and simply change their imagined viewing direction toward a named target object, they may study a map and imagine standing at one object facing another one, or they may look at actors or avatars and imagine what these might be seeing. In this latter case, perspective taking becomes an issue of social cognition where knowledge of the perceptions of others is studied as part of the theory of mind. Studies employing avatars or actors to mark the imagined viewpoint have been carried out, for example, by Vogeley et al. (2004) or Ward et al. (2020), but direct evidence for a superiority of social over nonsocial anchor-points for perspective taking seems to be missing. This does not exclude the possibility, however, that mechanisms of perspective taking support processes in social cognition and theory of mind.

Perspective taking may be different for imagined translations and imagined rotations. In a paradigm developed by Klatzky et al. (1998), subjects imagine walking two legs of a rectangular triangle and then indicate the direction from the imagined endpoint of the walk to the start point, which is still the actual position of the subject (figure 3.5). The walk may be performed with a turn after the first leg or by walking the second leg sideways. The pointing movement produced by the subjects in both conditions is approximately the same (in egocentric terms), even though this amounts to a vastly wrong pointing in the turning condition. This result indicates that imagined turning movements are neglected in this experiment, where turns are not defined by external landmarks ("imagine to be looking at object A ...") but by egocentric motor instructions ("imagine turning right..."). This may also be the reason why body turns are only poorly simulated in virtual environments presented on a screen, while immersive presentations allowing closed-loop body movements lead to much better perceptions.

Neuropysiological activity related to perspective taking has been found in various areas of the parietal cortex. In one study, Schindler and Bartels (2013) trained subjects in a virtual octagonal arena to remember the positions of eight objects placed in the corners of the octagon. In the function magnetic resonance imaging (fMRI) scanner, they then asked participants to take the perspective from the center of the arena to one of the corners as specified by the respective object. Next, participants attended to the direction of a second object in one of the remaining seven corners of the arena. fMRI signals from two parietal areas, the intraparietal sulcus (IPS) and the inferior parietal cortex, were recorded and fed into a support vector machine, together with the egocentric angle of the target to which the subject had been asked to attend. The support vector machine was able to learn this direction from the fMRI signals (representational similarity analysis). This indicates that the directional information about the attended target is indeed represented in the two parietal areas. However, the fact that a pattern recognition device such as the support vector machine is needed to find this representation may indicate that encoding is not made explicit in a maplike arrangement but remains implicit in the patterns of neural activity.

Perspective taking also plays a role outside peripersonal space, most notably in the context of navigational planning. Since this involves also other processes of spatial working and long-term memory, we will postpone the discussion of imagery of distant places until chapter 7.

3.4.3 Spatial Updating

Assume a cup of coffee is waiting for you on a table in front of you. If you turn away such that the cup gets out of sight, you will still be able to reach for the cup and grasp it. The positional information needed to control your arm movement is thought to be continuously held available in the representation of peripersonal space, which implies that this information is updated whenever you move. This process is known as spatial updating and occurs both in turning and in translational movements.

Spatial updating of represented object position occurs automatically upon observer motion, as was clearly demonstrated in the study by Farrell and Robertson (1998), that we mentioned already in the context of perspective taking (figure 3.4b). Subjects were seated in a revolving chair in the center of seven familiar objects placed equidistantly around the observer. The subjects first learned the arrangement of the objects; then they were blindfolded and asked to point at named objects. Before pointing, however, they underwent one of four treatments:

In the *control* condition, subjects were turned to a new position and then back to the original position; the errors and response times from the condition were taken as a baseline for the other conditions. In the *updating* condition, subjects were rotated to a new position and then asked to point from there. In this case, directional judgments were fairly correct and as fast as in the control condition. If, however, rotations had to be imagined without being actually performed (*imagination* condition), or if turns were performed but subjects were asked to ignore them and point as if they had not been turned (*ignoring* condition), large errors and increased reaction times were found. The results show that spatial updating is an automatic process that happens when the observer moves and cannot be suppressed at will. Updating does not occur during merely imagined turns. In the ignoring condition, subjects have to exert substantial cognitive effort to produce the requested pointings.

Spatial updating also occurs under translational movements of the observer that lead to novel viewpoints. One paradigm to study updating during observer locomotion is illustrated in figure 3.4d: an observer studies a scene, often a configuration of objects placed on a table, and then walks around the table to view it from the other side. While walking, the configuration may or may not be changed by relocating one object or by replacing it by another one. In a study by Wang and Simons (1999), subjects could view a round table from two windows in a circular curtain surrounding the table. Between two viewings, the windows were closed by blinds and the configuration of objects on the table could be changed by the experimenter. Also, the table could be turned by the same angle separating the two windows. In effect, four conditions arose in which the observer did or did not move from one window to the other and the table was or was not turned. In the stay–stay and walk–turn cases, the retinal image of the table would be unchanged. Interestingly, the observers' performance in detecting configurational changes of the objects on the table after walking to the other window was better if the table had not turned but appeared in the expected orientation. If the table had been turned with the observer, the retinal image was the same as before, but observers performed poorer in detecting configurational changes. The effect cannot be ascribed to a lack of information about the turning table, since table turning was disclosed to the observer by a handle that stuck out of the curtain and was always visible.

The detection of configuration change from novel viewpoints is thus supported by carrying out the related egomotion (i.e., by egocentric cues). In addition, environmental cues such as the layout of the room in which the experiment takes place have been shown to help recognition (Christou, Tjan, and Bülthoff 2003; Burgess, Spiers, and Paleologou 2004). Such environmental cues provide an allocentric reference frame that seems to interact with the updating process.

Spatial updating also underlies the walking without vision paradigm discussed already in section 3.3.2. In this case, the goal location is viewed and then updated during the observer's approach until its egocentric coordinates are reduced to zero. Similarly, in the experiment by Klatzky et al. (1998) (see figure 3.5), a condition with blindfolded walking was included, in which pointing was largely veridical. This indicates that the representation of the starting position had been successfully updated.

The task of remembering a location while walking was also used by Wolbers et al. (2008) to assess the neural correlates of spatial updating. In this study, subjects view a plain marked by random dots and experience egomotion from optic flow. At the time the experiment starts, a varying number of objects are visible that vanish after the motion onset. At the end of the experiments, subjects are asked to point to the original position of the objects. In a control condition, subjects stay at their starting point but still have to remember the object locations. In the fMRI data, the authors look for voxels whose activity depends on both the presence of egomotion and the number of objects to be remembered. Such voxels are found in the precuneus (medial parietal cortex) and in the supplementary motor area (dorsal precentral gyrus). These areas thus seem to be involved in the computations for spatial updating.

3.5 Recalibration in Peripersonal Space

The term "recalibration" was already introduced in section 2.6, where it referred to the readjustment of gain factors for the estimation of egomotion from proprioceptive and visual cues. In peripersonal space, similar calibrations are required for the visual control of motor actions such as grasping movements or approaches to a visually perceived goal.

3.5.1 Prisms, Inversion Goggles, and Varifocals

Recalibration has long been studied in eye–hand coordination. In a classical experiment going back to Helmholtz, subjects wearing prism glasses are asked to grasp an object under visual control (for a recent review, see Petitet, O'Reilly, and O'Shea 2018). In this situation, an object in the vertical midline is imaged at a retinal position displaced sideways in a direction and amount defined by the prism angle. Subjects will initially reach to a position that would be imaged at that retinal position, were the prism not in place. After a few trials, however, grasping error is reduced and eventually vanishes altogether. If the prism is removed, the error occurs in the opposite direction and is again compensated after a few trials. This is a clear recalibration between the visual and proprioceptive senses of position.

As an extreme version of this experiment, we discuss an old study by Kohler (1964) in which a subject was wearing inverting goggles that made them see the world upside down (see also Sachse et al. 2017). This is initially very confusing, but within a few days or weeks, the subject is able to interpret the scene correctly and eventually reports seeing everything upright as before. The turning upright of visual perception does not happen at once but gradually and may switch back and forth. Also, subjects report that individual objects that the observer attends to turn around while others remain inverted. In particular, objects touched with the hand or with a stick are more likely to turn upright than others left untouched. After two weeks, the subject is able to ride a bicycle in urban travel.

Initially, the subject also perceives illusory motion of visual targets upon turns of the head. This is due to the fact that the retinal movements predicted for that particular head movement are not the ones that actually happen. These illusory motion percepts also go away after a while. The effect is well known from people receiving new glasses, especially varifocals. When turning the head, strange distortions can be perceived that normally go away after some days or weeks. Again, this is an example of a recalibration of the retinal position of the stimulus and the perceived position of the object in peripersonal space. When the glasses are taken off, no aftereffect (as described above for the prism experiment) normally occurs. This indicates that the brain is able to store two calibration patterns and switches between them.

3.5.2 Spatial Presence

The perception (or awareness) of being present at a given place is considered an important part of consciousness. We already discussed perspective taking, in which that subjects imagine looking from elsewhere to the current scene. Of course, in perspective taking, subjects are only imagining to be elsewhere, while at the same time they are completely aware of their true location. Still, the feeling of being present at a certain location, or spatial presence, is a percept that can be subject to illusions, especially in neurological disorders.

In an experiment by Lenggenhager et al. (2007), subjects were equipped with video goggles showing the image of a camera placed behind them: that is, they would see a person looking like themselves standing a few meters in front of them. If they move their arm, the person would do the same thing, and so on. Subjects report thinking that they are looking at their own body from outside themselves. To increase this effect, the experimenter touches the back of the subject several times, and the subject feels the touch and at the same time sees the body in front also being touched. This touching procedure allows introducing a control condition: the visual presentation can be either immediate, in which case felt and seen touch are synchronous, or with a delay, so that synchrony is destroyed. After some time of stimulation, the subject is led out of the room, and the goggles are removed. They are then asked to walk back to the position where the stimulation occurred. In the synchronous condition, but not in the asynchronous one, subjects tend to walk to a position closer to the visually perceived "second body" and away from the true location. This also works if the second body is that of another person, but not with a geometric object. In conclusion, the results indicate that the position of perceived presence has been recalibrated from the standard at the center of view to a position about a meter in front of the center of view.

Key Points of Chapter 3

- The space and objects around us with which we immediately interact are represented as peripersonal space.
- It is constructed from sensory inputs such as visual depth cues but also from the integration of such inputs acquired during head and body movements. The relative impact and meaning of each cue can be changed in recalibration.
- Peripersonal space contains some metric information but is distorted by the vast discrepancy in the accuracy of measurements of depth and width. Its intrinsic geometry is non-Euclidean.
- Peripersonal space is an active representation that is continuously updated as we move and can be used to make predictions about the outcome of viewpoint changes.
- Peripersonal space also contains a representation of "self."

References

Blanz, V., and T. Vetter. 2003. "Face recognition based on fitting a 3D morphable model." *IEEE Transactions on Pattern Analysis and Machine Intelligence* 25:1063–1074.

Burgess, N., H. J. Spiers, and E. Paleologou. 2004. "Orientational manoeuvres in the dark: Dissociating allocentric and egocentric influences on spatial memory." *Cognition* 94:149–166.

Campagnoli, C., S. Croom, and F. Domini. 2017. "Stereovision for action reflects our perceptual experience of distance and depth." *Journal of Vision* 17 (9): 21.

Chen, J., I. Sperandio, M. J. Henry, and M. A. Goodale. 2019. "Changing the real viewing distance reveals the temporal evolution of size constancy in visual cortex." *Current Biology* 29:2237–2243.

Christou, C., and H. H. Bülthoff. 1999. "View dependence in scene recognition after active learning." *Memory & Cognition* 27:996–1007.

Christou, C., B. S. Tjan, and H. H. Bülthoff. 2003. "Extrinsic cues aid shape recognition from novel viewpoints." *Journal of Vision* 3:183–198.

Cumming, B. G., E. B. Johnston, and A. J. Parker. 1991. "Vertical disparities and perception of three-dimensional shape." *Nature* 349:411–413.

Diwadkar, V. A., and T. P. McNamara. 1997. "Viewpoint dependence in scene recognition." *Psychological Science* 8:302–307.

Erkelens, C. J. 2017. "Perspective space as a model for distance and size perception." *i-Perception* 8 (6): 2041669517735541.

Farrell, M. J., and I. H. Robertson. 1998. "Mental rotation and the automatic updating of body-centered spatial relationships." *Journal of Experimental Psychology: Learning, Memory and Cognition* 24:227–233.

Foley, J. M. 1966. "Locus of perceived equidistance as a function of viewing distance." *Journal of the Optical Society of America* 56:822–827.

Foley, J. M. 1972. "The size-distance relation and intrinsic geometry of visual space: Implications for processing." *Vision Research* 12:323–332.

Foley, J. M. 1978. "Primary distance perception." In *Perception. Vol. 8 of Handbook of sensory physiology,* edited by R. Held, H. W. Leibowitz, and H.-L. Teuber. Berlin: Springer Verlag.

Franz, V. H., K. R. Gegenfurtner, H. H. Bülthoff, and M. Fahle. 2001. "Grasping visual illusions: No evidence for a dissociation between perception and action." *Journal of Experimental Psychology: Human Perception and Performance* 27:1124–1144.

Frisby, J. P., and J. V. Stone. 2009. *Seeing. The computational approach to biological vision.* Cambridge, MA: MIT Press.

Georgopoulos, A. P., J. F. Kalasak, R. Caminiti, and J. T. Massey. 1982. "On the relation of the direction of two-dimensional arm movements and cell discharge in primate motor cortex." *The Journal of Neuroscience* 2:1527–1537.

Gilinsky, A. S. 1951. "Perceived size and distance in visual space." *Psychological Review* 58:460–482.

Goldstein, E. B., and J. Brockmole. 2016. *Sensation and perception.* 10th ed. Boston: Cengage Learning.

Haffenden, A. M., and M. A. Goodale. 1998. "The effect of pictorial illusion on prehension and perception." *Journal of Cognitive Neuroscience* 10:122–136.

Hegarty, M., and D. Waller. 2004. "A dissociation between mental rotation and perspective-taking spatial abilities." *Intelligence* 32:175–191.

Heller, J. 1997. "On the psychophysics of binocular space perception." *Journal of Mathematical Psychology* 41:29–43.

Helmholtz, H. 1876. "The origin and meaning of geometrical axioms." *Mind* 1:301–321. http://www.jstor.org/stable/2246591.

Helmholtz, H. von. 1909–1911. *Handbuch der physiologischen Optik.* 3rd ed. Hamburg: Voss.

Hillebrand, F. 1901. "Theorie der scheinbaren Grösse bei binocularem Sehen." *Denkschriften der Kaiserlichen Akademie der Wissenschaften, mathematisch-naturwissenschaftliche Classe* LXXII:255–307.

Hinterecker, T., C. Leroy, M. E. Kirschhock, M. T. Zhao, M. V. Butz, H. H. Bülthoff, and T. Meilinger. 2019. "Spatial memory for vertical locations." *Journal of Experimental Psychology: Learning Memory and Cognition* 45:1205–1223.

Howard, I. P., and B. J. Rogers. 1995. *Binocular vision and stereopsis.* Oxford Psychology Series 29. New York, Oxford: Oxford University Press.

Hunley, S. B., and S. F. Lourenco. 2018. "What is peripersonal space? An examination of unresolved empirical issues and emerging findings." *WIREs Cognitive Science* 9:e1472.

Indow, T. 1999. "Global structure of visual space as a united entity." *Mathematical Social Sciences* 38:377–392.

Kelly, J. W., L. A. Cherep, B. Klesel, Z. D. Siegel, and S. George. 2018. "Comparison of two methods for improving distance perception in virtual reality." *ACM Transactions on Applied Perception* 15 (2): 11.

Klatzky, R. L., J. M. Loomis, A. C. Beall, S. S. Chance, and R. G. Golledge. 1998. "Spatial updating of self-position and orientation during real, imagined, and virtual locomotion." *Psychological Science* 9:293–298.

Koenderink, J. J., A. J. van Doorn, and J. S. Lappin. 2000. "Direct measurement of the curvature of visual space." *Perception* 29:69–79.

Kohler, I. 1964. *The formation and transformation of the perceptual world.* Vol. III(4). Psychological Issues. New York: International Universities Press.

Lauer, J. E., E. Y. Yhang, and S. F. Lourenco. 2019. "The development of gender differences in spatial reasoning: A meta-analytic review." *Psychological Bulletin* 145:537–565.

Lenggenhager, B., T. Tadi, T. Metzinger, and O. Blanke. 2007. "Video ergo sum: Manipulating bodily self-consciousness." *Science* 317:1096–1099.

Logothetis, N. K., J. Pauls, H. H. Bülthoff, and T. Poggio. 1994. "View-dependent object recognition by monkeys." *Current Biology* 4:401–414.

Loomis, J. M., R. L. Klatzky, and N. A. Giudice. 2013. "Representing 3D space in working memory: Spatial images from vision, hearing, touch, and language." In *Multisensory imagery: Theory and applications,* edited by S. Lacey and R. Lawson, 131–156. New York: Springer.

Luneburg, R. K. 1950. "The metric of binocular visual space." *Journal of the Optical Society of America* 40:627–642.

Mallot, H. A. 2000. *Computational vision: Information processing in perception and visual behavior.* Cambridge, MA: MIT Press.

Marr, D. 1982. *Vision.* San Francisco: W. H. Freeman.

McKee, S. P., L. Welch, D. G. Taylor, and S. F. Bowne. 1990. "Finding the common bond: Stereoacuity and other hyperacuities." *Vision Research* 30:879–891.

Millard, A. S., I. Sperandio, and P. A. Chouinard. 2020. "The contribution of stereopsis in Emmert's law." *Experimental Brain Research* 238:1061–1072.

Milner, A. D., and M. A. Goodale. 1995. *The visual brain in action.* Oxford: Oxford University Press.

Mou, W. M., and T. P. McNamara. 2002. "Intrinsic frames of reference in spatial memory." *Journal of Experimental Psychology: Learning, Memory and Cognition* 28:162–170.

Nazareth, A., X. Huang, D. Voyer, and N. Newcombe. 2019. "A meta-analysis of sex differences in human navigation skill." *Psychonomic Bulletin & Review* 26:1503–1528.

Petitet, P., J. X. O'Reilly, and J. O'Shea. 2018. "Towards a neuro-computational account of prism adaptation." *Neuropsychologica* 115:188–203.

Philbeck, J. W., and J. M. Loomis. 1997. "Comparison of two indicators of perceived egocentric distance under full-cue and reduced-cue conditions." *Journal of Experimental Psychology: Human Perception and Performance* 23:72–75.

Pitts, W., and W. S. McCulloch. 1947. "How do we know universals. The perception of auditory and visual forms." *Bulletin of Mathematical Biophysics* 9:127–147.

Rieser, J. J., D. H. Ashmead, C. R. Taylor, and G. A. Youngquist. 1990. "Visual perception and the guidance of locomotion without vision to previously seen targets." *Perception* 19:675–689.

Sachse, P., U. Beermann, M. Martini, T. Maran, M. Domeister, and M. R. Furtner. 2017. "'The world is upside down'—The Innsbruck goggle experiments of Theodor Erismann (1883–1961) and Ivo Kohler (1915–1985)." *Cortex* 92:222–232.

Schindler, A., and A. Bartels. 2013. "Parietal cortex codes for egocentric space beyond the field of view." *Current Biology* 23:177–182.

Shelton, A. L., and T. P. McNamara. 2001. "Systems of spatial reference in human memory." *Cognitive Psychology* 43:274–310.

Shepard, R. N., and L. A. Cooper. 1982. *Mental images and their transformations.* Cambridge, MA: MIT Press.

Sperandio, I., and P. A. Chouinard. 2015. "The mechanisms of size constancy." *Multisensory Research* 28:253–283.

Srinivasan, M. V., S. B. Laughlin, and A. Dubbs. 1982. "Predictive coding: A fresh view of inhibition in the retina." *Proceedings of the Royal Society (London) B* 216:427–459.

Tanaka, K. 2003. "Columns for complex visual object features in the inferior temporal cortex: Clustering of cells with similar but slightly different stimulus selectivities." *Cerebral Cortex* 13:90–99.

Teufel, C., and P. C. Fletcher. 2020. "Forms of prediction in the nervous system." *Nature Reviews Neuroscience* 21:231–242.

Uexküll, J. von, and G. Kriszat. 1934. *Streifzüge durch die Umwelten von Tieren und Menschen. Ein Bilderbuch unsichtbarer Welten.* Berlin: Springer.

Vogeley, K., M. May, A. Ritzl, P. Falkai, K. Zilles, and G. R. Fink. 2004. "Neural correlates of first person perspective as one constituent of human self-consciousness." *Journal of Cognitive Neuroscience* 16:815–827.

Wallach, H., and D. N. O'Connell. 1953. "The kinetic depth effect." *Journal of Experimental Psychology* 45:205–217.

Wang, R. F., and D. J. Simons. 1999. "Active and passive scene recognition across views." *Cognition* 70:191–210.

Ward, E., G. Ganis, K. L. McDonough, and P. Bach. 2020. "Perspective taking as virtual navigation? Perceptual simulation of what others see reflects their location in space but not their gaze." *Cognition* 199:104241.

Wolbers, T., M. Hegarty, C. Büchel, and J. M. Loomis. 2008. "Spatial updating: How the brain keeps track of changing object locations during observer motion." *Nature Neuroscience* 11:1223–1230.

Yang, Z., and D. Purves. 2003. "A statistical explanation of visual space." *Nature Neuroscience* 6:632–640.

Zacks, J. M. 2008. "Neuroimaging studies of mental rotation: A meta-analysis and review." *Journal of Cognitive Neuroscience* 20:1–19.

Zajaczkowska, A. 1956. "Experimental test of Luneburg's theory. Horopter and alley experiments." *Journal of the Optical Society of America* 46:514–527.

4 In the Loop

The basic spatial behavior is to move around in an environment. The paths or trajectories thereby produced can be based on reflex-like stimulus–response mechanisms known as "taxes" (singular: taxis) or motor schemata, in which case path generation is basically a problem of feedback control. Control mechanisms of this type have been studied in simple animals but can also be found as components of human behavior. In this chapter, we will explore taxis as a powerful mechanism of navigation but also its limitations and their possible remedies. These will be addressed in a series of examples including (i) the role of eye movements to extract specific information from the environment, (ii) the generation of motion plans for extended episodes of movement, (iii) the usage of general world knowledge in path selection, and (iv) path planning in cluttered and dynamic environments. Taken together, these examples show how internal representations of goals, obstacles, and motor plans help to produce adaptive and flexible behavior. We conclude this chapter by discussing elaborated action–perception cycles that incorporate higher-level mechanisms and provide a framework for the evolution of spatial cognition from simple orientation reactions.

4.1 Directed Movement

The understanding of space depends not only on what we see or hear but also on how we move. The egocentric body axes forward/backward, left/right, and up/down not only provide a frame of reference for the perception of object locations but also, and maybe even primarily so, define directions and goals for bodily motion. Performances such as approaching a food item or mating partner, shying away from dangers, or following a guidance to explore new opportunities are at the basis of spatial behavior and exist at least as rudiments in all living beings.

For the description of directed movements, the following components can be identified. First, a structure or target in the environment needs to be present that can be sensed by the organism. Second, a sense organ or receptor has to generate an internal signal or code such

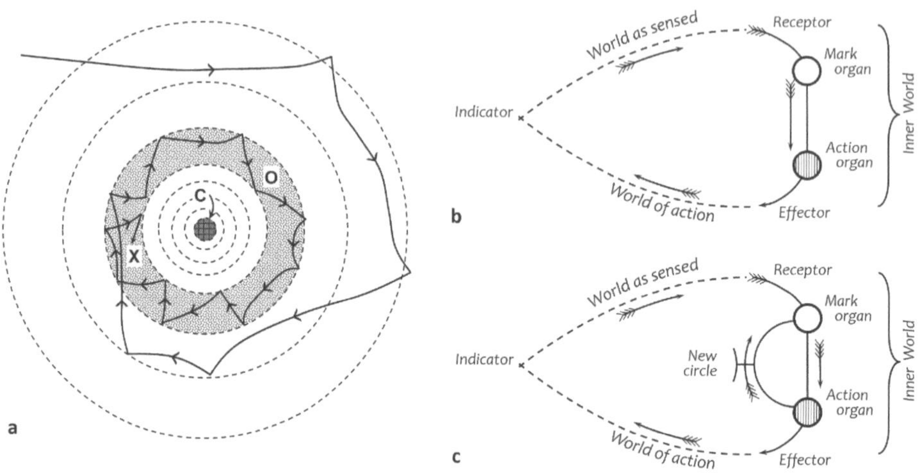

Figure 4.1
Simple behavior. (a) Trajectory of a *Paramecium* in a chemical gradient field generated by a central bubble of CO_2 (marked "C"). The concentration has an optimum marked by gray texture and the letter "O." The *Paramecium* passes by and performs an avoidance reaction (turn; kinks in trajectory) when sensing a change to the worse. As a result, it finds the optimum and stays within. "X" marks the endpoint of the observed trajectory redrawn from Kühn (1919) after Jennings (1906). (b, c) Functional circles for simple behaviors as conceived of by Uexküll (1926). The "indicator" is a feature in the environment for which the animal has a receptor or sensory filter. If recognized, it triggers an action that is executed by the "effector." The environment changes and the indicator may vanish. If it does, no further action is initiated. Panel (b) shows the simple functional circle as a feedback loop while panel (c) shows additional inner signals ("new circle") such as an efference copy from the action organ to the mark organ. This may be considered a simple type of representation. Panels (b) and (c) redrawn from Uexküll (1926).

as a nervous activity, which is the third element. The fourth component is the motor system to which the code is passed and then generates an action. In turn, the action changes the environment or the relation between the observer and the environment and therefore the sensory input, and so forth. This feedback circuit is thus a closed loop, not just an "arc" from perception to action.[1] It was first described as "functional cycle" by Jakob von Uexküll (1926) and is now known as the action–perception cycle. It is a basic idea

1. The idea of the reflex arc goes back at least to Descartes's (1686) famous illustration in his *Tractatus de Homine* (p. 58): In a man withdrawing his foot from a flame, "fire particles" cause "skin movement," which pull a string to a "ventricle" in the brain. This opens a valve from which "animate spirits" are released and flow into the muscles, where they cause contraction, presumably by some sort of hydraulic or pneumatic mechanism.

in many fields of psychology and neuroscience, including, for example, ecological psychology (Gibson 1950; Warren 2006), sensorimotor integration (Prinz 1997), behavioral neuroscience (Fuster 2001), and even music perception (Novembre and Keller 2014).

Simple examples of directed movements and perception–action couplings have been studied in unicellular organisms such as bacteria, protozoa, and algae, which orient in chemical gradients, light fields, or gravitation. An example is given in figure 4.1a. A *Paramecium*, a unicellular organism moving by the beat of a large number of "cilia" covering its surface, is swimming in a drop of water. In the example depicted in figure 4.1a, a bubble of carbon dioxide has been placed in the center of the image and a chemical gradient of carbonic acid will build around it. A *Paramecium* is passing the bubble without reaction, but as soon as it senses a reduction in H^+-concentration (increasing pH value), it performs an "avoidance reaction." This reaction is not directed but simply consists of a movement backward and a body turn by a small angle, after which the journey is continued in the new direction (Jennings 1906; Eckert 1972). If the unfavorable change is sensed again, the reaction will be repeated until a new path is found that avoids the hindrance. As a result of this trial-and-error procedure, the animal finds the optimum and stays within.

The trial-and-error behavior of *Paramecium* illustrates the basic perception–action approach as formulated by Uexküll (1926): see figure 4.1b. The "indicator" is the change of the H^+-concentration (pH) along the animal's path; the receptor is probably an H^+-channel in the cell membrane causing a depolarization of the cell. The "mark organ" is a calcium channel that opens up transiently upon depolarization and allows Ca^{2+}-ions from the surrounding medium to enter the cell. The calcium ions in turn cause the cilia to invert their beat direction and thus effect the backward movement (Eckert 1972). The backward motion ends when the transient opening of the calcium channels is finished.

This type of feedback behavior works without any form of "understanding" of what space or an acidity gradient is; it does not include an explicit "recognition" of objects or events or a "decision" about what is to be avoided. Rather, it is simply a control loop ensuring the "homeostasis" or steady state of the organism and starting to counteract as soon as a deviation from the homeostatic norm is detected. In the beginning of the twentieth century, scientists like Herbert S. Jennings (1906) and Jacques Loeb (1912) explored the idea that also more complex behavior can be described in this way, an idea that also influenced the early behaviorists (see Pauly 1981).

Stereotyped stimulus–response behaviors in freely moving organisms that result in a locomotion are known as taxes. Kühn (1919) and later Fraenkel and Gunn (1940) suggested taxonomies of taxis behavior distinguished by the physical nature of the orienting stimulus and by the control law applied. The classification by physical nature—for example, as phototaxis (orienting to light), chemotaxis[2] (orienting in a chemical gradient), geotaxis (orienting to gravity), thigmotaxis (wall-following), and so on—is still used today. With respect to the control laws, organisms may move so as to increase (or decrease) sensory input (positive or negative "telotaxis"), keep a fixed angle to a source ("menotaxis"), or balance the stimuli arriving at the left and right body side ("tropotaxis"). In realistic cases, however, the distinctions between these mechanisms are often not as clear as they may seem. We will discuss a number of examples below.

More appropriate explanations of orienting and other behaviors have to take into account a number of additional mechanisms that have also been pointed out by Uexküll (1926): first of all, the functional cycle may be improved by additional interior feedback loops allowing the system to predict the consequences of possible actions (figure 4.1c). A second mechanism is the integration of multiple functional circles and the selection of varying priorities in homeostasis (e.g., feeding vs. mating). Finally, the modification of functional circles by learning and memory will add to the flexibility and appropriateness of behavior. Of course, the investigation of such high-level amendments is the very focus of the cognitive sciences.

In the remainder of this chapter, we will study the relation between internal representations and the associated behavioral improvements in a number of examples.

4.2 Left–Right Balancing

In higher animals, a taxis can be defined as a reflex whose output is an orienting behavior. The notion of a reflex allows for signal transmission along nerve fibers and synapses, which obviously cannot exist in *Paramecium*, given that it is a unicellular organism. Still, it is interesting to note that the information processing allowing *Paramecium* to perform its avoidance behavior is based on biophysical processes that are surprisingly similar to those underlying neural signal propagation and processing (Eckert 1972).

In "bilaterian" animals, in which symmetric left and right body sides are anatomically defined, both sense and motor organs are generally organized in pairs. A common pattern of neural circuitry in this case is the coupling of the sensors from one body side to the effectors either on the same body side ("ipsilateral") or across the midline ("contralateral"). If motor force is stronger on one side, the organism will turn away from that side and toward the other side. This is the basic idea of "tropotaxis," which is thought to play a role in staying away from obstacles as well as in the approach of goals.

4.2.1 Braitenberg Vehicles

The concept has been elaborated by Braitenberg (1984) in a series of thought experiments with cybernetic "vehicles" (figure 4.2). Each vehicle has two lateralized frontal sensors

2. The behavior of *Paramecium* in the carbonic acid gradient as described above may also by classified as a "chemokinesis" rather than a "chemotaxis" because the reaction is triggered but not oriented by the stimulus; orientation results from the trial-and-error procedure.

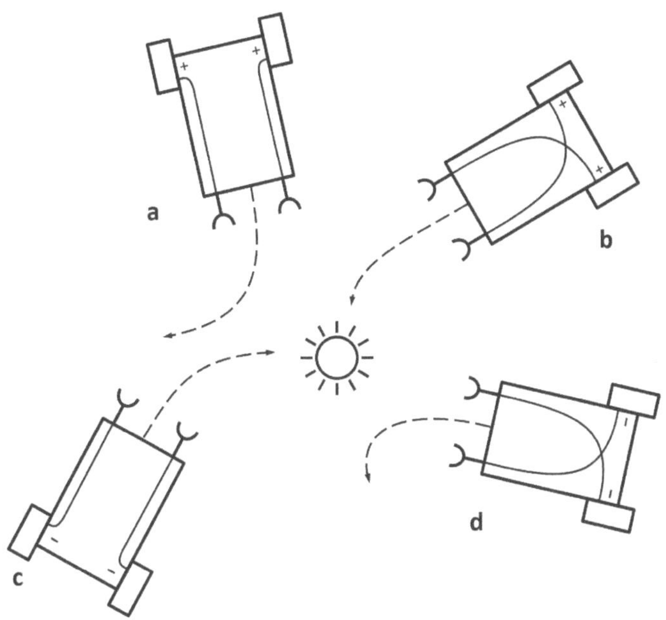

Figure 4.2
Braitenberg vehicles and the concept of tropotaxis. Four bilaterally symmetric vehicles are shown together with schematic paths in the vicinity of a light source. (a) Ipsilateral excitatory. (b) Contralateral excitatory. (c) Ipsilateral inhibitory. (d) Controlateral inhibitory. Modified and combined after Braitenberg (1984).

(i.e., a left and a right one) each coupled directly to a drive or motor in the rear. Sensor and drive are thus coupled in a "monosynaptic" reflex in which the thrust of the drive is proportional to sensor output or, in an inhibitory case, to some constant from which a quantity proportional to the sensor output is subtracted. Coupling can be ipsilateral, in which case each sensor controls the motor on the same side of the body, or contralateral, if the connections cross from one side to the other. In the case depicted in figure 4.2a, the agent encounters a source on its left: that is, the left sensor is closer to the source and therefore receives a stronger input. In the ipsilateral excitatory coupling of this agent, the left motor will then turn faster and the agent will perform a turn away from the source while speeding up in its vicinity and slowing down when the source is passed. In the case of figure 4.2b, coupling is contralateral excitatory, and an unequal stimulation of the two sensors will result in a turn toward the source. At the same time, the agent will speed up since the stimulation gets stronger the closer the agent gets to the source. In the end, the agent will hit the source with high speed. In ipsilateral inhibitory coupling (figure 4.2c), the agent will also turn toward the source but with decreasing speed, resulting in a kind of docking behavior. Finally, contralateral inhibitory coupling (figure 4.2d) leads to an avoidance of the source, this time, however, with a slow passage speed.

Behavioral complexity of Braitenberg vehicles can be increased in a number of ways. For example, a nonlinear transfer function may be included between the sensor and the motor such that the dependence of stimulation and thrust follows a maximum curve. The vehicle may then turn toward the source until a certain stimulus level is reached, at which time it will again turn away. As a result, it will start cycling about the source or perform other, more complex trajectories if other nonlinearities are used. Also, multiple sensor pairs could be included, each coupled to the motors in their own way. In this case, the vehicle will react differently to different stimuli in the sense that it might approach one stimulus while avoiding another.

In robotics and computer science, the idea of the Braitenberg vehicle has been influential in many fields, including, for example, embodied cognition and artificial life (e.g., Brooks 1986; Pfeifer and Scheier 2001) or biomorphic robotics (Webb 2020). Minimal robots in the style of Braitenberg vehicles also lend themselves for the exploration of evolutionary strategies in robot design (Floreano and Keller 2010) and collective behavior of robot swarms (Kube and Zhang 1994).

Braitenberg (1984) himself discusses the interpretations that these behaviors may elicit in the eyes of an observer, ranging from aggression in the accelerated approach and hitting behavior of the contralateral excitatory vehicle, curiosity in the slow passages of the ipsilateral excitatory vehicle, or attraction and even love in the docking behavior of the inhibitory ipsilateral one. While this may seem a bit far-fetched at first glance, it should remind us of a more important question: if we ascribe aggression or attraction to an animal or human, how do we know that it is not just stimulus–response behavior?

4.2.2 Trail Following

Many ant species use pheromone[3] trails that mark paths from the nest to a food source. Trail following is based on chemical sensors localized in the ant's antennae. When the ant walks along a path, it receives odor input on both antennae and can keep track by balancing the input on both sides or, more correctly, by continuously making small turns toward the side where the stronger stimulus is sensed. The result is overall path following with slight oscillations to the left and right, as illustrated in figure 4.3a (Hangartner 1967). This looks like a clear example of a tropotaxis: that is, balancing reaction as would also be seen in a Braitenberg vehicle with contralateral coupling.

One way to test this assumption is the removal of one antenna. In this case, the input from the lesioned side is always zero while the unlesioned side continues to produce some signal as long as the ant is in the vicinity of the trail and the odor can at all be sensed.

3. A pheromone is a substance functioning as a chemical signal. Unlike hormones, pheromones act between different individuals.

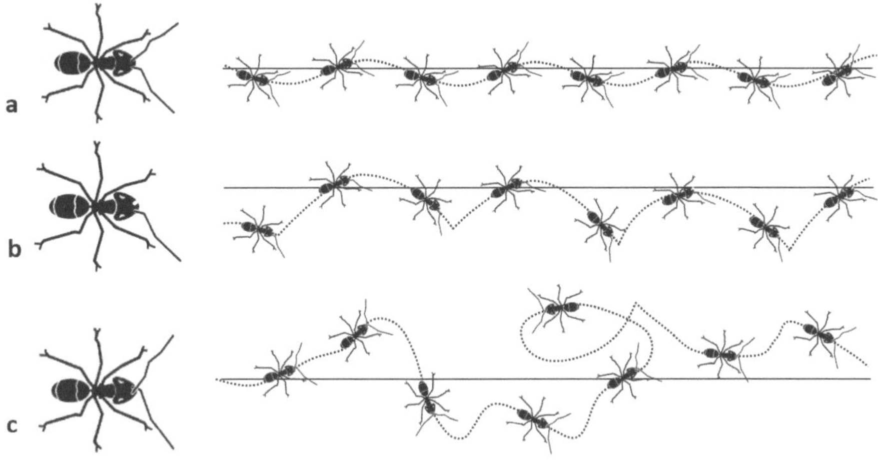

Figure 4.3
Following of a pheromone trail by tropotaxis in the jet ant *Lasius fuliginosus*. The solid line marks the pheromone trail; the dotted line is the trajectory of the ant. (a) Normal ants follow the trail in an oscillating pattern resulting from turns toward the side of stronger stimulation. (b) If the left antenna is removed, tropotaxis is no longer possible. However, the ant can still sense reductions in pheromone concentration, presumably by monitoring changes over time. (c) If the left antenna is bent to the right and vice versa, tropotaxis becomes misleading. The trajectory gets more curved, but the trail is not completely lost. Adapted from Hangartner (1967), by permission.

Tropotaxis then predicts a continuous circling movement in the direction of the unlesioned side. Such circling movements have indeed been described, for example, in hoverflies of the genus *Eristalis* moving in homogeneous light after one eye has been covered with black lacquer. In other examples of unilateral blinding, however, no circling behavior has been observed, which indicates that other mechanisms may also be present (Fraenkel and Gunn 1940).

In the case of *Lasius fuliginosus* trail following, removal of one antenna does not result in the expected pattern. Rather, the ant is still able to follow the trail, albeit with larger error (figure 4.3b). Path deviations to the lesioned side (left of the ant in figure) result in smooth correcting turns as occur also in the intact ant. Larger path deviations occur toward the unlesioned side, where the backward pulling action of the contralateral antenna is missing. Detours to this side end in abrupt turns back, which look like an avoidance reaction. Trail following with one antenna cannot be explained by left–right balancing but requires a comparison of sensory stimuli over time, as in a sequence of peering movements. This is sometimes called a "klinotaxis."

In a third experiment, the two antennae are crossed over and glued in this unnatural position such that tropotaxis would be missleading. Even then, the ant is not completely lost but can still follow the path to some extent (figure 4.3c). This example shows that while tropotaxis may play a role in path following, other mechanisms are also present. The idea of taxes, attractive as it is, must therefore not lead to oversimplified models of animal behavior.

Despite their unquestionable usefulness in many domains of behavior, taxes such as trail following also have obvious limitations, which result from their stereotyped nature and eventually are the driving forces for the evolution of more cognitive mechanisms. An instructive example illustrating these limitations is the phenomenon of milling in ants (Schneirla 1944). Army ants such as *Labidus praedator* (formerly called *Eciton p.*) raid in large colonies, swarming over wide areas with little visual control by or interaction between the individual raiding ants. Traffic is therefore largely controlled by stereotyped route following with the neighbor ants as main sensory input. When a part of the swarm gets detached from the main colony, it can happen that the raiding movement runs into a loop in which the ants continue to go in circles for hours and days and presumably until starvation. The stimulus–response scheme guaranteeing coordinated traffic in normal situations now forces the ants to keep running in the ant mill, and no higher-level mechanisms seem to exist that would be able to interrupt the flow.

For further examples of orienting behavior in animals, as well as their underlying mechanisms, see, for example, Schöne (1984) and Wehner (2020). Following a pheromone trail is also an example of a *guidance*, which is an important mechanism in route navigation. Guiding structures can be well-worn paths, walls, rivers, or even a personal guide who leads the way by walking in front of us (see O'Keefe and Nadel 1978, 80ff, for an elucidating discussion). All we would need to do in this situation is to keep an eye on the guide and go ahead. In spatial cognition, representations of guidances and the goals they lead to are important parts of spatial knowledge that play a role in route following, planning, and wayfinding. For behaviors consisting of a taxis and a memory component (such as recognizing a guidance), O'Keefe and Nadel (1978) coined the term "taxon," which is basically a stimulus–response schema involving the attraction to a goal (see also Redish 1999). The taxon system was contrasted to a "locale" system in which behavior depends on some sort of representation of the current position; the simplest part of the locale system is path integration.

4.2.3 A Digression on Collective Building

The formation of the pheromone trail is an example of the collective and self-organized building of a structure that forms the environment and makes it "navigable" for the ants. The trail pheromones are produced in special glands and are first laid out by scout ants on their way back from profitable food sources. When they arrive in the nest, the scouts recruit more foragers, who will follow the existing trail to the source and renew it with their own pheromone placements when returning to the nest. Since earlier markings quickly evaporate, the trail is a dynamic structure, which converges to a near-optimal course between the food-source and the nest and vanishes when the food source is depleted (Deneubourg et al. 1990). The collective behavior of many ants can thus be said to solve the problem of path optimization without mechanisms of planning or memory involved (see, e.g., Bonabeau, Dorigo, and Theraulaz 2000).

Trail formation is the result of an automatic behavior that can be characterized by two rules: (i) if you go out foraging and encounter a pheromone marker, follow the marker, and (ii) if you are returning from a successful foraging trip, lay out pheromone markers along your way. In a large ant colony, these rules suffice to produce the trail. The idea of an automatic behavior to be triggered by a marker that has been generated or placed by other members of the hive as a part of this very behavior has been called "stigmergy" by Grassé (1967). It is the basis not only of path formation but also of building impressive nest structures of social insects, which are among the most complex productions found in nature. For example, Perna and Theraulaz (2017) have investigated the connectivity of the chambers and galleries of a termite nest and found a surprisingly well-organized structure with few through-running highways and many short dead ends each connecting just a few chambers. Stigmergy thus allows the animals to structure space in a way appropriate for their navigational skills.

Self-organized formation of trails is not limited to insects but also occurs in many other cases ranging from grazing animals to human crowds. If many people have to cross a lawn, say, they will tend to step into the footprints of earlier walkers where the grass is low. They will do so even if their goal is not quite in the same direction. As a result, a "desire line" will develop, which can be modeled as an optimal combination of the comfort of walking and the distance to the goal (Helbing et al. 2001).

4.2.4 Centering in Corridors

If a honey bee is flying in a corridor with vertical stripes covering the corridor walls, it tends to stay in the center, keeping equal distance to either side. In a cleverly devised experimental setup, Srinivasan et al. (1996) showed that the distance estimate required for this behavior is based on optic flow. Figure 4.4a shows a top view of the flight corridor with the flight path taken by the bees. In the experiment, the sides of the corridor were made movable either against to or along with the flight direction of the bee. If the wall pattern moves with the bee (figure 4.4b), the optic flow perceived on the moving side is reduced, as if the structure generating the flow would be far away. In this situation, the bee approaches the moving side, such that the flow from this side is again increased while the flow on the far (stationary) side is reduced. Vice versa, if the wall pattern moves toward

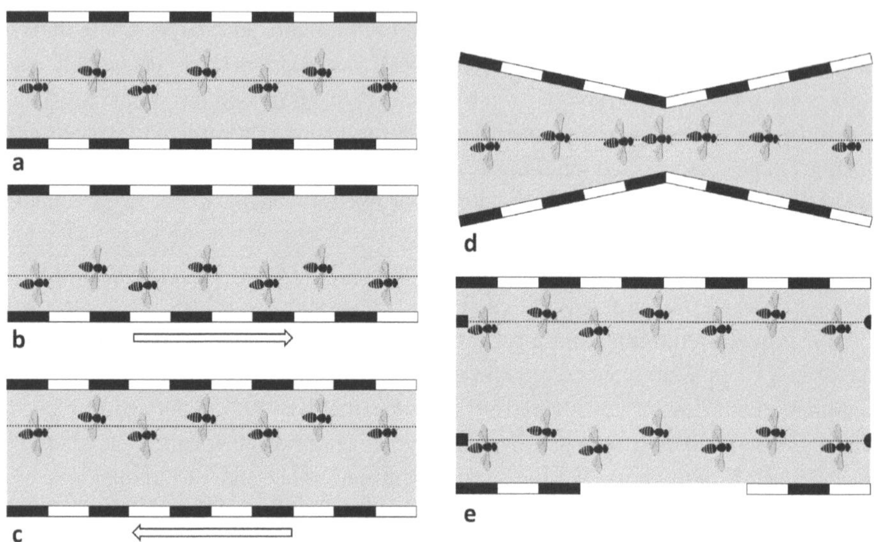

Figure 4.4
Centering behavior in honey bees. (a) Top view of the corridor with striped walls and the flight trajectory of the bees in the center (schematic). (b) If the right wall is moving with the bees, optic flow on this side is reduced and the bees choose a path to the right. (c) If the wall is moving toward the flies, optic flow is increased and the bees move away from this side. (d) In a constricted corridor, bees slow down and speed up when the corridor widens again. (e) If released on the side of a wider corridor and heading toward a feeder straight ahead, the bees can follow a straight line off the corridor center. Removal of part of the corridor wall does not lead to path deviations. Panels (a) to (d) after Srinivasan et al. (1996), panel (e) after Serres et al. (2008). Figures not to scale.

the bee, it adjusts its flight path to the stationary side. In both cases, the flight position is chosen such as to equalize the optic flow perceived in the left and right eye.

If the width of the corridor is changing (figure 4.4d), centering still works. In addition, the bee slows down in the narrow section and speeds up after passing it. Srinivasan et al. (1996) showed that the speed is chosen such that optic flow perceived by the bee stays constant while passing the constriction. Both behaviors, centering and slowing down in a constriction, are exactly as would be expected for a Braitenberg vehicle with contralateral inhibitory coupling, which turns away from the side with larger flow and slows down if overall flow increases.

However, as in the case of following chemical trails, tropotaxis or left–right balancing does not seem to be the end of the story. In a setup with a wider corridor, Serres et al. (2008) had bees fly from a release site (black square in figure 4.4e) to a feeder (half circle in the figure). If both release site and feeder are sideways displaced from the midline of the corridor, the bees still take a straight flight path: that is, they do not show the centering behavior but keep the distance to the wall constant. This wall-following behavior is also known as "thigmotaxis." If one side of the corridor is removed over a distance of 1.5 m, such that no optic flow signal is coming from this side at all, the bees still keep their straight flight line. Indeed, if tropotaxis were strictly operating, the bee should leave the corridor as soon as it reaches the gap in the wall. Likewise, tropotactic animals should not be able to fly by a fence with constant distance but should turn away from it into the open space. Serres et al. (2008) show that this is not the case. As an alternative explanation, they suggest a lateral optic flow regulator that keeps the flow on one side constant.

In conclusion, animal behavior does not conform with monofactorial explanations but is based on the interaction of multiple mechanisms of which tropotaxis is likely to be one. These mechanisms use simple principles of information processing, including optic flow, feedback control in the action–perception cycle, left–right balancing, wall following, and so on. Dickinson (2014) describes these mechanisms as the "Devonian toolkit" which is thought to have been part of the original insect "bauplan" when they first appeared in evolution. In order to understand how the elements of this toolkit interact to control behavior in each species, further investigations into the neurophysiological mechanisms are needed (Webb and Wystrach 2016; Mauss and Borst 2020).

4.2.5 Beyond Insects

Stimulus–response behavior in a feedback loop with the environment is not limited to insects but occurs also in humans. One example is the braking performance based on optical flow ($\dot{\tau}$) as discussed in section 2.5.2. Another example concerns the role of optic flow in the control of egomotion speed. Snowden, Stimpson, and Ruddle (1998) instructed drivers in a driving simulator to keep a constant speed while the visibility of the scene changed from clear to foggy conditions. In fog, the density of the optic flow field is reduced since fewer features can be distinguished that would provide reliable motion information. As a result, the drivers sped up, presumably in an attempt to keep total flow constant. Of course, this behavior is not advisable in real-world driving situations and could not be reproduced in that situation (Owens, Wood, and Carberry 2010) where drivers acted more wisely by slowing down during fog. Still, speeding on highways does seem to increase as lighting conditions deteriorate (de Bellis et al. 2018). Driving speed also varies with the width of a corridor, that is, subjects tend to slow down when passing a constriction and speed up afterward, just as the honey bees shown in figure 4.4d (see, e.g., Ott et al. 2016). The implications of these results for the template-matching approach to optic flow analysis have been discussed in section 2.5.3.

The presence of simple stimulus–response behavior can also be tested in open-loop situations where the feedback from the environment is missing. In a virtual reality experiment by Wallis, Chatziastros, and Bülthoff (2002), subjects driving on a straight four-lane road were asked to change the lane. With visual feedback, this is no problem. For changing to the left lane, subjects first turn the steering wheel to the left and, when the left lane is reached, the car has to be turned back to the forward direction, which is achieved by turning the steering wheel to the right. When heading is again parallel to the street, the steering wheel is turned back to the forward direction. Thus, the steering sequence is *straight–left–right–straight*. In a second condition, lane change occurs within a tunnel without light (i.e., visual feedback is taken away). In this condition, subjects do turn the vehicle to the left, but then immediately turn back to straight forward so that the car will keep an angle to the street direction; the steering sequence is *straight–left–straight*. Of course, the car will then leave the street. If vision is reinstalled, subjects realize their mistake and readjust their heading to the street direction. This result indicates that visual feedback is needed in street following and that internal monitoring, if at all present, is not sufficient. The effect persists even if clear egomotion cues are provided without showing the streets margins, for example, by adding the optic flow of a cloud of random dots (Xu and Wallis 2020).

Insect-like mechanisms of course control have also been explored in technical applications in the field of biomorphic robotics. For example, Argyros, Tsakiris, and Groyer (2004) have tested left–right balancing of optic flow in a mobile robot navigating a corridor. This algorithm works well but has the disadvantage that the robot has to move in order to acquire obstacle information. This problem does not arise in classical engineering approaches where the visual scene is analyzed in all available detail and represented as a map in which motion is then planned. Biomorphic course control is applied in robotics in special cases where processing capacity is limited and the robot is moving anyway. Examples include obstacle avoidance in small aerial robots (drones, e.g., Beyeler, Zufferey, and Floreano 2009) or in the design of a lunar landing device (Sabiron et al. 2015).

4.3 Cognitive Components

The behaviors discussed so far are of reflexive nature: that is, perception is coupled directly to action. However, feedback loops from perception to action are not restricted to this simple case but may include additional components or inner states of the agent. Here we discuss a number of examples that occur also in the absence of an explicit long-term memory of a specific spatial environment. These examples include the control of information intake via eye movements, motor plans for extended trajectories or maneuvers, the knowledge of general rules for successful navigation, the representation of goals and obstacles, and strategies for optimal foraging. Elaborations based on memory functions will be discussed in later chapters.

4.3.1 Eye Movements

A task similar to centering in a corridor is street following in a car. We have already seen in the previous section that this requires visual feedback since turns performed for a lane change are not automatically compensated if feedback is missing.

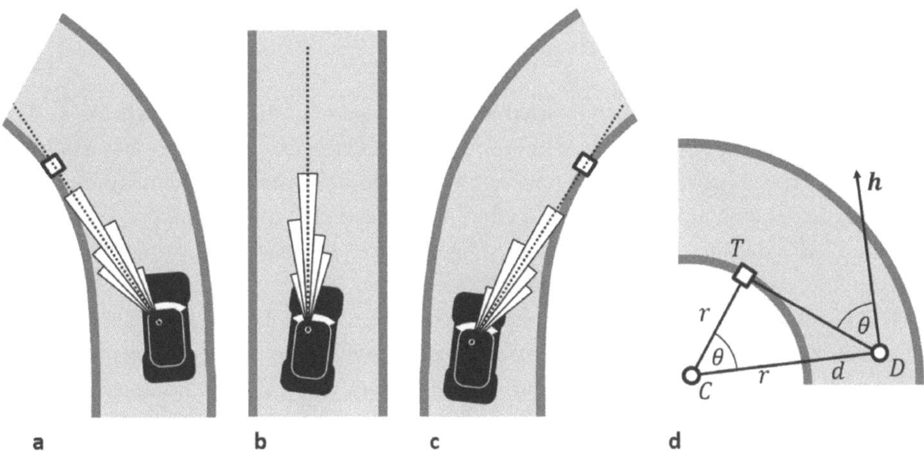

Figure 4.5
Eye movements during steering in curves. (a-c) When driving in a curve, drivers tend to fixate a point on the inner road margin where the viewing direction is tangent to the curve (diamond mark). The dotted line marks the average viewing direction while the fan shows the distribution of viewing directions (schematically, after data from Land and Lee 1994). (d) Geometry of curve driving. *D*, driver; *C*, center of curvature (midpoint of a curve following a circular arc); *T*, tangent point from *D* to inner street margin; ***h***, current heading or forward direction on road; *r*, radius of curvature; *d*, driver distance to inner road margin; θ, splay angle. The required setting of the steering wheel corresponds to the curvature $1/r$ and can be calculated from θ and *d*.

An important factor in street following is the control of eye gaze, which is not fixed to the heading direction (which might be helpful for optic flow analysis) but moves around in the visual field. The presence of eye movements also rules out simple left–right balancing as a mechanism of street following in humans. Land and Lee (1994) filmed car drivers on curved narrow roads in the Scottish Highlands and analyzed their gaze direction in dependence on upcoming steering needs. One strategy used by the observers is to fixate a point on the margin of the road on the inner side of a curve where the line from the observer is tangent to the road margin (figure 4.5). As a point in the outside world, this point moves with the car, but its projection in the image is stable when navigating a circular curve. In a left curve, it would be the rightmost image point depicting the left roadside. The angle between the current heading and the tangent point can easily be determined from the image; it is also known as the "splay" angle. Note that the "heading" direction can be replaced by the projection of the image vertical to the ground plane. It can thus be obtained without optic flow evaluation. Simple geometry shows that the local curvature of the road can be calculated from the splay angle and the distance of the driver from the roadside (see figure 4.5d) by solving the relation

$$(r+d)\cos\theta = r \tag{4.15}$$

for r. The curvature $1/r$ can then be used to set the position of the steering wheel.

In figure 4.5d, the street is assumed to follow a circular arc and therefore has a constant curvature. This is not true for most curves, which are built with continuously increasing and then decreasing curvature, so that a continuous rotation of the steering wheel would be required.[4] Besides the splay angle itself, splay rate (i.e., its change over time) is therefore also an important factor in driving control.

Unlike the approach-under-drift experiment shown in figure 2.5, the splay angle account of curve steering works without the use of optic flow but relies on the position of the tangential point as an environmental cue. Recent experiments with intermittent image presentations, where optic flow is absent, support the relevance of the splay angle in steering (Macuga et al. 2019). In any case, instantaneous measurements of splay angle, heading from optic flow, and motor action such as turning the steering wheel have to be considered as parts of a larger control loop to obtain a complete model of human driving. For a recent review, see El et al. (2020).

4.3.2 Motor Plans for Extended Maneuvers

Human walking trajectories are composed of steps during which the center of mass of the body is oscillating from left to right while the feet are placed so as to stabilize gait. When walking on a treadmill (i.e., on a flat surface and with no particular goal) foot placements can be nicely described by biomechanical models (Bruijn and Dieën 2017). When walking freely in outdoor environments, however, foot placement is additionally constrained by the terrain. Matthis, Yates, and Hayhoe (2018) showed that in this condition, eye movements tend to fixate the next stepping point. As in curve following discussed above or in visually controlled manual behavior, fixations thus seem to anticipate the next action step, which is guided by or directed toward the fixated target.

This does not mean, however, that food placement is driven by the envisaged goal in a simple stimulus–response or taxis-like manner. Rather, an internal representation of planned motion trajectories seems to exist, as was demonstrated in a study by Hicheur et al. (2007). Subjects standing at a starting point were shown a gateway a few meters away. The gateway is basically the frame of a door, which marks not just a goal point to walk to but also a body orientation or heading with which the door has to be passed. If the door is turned, subjects would have to walk in an arc to arrive at the door in a frontal orientation. When subjects repeat this task several times, the body trajectories produced in each trial show but a small variation while the placement of the feet differs substantially across

4. A curve whose curvature is linearly changing with length is the clothoid or Euler spiral. It is used to design curved lanes, for example, in highway interchanges. It is also used as a model of the racing line in circular curves.

trials. This indicates that the next motion step is not simply determined by the view of the goal and the current pose, as one might expect in a stimulus–response model. Rather, the participants seem to generate a motion plan such as a smooth body trajectory leading to the goal. Foot placement may vary due to noise or additional constraints.

4.3.3 General Knowledge and the "Legibility" of an Environment

In humanmade environments such as cities or airports, navigators are able to arrive at their goal via complex paths, even if they visit the area for the first time and therefore cannot rely on spatial memory. This ability is a result of the interaction between the structure of built environments, possibly including signage displayed to support wayfinding, general knowledge of the navigators about the structure of the type of environment, and simple behaviors such as following the guidances provided by salient landmarks or passages. In architecture and urban planning, these structures and mechanisms are said to constitute the "legibility" of the environment. In his influential book *The Image of the City*, Kevin Lynch (1960) listed a number of elements used to structure a city. These are (i) *paths*, or channels along which people usually move; (ii) *edges* such as barriers or perceived boundaries (e.g., between a front of houses and a park); (iii) *districts*, which may be delineated by edges and can be recognized by their character or building style; (iv) *nodes*, which are approached as hubs or intersections of paths; and (v) *landmarks*, such as a tower in a neighborhood of small houses, which, like nodes, are point-like structures. They are, however, usually not entered by the observer but only viewed from a distance.

Of these structuring elements, paths and landmarks can be thought of as guidances that direct the movement of a navigator. The overall "image" of the city that Lynch is talking about is, however, a spatial memory in which the named elements are the categories. Indeed, the question of spatial "ontology" (i.e., of the categories or types of objects that are represented and stored in spatial memories) is fundamental for the study of spatial cognition. The understanding that unknown cities can be structured along these lines is helpful for navigation, even if no specific knowledge of landmarks or other spatial memory has yet been acquired. For example, when entering an unknown city and trying to reach its center, it is usually a good idea to follow a riverfront or a wide avenue instead of crossing a labyrinthine neighborhood in an attempted straight line. An architectural environment is said to be "legible" if the way choices that the navigator intuitively tends to make are indeed suitable to reach the goal.

Spatial behavior in real cites is based on the interplay of both reactive guidance behavior and spatial memory. A case where reactive behavior is the dominant part is collective behavior of large groups of people, which can often be explained by simple, reactive rules without considering spatial memory or other cognitive mechanisms (agent-based modeling; see, e.g., Moussaid, Helbing, and Theraulaz 2011). Understanding traffic flow in large cities, however, has to take into account the actual street network as well as the travelers' perceptions and knowledge thereof. In urban planning, models of such behavior are studied in the field of space syntax pioneered by Hillier and Hanson (1984).

Indoor navigation is similar to outdoor navigation in many respects (Carlson et al. 2010) but differs by the possible presence of multiple floor levels and the significance of staircases as another structuring element (Hölscher et al. 2006). Also, metric information may be more important than in outdoor environments, possibly due to the smaller overall distances (Hölscher et al. 2006). Another important concept in indoor environments is the "isovist" (Benedikt 1979), which is defined as the portion of the ground floor visible from a given point of view (see figure 6.8e). In rooms with a convex geometry, the isovist is the same everywhere in the room and coincides with the floor plan. In corridors, lobbies, or the halls of a museum, the isovist can become quite complex especially if occluding walls are standing isolated in an otherwise free space. Wiener et al. (2007) have shown that the complexity of the isovist correlates with affective experiences of observers in a room and with their wayfinding performance. Also, in a search task in an unknown virtual environment composed of multiple rooms among which subjects had to choose, gaze pattern and search preferences were partially predicted by the isovist at each decision point (Wiener et al. 2012).

4.3.4 Goals and Obstacles

Path finding Real paths are not controlled by just one factor but depend on multiple constraints that are taken into account by the navigator. For example, an agent may be attracted by a distant light but turns away from optic flow to avoid obstacles. If an obstacle blocks the direct approach to the goal, the combination of attraction and repellence will lead to a curved detour around the obstacle and onto the goal. Obstacles are passed with a margin, even if they are not actually blocking the direct approach, but occur with slight sideways displacement. This behavior is indeed observed in human navigators (Warren and Fajen 2002) but would not be expected if path planning would proceed by finding the shortest collision-free path in a map of the environment. Such algorithms are well known in robotics (configuration space approach; Lozano-Perez 1983) but will not be discussed here.

In the logic of taxis and simple behavior, goals and obstacles can be treated as points to which the travelers are either attracted or from which they are repelled. Each point would then be the target of a positive or negative taxis, which exerts a "force" on the navigator. For a quantification of this idea, each taxis (or "motor schema") can be modeled by a radial potential function with a positive or negative sign, which is centered at the target and superimposed with the potential functions generated by other targets. The potentials of obstacles are peaks from which the agent will move away while the potential of the goal is a trough or basin to which the agent will be attracted. The agent would then be moving downhill on the combined potential surface (Arkin 1989).

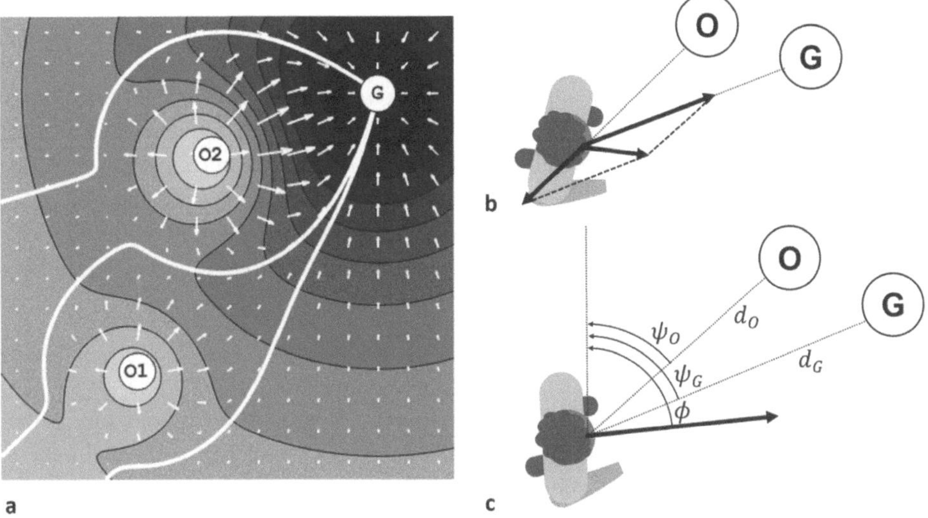

Figure 4.6
Path planning by combined attracting and repelling "forces." (a) In the potential field approach, a path is planned in a local map (peripersonal space) in which the goal (G) and the obstacles (O1, O2) are represented. The goal is surrounded by a wide negative potential forming a "basin of attraction," while obstacles are surrounded by narrower, positive potentials. The white arrows show the local downhill direction in the combined potential. Three paths from different starting points to the goal are shown. They are calculated for a navigator without inertia: that is, they tightly follow the local downhill direction. (b) Combination of attraction by the goal G and repellence from an obstacle O by vector summation. This mechanism underlies the trajectories shown in subfigure (a). (c) Angle-based model. Attraction and repellence are modeled as turns toward or away from the source. The model also includes a "stiffness" of the trajectory (i.e., a type of directional inertia in walking).

The idea is illustrated in figure 4.6a. The agent will move downhill on the potential landscape, and the lowest point is the location of the goal. Obstacles are modeled as local peaks from which the trajectory will automatically stay away. If this mechanism is used for planning, a local map is needed on which the potential is computed. Alternatively, the potential could simply be measured as a function of the nearness to the source (goal or obstacle). Together with a measurement of the bearing of these sources, the agent would then simply determine the vector resultant of the attracting and repelling forces, as shown in figure 4.6b. Indeed, vector summation of competing behaviors is a general scheme in action selection; for an overview, see Crabbe (2007).

Note that the trajectories shown in figure 4.6 contain rather abrupt turns. This is a result of the tight following of the local gradient of the potential function. In mechanics, such curves would occur if a body with zero inertia would be moving downhill in the potential surface. With inertia, the downhill vector will gradually accelerate the body in its direction as described by Newton's second law of motion, which results in smoother trajectories but also in occasional uphill sections.

More realistic models based on angular acceleration have been developed, for example, by Schöner, Dose, and Engels (1995) and Fajen et al. (2003); see figure 4.6c. The state variable is the heading ϕ of the navigator measured with respect to some arbitrary reference direction. At each instant in time, the bearings of the goal (ψ_G) and the obstacles (ψ_{Oi}) are measured together with the respective distances d_G and d_{Oi}. The potentials are replaced by attraction and repellence functions f_G and f_O, respectively. Motion is then modeled by a second-order differential equation as

$$\ddot{\phi} = -f_d(\dot{\phi}) - f_G(\phi - \psi_G, d_G) + \sum_i f_O(\phi - \psi_{Oi}, d_{Oi}). \tag{4.16}$$

The equation contains a damping term f_d, which could be proportional to angular velocity $\dot{\phi}$, as in ordinary friction, or could be made nonlinear to model experimental data. If it is large, the trajectory will be smooth, or "stiff," whereas small damping and strong attraction and repellence result in more kinky trajectories. The repellence function f_O will decrease with angular offset and distance, whereas the attractor function will stay high to ensure attraction to the goal also from large distances. The model has been tested with subjects passing a couple of obstacles while approaching a goal and produces realistic results (Warren and Fajen 2002).

Note that the reference direction shown in figure 4.6c is needed only for the graphical convenience. Equation 4.16 contains only angular differences in which the actual choice of the reference direction cancels out. The model does assume, however, some knowledge about the distance of obstacles and goals that have to be extracted from the image.

Path finding based on potential fields or the interaction of attracting and repelling forces is generally subject to the problem of dead-locking: that is, of getting trapped in local minima of the potential field. For example, if an animal would have spotted some prey behind a fence but would be unable to cross the fence, the sketched path finding procedures might force the animal to keep standing in front of the fence, watching the prey without being able to reach it. Detours around the fence would be impossible since they would require initially going away from the goal (i.e. uphill in the potential landscape). Clearly, most animals would do better than that; for a well-studied example in the European green toad *Bufo viridis* see Collett (1982). As in the ant milling example discussed above, a higher-level mechanism is needed that is able to interrupt the simple behavioral loop. For example, the repellence of the obstacle in front (i.e., the fence) might be increased until the local minimum in the potential landscape is filled. The downhill path in the corrected potential landscape might than be able to reach the goal. It may, however, also get stuck in the next local minimum. This seems to be the case in the toad, as was studied in experiments with double fences by Collett (1982). A general solution requires a working memory of the local environment together with a path-planning mechanism, as will be discussed in section 7.3.2.

Moving obstacles and goals If obstacles or goals are moving, the path-planning procedures discussed above will still work but may produce suboptimal results. If a moving target is pursued by simply following its shifting potential, it might in fact never be reached. Instead, the navigator should set a course with a lead ahead of the target, so that the trajectories will eventually intercept (see, e.g., Steinmetz et al. 2020). Control laws for interception tasks can be derived from optic flow, based on variables derived from time to collision (cf. section 2.5.2). Zago et al. (2009) review the literature on interception (including three-dimensional movements such as flying balls) and conclude that control laws alone are not sufficient to explain the experimental data but that internal models and representations are likely to play a role.

As an example of memory involvement in course control, consider an experiment on crossing a street with interfering traffic (Hardiess, Hansmann-Roth, and Mallot 2013). In a driving simulator, subjects approach a crossing street with traffic in both directions. Egomotion speed is fixed and cannot be changed. Shortly before the crossing, the motion stops and the subjects have to decide whether or not they would have hit one of the crossing cars in case the motion had continued.

During the approach, subjects tend to follow individual cars on the crossing street with their eye gaze. From the many cars present in every trial, fixation often goes to the one that is threatening a collision. Indeed, this car can be detected with a simple strategy not requiring memory involvement: if the bearing angle (i.e., the angle between the driver's heading and the approaching car) does not change in time, this car will be hit at the intersection.

After the trial stops ahead of the crossing, the cars on the crossing street are deleted from the image and the subject is asked to pick identical cars from a reservoir area with the computer mouse and place them at the positions where the cars from the approach phase vanished, thus reconstructing the image as it looked before the deletion of the cars. Subjects are generally able to do this, but they tend to leave out those cars that already passed the crossing and are thus no longer threatening a collision. This indicates that subjects maintain a memory of the car position together with their expected movement and the associated risk of collision. In real life, where subjects are able to control their approach speed to find a gap in the crossing traffic, this information will be very helpful.

4.3.5 Optimal Search and Foraging

One important reason for moving in space is the need to forage for food. Taxis may direct the search for promising locations such as in unicellular algae, whose positive phototaxis is clearly related to the need to run photosynthesis. If promising directions toward a food item cannot be told from a larger distance, however, the agent has to resort to search. This can be done in different ways; for example, the agent may simply wander around until it

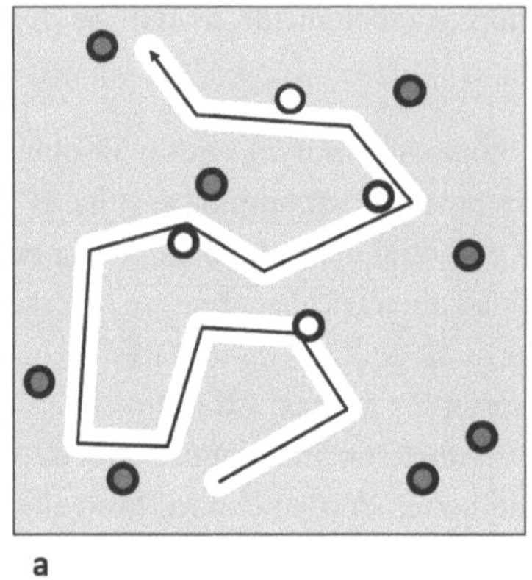

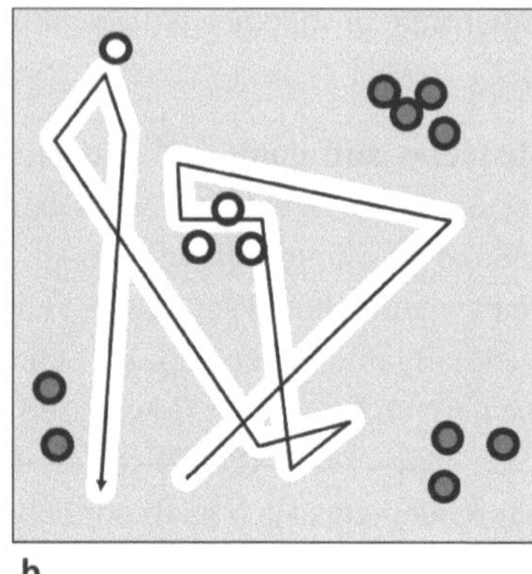

Figure 4.7
Optimal foraging strategies depend on the spatial distribution of food items. Filled dots: food items; open dots: depleted food sites. Black line: random walk trajectory. A food item is found if the trajectory passes by with a margin indicated by the white trace. (a) If food items are distributed homogeneously, a random walk search strategy with normally distributed steps is adequate. (b) If food items tend to occur in clusters, random walk with occasional long-distance steps ("Lévy flights") may be more appropriate. Animals such as predatory fish adapt their search strategy to the distribution of their prey. This implies that the according knowledge is represented in their brains.

hits a food item, or it may search a patch exhaustively and then jump to another patch. For a review of animal search strategies, see Bell (1991).

One distinction between different search strategies is illustrated in figure 4.7. If food items occur with a homogeneous spatial distribution, searching with a simple random walk[5] strategy may be sufficient, as shown in figure 4.7a. However, in the case of patchy distributions of food, as shown in figure 4.7b, a random search pattern with occasional larger movements to enter a new foraging range (Lévy flights) is more appropriate. In a study on marine predatory fish (sharks, tuna, etc.), Humphries et al. (2010) analyzed a large number of telemetrically obtained movement data and fitted the underlying distribution of the steps. The interesting parameter is the strength of the tail of the distribution of movement steps. If it is large, the agent will occasionally switch to a new foraging range as is optimal in patchy prey distributions, whereas short tails should be used for homogeneous distributions of prey. The authors show that sharks and other fish adapt the distribution to the environment and the likely occurrence pattern of their prey. This exceeds plain stimulus–response behavior since the environment type has to be represented in order to generate the behavioral switch.

5. A random walk in two or more dimensions is a vector-valued random variable where each vector is obtained from the previous one as $x_{t+1} = x_t + v$ and the increments v are drawn independently from a fixed random distribution. Random walk is also known as Brownian motion.

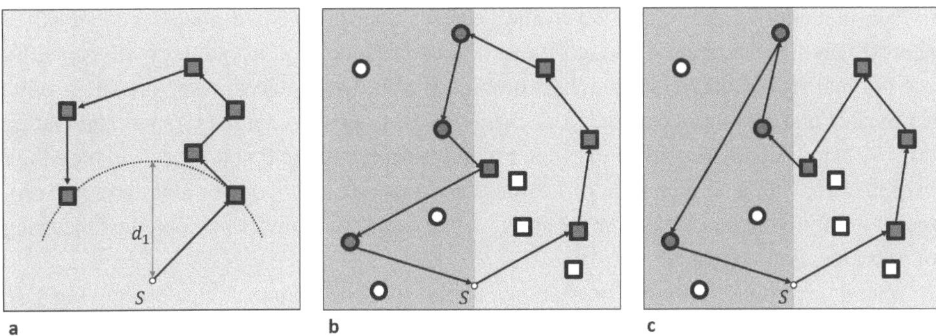

Figure 4.8
Traveling salesperson problem (TSP). (a) Array of six bait locations used by Cramer and Gallistel (1997) to study TSP performance in monkeys. From the start point *S*, two baiting sites have the same distance d_1. When using the nearest-neighbor strategy, they should therefore be chosen with equal probability. However, the monkeys prefer the rightward path, indicating that they think ahead, considering the reward gained in further steps of their trip, or that they plan paths to bait clusters. (b, c) Hierarchical representations in TSP performance. Human subjects visit a set of locations that form two regions containing different classes of goals (circles to the left and squares to the right of the figure). The open symbols are distractors. (b) Trajectory generated with the nearest neighbor strategy. (c) Trajectory observing region boundaries. Human subjects prefer this type of strategy even if it generates longer paths than the nearest-neighbor strategy (Wiener, Ehbauer, and Mallot 2009).

More information about the environment needs to be represented in order to optimize a path to multiple targets. The minimization of path length in this situation is known as the traveling salesperson problem (TSP) and has become paradigmatic of the study of optimization in both computer science and behavior. Cramer and Gallistel (1997) had vervet monkeys watch a scene in which a human experimenter walked around in a zigzagging course and placed food items along the way. The food items themselves were concealed from viewing. When baiting was completed, the monkey was allowed into the environment and started to collect the food. The paths produced by the monkey turned out to be much shorter than the baiting path previously watched and were in fact close to the shortest possible path. This indicates that the monkey did not simply remember and reproduce the baiting path but was able to plan a new path connecting the food locations in a novel sequence resulting in a shorter path.

In additional experiments reported in the same paper, Cramer and Gallistel (1997) were also able to show that the underlying planning strategy is taking into account more than just one future step (see figure 4.8a). If two food sites occur equally close to the start position but belong to two larger food clusters behind them, monkeys prefer the goal that is part of the larger cluster of resources, indicating that they think beyond their next step.

Path planning based on moving to the nearest hitherto unvisited goal is known as the nearest-neighbor strategy. Although the solutions produced by this strategy are generally not optimal (particularly in large TSP problems with many places), they are often quite reasonable and may suffice in everyday behavior. However, the data of Cramer and Gallistel (1997) indicate that monkeys do better than the nearest-neighbor strategy and base their path decisions on a combination of nearness and reward. This requires a hierarchical representation in which not only the location of individual food items but also the clustering of these locations is available.

Spatial hierarchies have also been shown to play a role in human navigators solving TSP problems. Wiener, Ehbauer, and Mallot (2009) designed TSP configurations in environments composed of a number of disjoint regions (figure 4.8b,c). In these environments, an additional strategy is possible in which subjects would visit first the goals of one region and then switch to the next region. This region-based strategy was indeed used by the subjects even if it led to poorer results (longer overall path length) than the simple nearest-neighbor search. This indicates that TSP planning is performed in a representation sensitive to spatial hierarchies.

4.4 Augmented Action–Perception Cycles

The examples discussed in the previous section suggest that the classical idea of the action–perception cycle needs to be generalized in various ways. With respect to the architecture, or modularization, of the system, the following extensions can be made (see figure 4.9).

Acquisitive behavior Sensory information is not only impinging on the receptors and used as it appears but may also actively be searched for. This is most obvious in eye movements where gaze is directed in a way as to optimize the information intake for a given behavioral situation. Similarly, directional turns of the pinna of the ear or peering movements of the head and body are examples of acquisitive or sensorimotor behavior. In the action–perception loop, acquisitive behavior differs from other behaviors in that the feedback from effector (e.g., the eye muscle plant) to the sensed information (e.g., the retinal image) is not mediated by the environment (figure 4.9b). Its purpose is indirect and the information obtained has to be made available to other, environment-oriented behaviors pursued by the agent.

Subsumption Taxis is a basic and powerful mechanism of spatial (and other) behavior but with low flexibility. It forms a default level that can run if no higher-level mechanisms interfere. The idea of a hierarchy of control layers was suggested in robotics as the "subsumption architecture" by Brooks (1986); see figure 4.9c. It contrasts previous top-down approaches of robot control that start with high-level goals and develop elementary action steps only after that. The idea that simple behaviors can run independently and higher-level

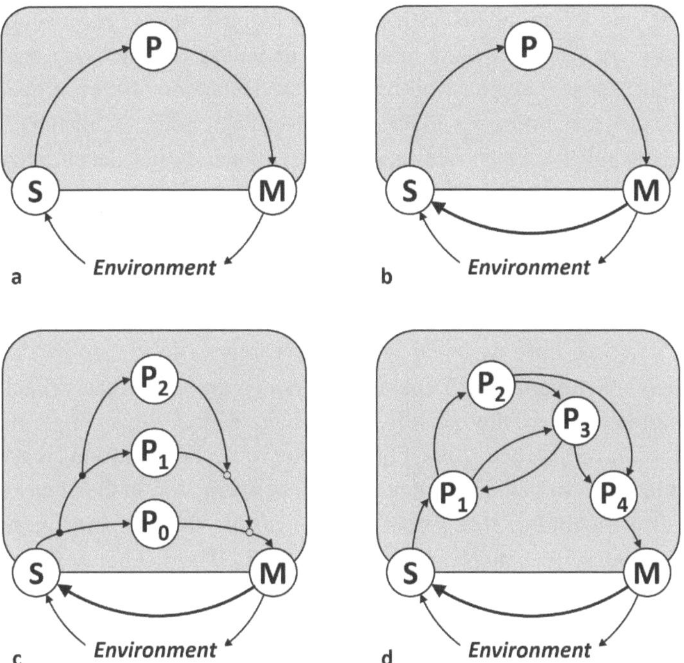

Figure 4.9
Augmented action–perception cycles. (a) Simple cycle with sensor (S), processor (P), and motor plant (M). (b) Acquisitory behavior such as eye movement forms an additional feedback loop bypassing the environment. (c) Subsumption. The default processing mode P_0 can be interrupted or modulated by higher-order modes P_1 or P_2. (d) General modularizations may evolve from the basic cycle by splitting existing modules or adding new ones. For example, the structure depicted in the figure may have developed from a simpler loop S–P_1–$P_{3/4}$–M by adding a control layer P_2 and splitting up P_3 and P_4.

control is added only as a secondary mechanism is also attractive as an approach to the evolution of behavior. Higher-level control may involve more elaborate data processing, cue integration, and memory.

Instincts When we talk of "behaviors" as a plurality that comprises a "repertoire," we assume that the inner part of the action–perception cycle can be subdivided into meaningful components. One way to do this is the subsumption architecture discussed above. Another idea is the breakdown into Freudian "drives": that is, competing motivations that lead to the execution of one or another behavior. Unlike subsumption, this does not result in a fixed hierarchy but allows for motivational changes between feeding, mating, looking for shelter, and so on, as are elementary in living beings. In the action–perception view, behaviors are "arcs" of the control loop including both sensory and motor components, a point that was emphasized by Nikolaas Tinbergen (1951) in his theory of the "instinct." The process of decision-making, or selection of a specific instinct appropriate to deal with a given situation, has been modeled in various, mostly decentralized ways: that is, by some sort of competition or voting based on the motivational components intrinsic to each behavior. Interesting examples include the hydraulic model of motivation and drive reduction by Lorenz (1978), winner-take-all neural networks (Arbib and Lieblich 1977), or the "societies of the mind" of Minsky (1986).

Modules So far, we have assumed that subdivisions of the inner part of the action–perception loop are "longitudinal": that is, they are processing streams reaching from the sensory to the motor side. Alternatively, processing units or steps can be delineated that may be involved in various behaviors. For example, "object recognition" may be a separate unit involved in different behaviors such as feeding, mating, or, in the context of landmark recognition, also wayfinding. On a coarser level, stages such as perception, memory, or planning can be named as examples of processing steps. The distinctions between processing streams supporting behaviors, on the one hand, and processing steps potentially useful in multiple behaviors, on the other, emphasize different dimensions that have been discussed as "horizontal" and "vertical" modules by Fodor (1983). This terminology suggests a matrix-like architecture where object recognition (to stay with this example) is the same irrespective of whether it is used in the context of mating or wayfinding. Indeed, this is how an engineer would probably design a behaving system: each processing step is the responsibility of one module that is used in all processing streams in which its computation is required. Modules should not be "committed" to just one behavior but can be freely reused wherever their computation is needed. David Marr (1982) formulated this requirement for an optimal modularization as his "principle of least commitment."

In the neuroscientific literature, "modules" are usually considered anatomically defined parts of brain, such as Brodmann's areas of the cerebral cortex. At the same time, these modules are parts of processing streams, as is discussed for the prefrontal cortex by Fuster (2001). Cortical areas can often be associated with well-defined processing steps, leading to a division of labor between areas for motion vision, place recognition, or language parsing and production. A simple example of a modularization that partially reflects both processing streams and steps is shown in figure 4.9. The principle of least commitment, and with it the distinction of horizontal and vertical modules, however, is not generally realized in the brain. Violations occur if a seemingly identical task occurring in different behavioral contexts activates different brain areas in fMRI or is differentially affected by lesions. Such cases are studied as "dissociations" in neuropsychology; they allow a better understanding of how the brain's division of labor works in detail. One important example of a dissociation that we will discuss at length in chapter 7 is that between the route and the map type of spatial knowledge.

We conclude this chapter with a brief note on cognitive evolution: cognition has evolved under the need to produce successful behavior. In its simplest case, this may have been taxis behavior in a basic action–perception cycle, but environmental needs and competition with other agents have led to more complex cognitive architectures. These are gradual modifications and extensions of the basic action–perception cycle in which new modules must have had an adaptive value right from the time of their first appearance. The result is the architecture of modern brains and of the cognitive apparatus, in which the action–perception structure, though always present as an underlying principle, may occasionally be hard to discover.

Key Points of Chapter 4

- Behavior goes on in a feedback loop of perception, internal processing, motor action, and the environment. It aims at the maintenance of an inner steady state ("homeostasis").
- Simple stimulus–response behaviors in which the motor action is a bodily movement are called taxes.
- Taxis behavior is rarely observed in isolation but is often combined with additional mechanisms such as task-specific intake of information, temporally extended motor plans, previously acquired knowledge, or the representation of goals and obstacles.
- In the evolution of complex behavior, simple action–perception cycles have developed into interacting networks of modules. These modules can be distinguished according to control level, processing steps performed, and the behavioral competences they subserve.

References

Arbib, M. A., and I. Lieblich. 1977. "Motivational learning of spatial behavior." In *Systems neuroscience,* edited by J. Metzler, 221–239. New York: Academic Press.

Argyros, A. A., D. P. Tsakiris, and C. Groyer. 2004. "Biomimetic centering behavior—Mobile robots with panoramic sensors." *IEEE Robotics & Automation Magazine* 11 (4): 21–30.

Arkin, R. C. 1989. "Motor schema-based mobile robot navigation." *The International Journal of Robotics Research* 8:92–112.

Bell, W. J. 1991. *Searching behaviour. The behavioural ecology of finding resources.* London etc.: Chapman & Hall.

Benedikt, M. L. 1979. "To take hold of space: Isovists and isovist fields." *Environment and Planning B* 6:47–65.

Beyeler, A., J.-C. Zufferey, and D. Floreano. 2009. "Vision-based control of near-obstacle flight." *Autonomous Robots* 27:201–219.

Bonabeau, E., M. Dorigo, and G. Theraulaz. 2000. "Inspiration for optimization from social insect behavior." *Nature* 406:39–42.

Braitenberg, V. 1984. *Vehicles. Experiments in synthetic psychology.* Cambridge, MA: MIT Press.

Brooks, R. A. 1986. "A robust layered control system for a mobile robot." *IEEE Journal of Robotics and Automation* RA-2:14–23.

Bruijn, S. M., and J. H. van Dieën. 2017. "Control of human gait stability through foot placement." *Journal of the Royal Society Interface* 15:20170816.

Carlson, L. A., C. Hölscher, T. F. Shipley, and R. Conroy Dalton. 2010. "Getting lost in buildings." *Current Directions in Psychological Science* 19:284–289.

Collett, T. S. 1982. "Do toads plan routes? A study of the detour behaviour of *Bufo viridis*." *Journal of Comparative Physiology A* 146:261–271.

Crabbe, F. L. 2007. "Compromise strategies for action selection." *Philosophical Transactions of the Royal Society B* 362:1559–1571.

Cramer, A., and C. Gallistel. 1997. "Vervet monkeys as travelling salesmen." *Nature* 387:464.

de Bellis, E., M. Schulte-Mecklenbeck, W. Brucks, A. Herrmann, and R. Hertwig. 2018. "Blind haste: As light decreases, speeding increases." *PLoS ONE* 13:e0188951.

Deneubourg, J. L., S. Aron, S. Goss, and J. M. Pasteels. 1990. "The self-organizing exploration pattern of the argentine ant." *Journal of Insect Behavior* 3:159–168.

Descartes, R. 1686. *Tractatus de homine.* Amsterdam: Typographia Blaviana.

Dickinson, M. H. 2014. "Death Valley, *Drosophila*, and the Devonian toolkit." *Annual Review of Entomology* 59:51–72.

Eckert, R. 1972. "Bioelectric control of ciliary activity." *Science* 176:473–481.

El, K. van der, D. M. Pool, M. R. M. van Paassen, and M. Mulder. 2020. "A unified theory of driver perception and steering control on straight and winding roads." *IEEE Transactions on Human-Machine Systems* 50:165–175.

Fajen, B. R., W. H. Warren, S. Temizer, and L. P. Kaelbling. 2003. "A dynamical model of visually-guided steering, obstacle avoidance, and route selection." *International Journal of Computer Vision* 54:13–34.

Floreano, D., and L. Keller. 2010. "Evolution of adaptive behaviour in robots by means of Darwinian selection." *PLoS Biology* 8 (1): e1000292.

Fodor, J. A. 1983. *The modularity of the mind.* Cambridge, MA: MIT Press.

Fraenkel, G. S., and D. L. Gunn. 1940. *The orientation of animals. Kineses, taxes, and compass reactions.* Oxford: Clarendon.

Fuster, J. M. 2001. "The prefrontal cortex—An update: Time is the essence." *Neuron* 30:319–333.

Gibson, J. J. 1950. *The perception of the visual world.* Boston: Houghton Mifflin.

Grassé, P.-P. 1967. "Novelles experiénces sur le termite de Müller (*Macrothermes mülleri*) et considérations sur la théorie de la stigmergie." *Insects Sociaux* 14:73–102.

Hangartner, W. 1967. "Spezifität und Inaktivierung des Spurpheromons von *Lasius fuliginosus* Latr. und Orientierung der Arbeiterinnen im Duftfeld." *Zeitschrift für vergleichende Physiologie* 57:103–136.

Hardiess, G., S. Hansmann-Roth, and H. A. Mallot. 2013. "Gaze movements and spatial working memory in collision avoidance: A traffic intersection task." *Frontiers in Behavioural Neuroscience* 7 (32): 1–13. doi:10.3389/fnbeh.2013.00062.

Helbing, D., P. Molnár, I. J. Farkas, and K. Bolay. 2001. "Self-organizing pedestrian movement." *Environment and Planning B: Planning and Design* 28:361–383.

Hicheur, H., Q.-C. Pham, G. Arechavaleta, J.-P. Laumond, and A. Berthoz. 2007. "The formation of trajectories during goal-oriented locomotion in humans. I. A stereotyped behaviour." *European Journal of Neuroscience* 26:2376–2390.

Hillier, B., and J. Hanson. 1984. *The social logic of space.* Cambridge: Cambridge University Press.

Hölscher, C., T. Meilinger, G. Vrachliotis, M. Brösamle, and M. Knauff. 2006. "Up the down staircase: Wayfinding strategies in multi-level buildings." *Journal of Environmental Psychology* 26:284–299.

Humphries, N. E., N. Queiroz, J. R. M. Dyer, N. G. Pade, M. K. Musyl, K. M. Schaefer, D. W. Fuller, et al. 2010. "Environmental context explains Lévy and Brownian movement patterns of marine predators." *Nature* 465:1066–1069.

Jennings, H. S. 1906. *Behavior of the lower organisms.* New York: Columbia University Press.

Kube, C. R., and H. Zhang. 1994. "Collective robotics: From social insects to robots." *Adaptive Behavior* 2:189–218.

Kühn, A. 1919. *Die Orientierung der Tiere im Raum.* Jena: Gustav Fischer Verlag.

Land, M. F., and D. N. Lee. 1994. "Where do we look when we steer?" *Nature* 369:742–744.

Loeb, J. 1912. *The mechanistic conception of life.* Chicago: University of Chicago Press.

Lorenz, K. 1978. *Vergleichende Verhaltensforschung. Grundlagen der Ethologie [Foundations of ethology].* Wien: Springer Verlag.

Lozano-Perez, T. 1983. "Spatial planning—A configuration space approach." *IEEE Transaction of Computers* 32:108–120.

Lynch, K. 1960. *The image of the city.* Cambridge, MA: MIT Press.

Macuga, K. L., A. C. Beall, R. S. Smith, and J. M. Loomis. 2019. "Visual control of steering in curve driving." *Journal of Vision* 19 (5): 1.

Marr, D. 1982. *Vision.* San Francisco: W. H. Freeman.

Matthis, J. S., J. L. Yates, and M. M. Hayhoe. 2018. "Gaze and the control of foot placement when walking in natural terrain." *Current Biology* 28:1224–1233.

Mauss, A. S., and A. Borst. 2020. "Optic flow-based course control in insects." *Current Opinion in Neurobiology* 60:21–27.

Minsky, M. 1986. *The society of mind.* New York: Simon & Schuster.

Moussaid, M., D. Helbing, and G. Theraulaz. 2011. "How simple rules determine pedestrian behavior and crowd disasters." *PNAS* 108:6884–6888.

Novembre, G., and P. E. Keller. 2014. “A conceptual review on action-perception coupling in the musicians brain: What is it good for?” *Frontiers in Human Neuroscience* 8:603.

O’Keefe, J., and L. Nadel. 1978. *The hippocampus as a cognitive map.* Oxford: Clarendon.

Ott, F., L. Pohl, M. Halfmann, G. Hardiess, and H. A. Mallot. 2016. “The perception of ego-motion change in environments with varying depth: Interaction of stereo and optic flow.” *Journal of Vision* 16 (9): 4.

Owens, D. A., J. Wood, and T. Carberry. 2010. “Effects of reduced contrast on perception and control of speed when driving.” *Perception* 39:1199–1215.

Pauly, P. J. 1981. “The Loeb-Jennings debate and the science of animal behavior.” *Journal of the History of Behavioral Sciences* 17:504–515.

Perna, A., and G Theraulaz. 2017. “When social behaviour is moulded in clay: On growth and form of social insect nests.” *Journal of Experimental Biology* 220:83–91.

Pfeifer, R., and C. Scheier. 2001. *Understanding intelligence.* Cambridge, MA: MIT Press.

Prinz, W. 1997. “Perception and action planning.” *European Journal of Cognitive Psychology* 9:129–154.

Redish, A. D. 1999. *Beyond the cognitive map: From place cells to episodic memory.* Cambridge, MA: MIT Press.

Sabiron, G., T. Raharijaona, L. Burlion, E. Kervendal, E. Bornschlegl, and F. Ruffier. 2015. “Suboptimal lunar landing GNC using nongimbaled optic-flow sensors.” *IEEE Transactions on Aerospace and Electronic Systems* 51:2525–2545.

Schneirla, T. C. 1944. “A unique case of circular milling in ants, considered in relation to trail following and the general problem of orientation.” *American Museum Novitates* 1253:1–26.

Schöne, H. 1984. *Spatial orientation. The spatial control of behavior in animals and man.* Princeton: Princeton University Press.

Schöner, G., M. Dose, and C. Engels. 1995. “Dyanmics of behavior: Theory and applications for autonomous robot navigation.” *Robotics and Autonomous Systems* 16:213–245.

Serres, J. R., G. P. Masson, F. Ruffier, and N. Franceschini. 2008. “A bee in the corridor: Centering and wall-following.” *Naturwissenschaften* 95:1181–1187.

Snowden, R. J., N. Stimpson, and R. A. Ruddle. 1998. “Speed perception fogs up as visibility drops.” *Nature* 392:450.

Srinivasan, M. V., S. W. Zhang, M. Lehrer, and T. S. Collett. 1996. “Honeybee navigation *en route* to the goal: visual flight control and odometry.” *The Journal of Experimental Biology* 199:237–244.

Steinmetz, S. T., O. W. Layton, N. V. Powell, and B. R. Fajen. 2020. “Affordance-based versus current-future accounts of choosing whether to pursue or abandon the chase of a moving target.” *Journal of Vision* 20 (3): 8.

Tinbergen, N. 1951. *The study of instinct.* Oxford: Clarendon Press.

Uexküll, J. von. 1926. *Theoretical biology.* New York: Hartcourt, Brace & Co.

Wallis, G., A. Chatziastros, and H. H. Bülthoff. 2002. “An unexpected role for visual feedback in vehicle steering control.” *Current Biology* 12:295–299.

Warren, W. H. 2006. “The dynamics of perception and action.” *Psychological Review* 113:358–389.

Warren, W. H., and B. R. Fajen. 2002. "Behavioral dynamics of human locomotion." *Ecological Psychology* 16:61–66.

Webb, B. 2020. "Robots with insect brains." *Science* 368:244–245.

Webb, B., and A. Wystrach. 2016. "Neural mechanisms of insect navigation." *Current Opinion in Insect Science* 15:27–39.

Wehner, R. 2020. *Desert navigator: The journey of an ant.* Cambridge, MA: Harvard University Press.

Wiener, J. M., N. N. Ehbauer, and H. A. Mallot. 2009. "Planning paths to multiple targets: Memory involvement and planning heuristics in spatial problem solving." *Psychological Research* 73:644–658.

Wiener, J. M., G. Franz, N. Rosmanith, A. Reichelt, H. A. Mallot, and H. H. Bülthoff. 2007. "Isovist analysis captures properties of space relevant for locomotion and experience." *Perception* 36:1066–1083.

Wiener, J. M., C. Hölscher, S. Büchner, and L. Konieczny. 2012. "Gaze behaviour during space perception and spatial decision making." *Psychological Research* 76:713–729.

Xu, X., and G. Wallis. 2020. "When flow is not enough: Evidence from a lane changing task." *Psychological Research* 84:834–849.

Zago, M., J. McIntyre, P. Senot, and F. Lacquaniti. 2009. "Visuo-motor coordination and internal models for object interception." *Experimental Brain Research* 192:571–606.

5 Path Integration

Path integration is the computation of the current position from the start point of a journey based on egomotion measurements, usually forward velocity and turn. These are obtained from the vestibular, proprioceptive, or optic senses as discussed in chapter 2. Landmark information is treated separately from path integration but may enter as a special case of compass information. Path integration is probably the most widespread navigational mechanism in the animal kingdom and an important source of metric information about space.

In this chapter, we discuss the basic computational approach characterized by the home vector that has been studied intensively in insects, most notably desert ants of the genus *Cataglyphis*. The underlying models are minimal in the sense that they assume the representation of the instantaneous home vector only, expressed by two or three scalar variables. We will then turn to path-integration behavior in humans and discuss evidence for richer representations. Finally, we present the computational neuroscience of path integration based on the grid and head direction cells found in the rodent brain. These models have greatly expanded the understanding of how the brain deals with space.

5.1 Dead Reckoning

Navigation, as opposed to mere course control, is a means to reach distant goals that are out of sight when the journey starts and may be located in the middle of completely unknown terrain. In this chapter, we assume that the goal is characterized by its distance and direction from the start point and ignore the role of landmarks marking the goal or important waypoints. The agent is then dependent on measurements of egomotion, which can be coupled or integrated to produce an estimate of the current position. The situation is simplified if a compass direction is available allowing the agent to correctly add each step to the estimate. In principle, however, it is also possible to replace the compass by a second integration process estimating the current heading by integration of former heading changes.

Path integration has long been used in ship navigation, where the speed may be measured with a log, magnetic north with a compass, and travel time with a clock or hourglass. At regular times during their watch, the helmsmen would check the current speed and compass direction and record it by inserting a pin into a so-called traverse board or pin compass from which the path can be approximately reconstructed by vector addition. In ship navigation, this process is also known as dead reckoning. If egomotion measurements are mostly based on counting the strides of the legs or the rotations of the wheels of a vehicle, the term "odometry" is also used.

Path integration can be carried out with easily obtained measurements of egomotion. It has, however, the disadvantage that errors once committed cannot be corrected later and will therefore accumulate. The only remedy for this is the use of landmarks and compasses. Even then, the distance up to which path integration reliably works is limited.

5.2 The Home Vector

5.2.1 Desert Ant Homing

Animals building nests and making feeding excursions from there ("central place foragers") need to be able to find back after the excursion. This behavior is known as "homing" and occurs in insects, spiders, birds, mammals, and many other groups across the animal kingdom; for an overview, see Papi (1992). Homing can be based on many cues, including, for example, previously laid pheromone trails, landmark knowledge, or the combination of multiple cues in a map of the environment. The term "path integration" refers to homing by the accumulation of egomotion measurements, which are stored as a "home vector": that is, as a minimal piece of memory specifying nothing but the direction and distance of the goal.

In naturally behaving animals, it is often difficult to separate path integration from other mechanisms, most notably from landmark usage. One group of model animals where path integration can be studied in isolation are desert ants living in environments where visual landmarks are rare and where the ground temperature is so high that pheromone trails would quickly evaporate. The best known of these are several species of the ant genus *Cataglyphis*, which has been studied intensively in the Tunisian Sahara desert by Rüdiger Wehner (2020) and his collaborators. Path integration does, however, also occur in ants of more temperate climates, but interactions with landmarks and self-generated trail markings make the analysis harder in these cases (Brun 1914; Maurer and Séguinot 1995).

Some of the behavioral findings are summarized in figure 5.1. *Cataglyphis* lives in underground nests, of which only the entrance is visible as a hole in the sand. A forager ant emerging from the hole will start a foraging excursion on a meandering path, keeping a rough overall direction. When a food item is found, for example, another insect, the forager will take it and move back to the nest in a straight path (Müller and Wehner 1988). In one example, the outbound path was 350 m long and the farthest distance from the nest was 113 m. The home vector needed to find back is symbolized in the figure by a black arrow. This home vector is continuously updated whenever the ant moves. Outward and

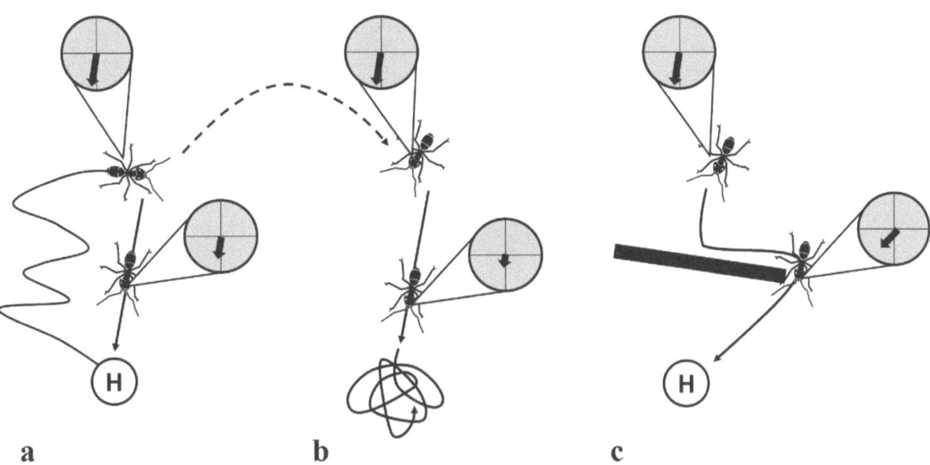

Figure 5.1
Vector navigation in ants. (a) An ant appears at the nest entrance ("home," H) and takes a curved excursion until it encounters a food item. During the excursion, the distance and direction to the nest are stored and continuously updated. Once a food item is found, the ant's motivation switches to returning to the nest, which is achieved by walking so as to reduce the stored home vector (i.e., in a straight line). (b) If the ant is picked up from its return run and displaced to a distant, inconspicuous place (dashed arrow in figure), it continues to walk in the assumed geocentric direction, producing roughly the correct distance: that is, until the home vector is reduced to zero. It then starts a search routine to find the nest entrance. (c) If an obstacle causes the ant to detour on its way back to the nest, the continuous updating of the home vector allows a direct approach to the nest after the obstacle is passed, without return to the initial course. For details, see Wehner (2020).

return paths differ by the motivational state of the ant. When this switches from foraging to return, the home vector is not only updated but also used to control the movement direction. When its length is reduced below some threshold, the ant will start to search for the nest entrance. A complete reset of the vector occurs only when the ant actually enters the nest.

If an ant is picked up during the return and displaced to a new location, the home vector is still present and the ant will run in the same geocentric direction as before until the vector is reduced to zero (figure 5.1b). It then starts a search routine that consists of loops of increasing size and in different directions but regularly coming back to the endpoint of the vector navigation, which marks the most likely location of the nest (Müller and Wehner 1995).

The continuity of home vector update is demonstrated in an experiment illustrated in figure 5.1c. In this case, a barrier is placed in the way of the returning ant such that it is forced to take a detour in its path. When the ant reaches the end of the barrier, it continues

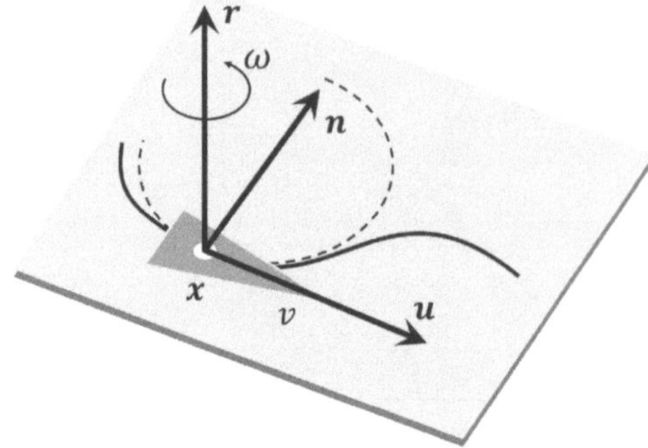

Figure 5.2
Path-integration geometry for planar movements. The agent is moving along a trajectory $\boldsymbol{x}(t)$ shown as a solid line. The unit vectors $\boldsymbol{u}$, $\boldsymbol{n}$, and $\boldsymbol{r}$ mark the local Frenet–Serret frame moving with the agent. $\boldsymbol{u}$ is the current forward direction. $\boldsymbol{n}$ is the "normal" vector pointing to the center of the local osculating circle, which is shown as a dashed line. The vector $\boldsymbol{r}$ is the current axis of rotation, which in the planar case coincides with the "binormal" vector of the Frenet–Serret theory. The scalar v and ω are the velocities of forward movement and rotation measured by the agent. Notations are as in equation 2.2.

its path directly to the nest without returning to the initial course. This indicates that path integration was continuing during the return and forced detour.

Maintaining precise home directions for distances of hundreds of meters by an animal of just 1 cm in length is an amazing ability. In doing so, ants make extensive use of the position of the sun, whose azimuth,[1] together with the time of day, provides a reliable compass information. Like many other animals with rhabdomic[2] photoreceptors, ants can see the polarization pattern of the sunlit sky. Since this pattern is symmetric to the meridian from the zenith passing through the sun, it can be used for precise measurement of the solar azimuth even when the sun itself is covered by clouds. While this may be a rather rare condition in the Sahara desert, it allows honey bees in central Europe to use the compass as long as a small spot of blue sky is visible (Frisch 1950). For further details on the polarization compass, see Rossel and Wehner (1986) and Labhart and Meyer (1999).

1. The solar azimuth angle is the compass direction of the sun, projected to the ground plane. For example, at places north of the Tropic of Cancer, it is strictly south at 12-o'clock noon.

2. Animal eyes generally use one of two types of photoreceptor cells called rhabdomic and ciliar. Among other things, they differ in the geometry of the membrane structure in which the rhodopsin (i.e., the light-absorbing protein) is located. In insects and other arthropods, these are microvilli: that is, tube-like protrusions of the receptor cells extending orthogonally to the direction of the incident light in a fixed angle for each receptor cell. Being attached to the cell membrane of these tubes, the rhodopsin has to align with the length direction of the tube. A given receptor cell therefore absorbs light only if the polarization plane aligns with its microvilli angle. A group of receptor cells can thus encode the direction of polarization of incident light. In contrast, the rhodopsin in the ciliar photoreceptor cells of vertebrates is located in "discs," in which it may take any angle. These receptors can therefore not distinguish different directions of light polarization (see, e.g., Land and Nilsson 2012).

5.2.2 The Geometry of Idiothetic Path Integration

Path integration and the differential geometry of curves The computational problem of idiothetic path integration is to update the current vector from home to the agent (or vice versa) from the instantaneous measurements of egomotion speed v and turning speed (heading change) ω. These measurements are "idiothetic" in the sense that they can be obtained without reference to external, environmental information such as compass directions or landmarks (see section 2.2.2). Mathematically, this problem is best stated with the tools of the differential geometry of curves; see figure 5.2 and do Carmo (2014). An agent is moving along a trajectory $\boldsymbol{x}(t) = (x_1(t), x_2(t), 0)$ in the ground plane. Speed and forward direction are then given by the derivative of $\boldsymbol{x}(t)$ with respect to time,

$$v = \|\dot{\boldsymbol{x}}\| \quad \text{and} \quad \boldsymbol{u} = \frac{\dot{\boldsymbol{x}}}{v}. \tag{5.17}$$

Since the length of $\boldsymbol{u}$ is fixed to 1, the direction of change of $\boldsymbol{u}(t)$ (i.e., its derivative) is orthogonal to $\boldsymbol{u}$; the unit vector in this direction is therefore called the normal vector, $\boldsymbol{n} = \dot{\boldsymbol{u}}/\|\dot{\boldsymbol{u}}\|$. In the planar case studied here, both $\boldsymbol{u}$ and $\boldsymbol{n}$ are confined within the ground plane. We may therefore write $\boldsymbol{u} = (\cos\eta, \sin\eta, 0)$, where η denotes the heading angle. In the general case (i.e., for space curves $\boldsymbol{x}(t)$), the vectors $\boldsymbol{u}$ and $\boldsymbol{n}$ define the so-called osculating plane, which is tangent to the trajectory. Finally, the vector orthogonal to both $\boldsymbol{u}$ and $\boldsymbol{n}$ is known as the binormal vector; in the planar case, it is the axis of instantaneous rotation of the agent and therefore denoted $\boldsymbol{r}$ in the figure. Together, the three vectors $\boldsymbol{u}$, $\boldsymbol{n}$, and $\boldsymbol{r}$ form the Frenet–Serret frame of the agent's trajectory. The speed of rotation about $\boldsymbol{r}$ is called ω and equals the change of heading over time. As an angular quantity with positive angles describing counterclockwise turns, it can be written as

$$\omega = \frac{\dot{x}_1\ddot{x}_2 - \ddot{x}_1\dot{x}_2}{\dot{x}_1^2 + \dot{x}_2^2}. \tag{5.18}$$

If the agent is moving with constant speed $v \equiv 1$, ω is the local curvature of the trajectory. Note that any time-dependent trajectory can be transformed into a curve with $v \equiv 1$ by a process known as rectification or parameterization by path length. While both descriptions are mathematically equivalent, the question whether ant path integration works with time-dependent or parameterized curves has received some attention in the experimental literature (Müller and Wehner 1988; Maurer and Séguinot 1995).

The geocentric cartesian solution The fundamental theorem of the local theory of curves (do Carmo 2014) implies that a planar trajectory $\boldsymbol{x}(t) = (x_1(t), x_2(t), 0)$ is completely defined by the measurements of $v(t)$ and $\omega(t)$ together with initial values for position $\boldsymbol{x}(0)$ and heading $\eta(0)$. As before, $(\cos\eta, \sin\eta) = (u_1, u_2)$: that is, η can be expressed via the two-argument arctangent function as $\eta = \text{atan2}(u_2, u_1)$. The curve is given by the solution of the following set of ordinary differential equations:

$$\begin{aligned}\dot{\eta} &= \omega \\ \dot{x}_1 &= v \cos \eta \\ \dot{x}_2 &= v \sin \eta\end{aligned} \tag{5.19}$$

This solution to the path-integration problem based on the classical Frenet–Serret equations was first proposed by Mittelstaedt and Mittelstaedt (1982). Since the variables on the left side appear only as derivatives, the equations can immediately be integrated to

$$\begin{aligned}\eta(t) &= \eta(0) + \int_0^t \omega(t')dt' \\ x_1(t) &= x_1(0) + \int_0^t v(t') \cos \eta(t')dt' \\ x_2(t) &= x_2(0) + \int_0^t v(t') \sin \eta(t')dt'\end{aligned} \tag{5.20}$$

where $\eta(0)$ and $(x_1(0), x_2(0))$ are the initial values of heading and position. If we set them to zero, "home" will be at the coordinate origin, and the x_1-direction of the coordinate system is the direction of first departure from home. Alternative formulations with discrete motion steps instead of continuous velocities have been given by Mittelstaedt and Mittelstaedt (1982) and Cheung and Vickerstaff (2010).

It is interesting to note that the fundamental theorem also applies to curves in three-dimensional space, in which case the "torsion," or roll, of the agent would have to be measured and integrated as a third quantity. This would allow idiothetic path integration also for nonplanar trajectories in the three-dimensional world. Using a sequence of elevated ramps zigzagging up and down, Wohlgemuth, Ronacher, and Wehner (2002) showed that *Cataglyphis* ants have correct path integration when returning from the ramps in the ground plane. This would be expected from three-dimensional path integration but could also be explained by two-dimensional path integration if the ants would be able to measure the horizontal component of their forward speed.

The procedure based on equation 5.19 is illustrated in figure 5.3a. The agent's position is represented as a Cartesian vector (x_1, x_2) with the starting point (marked "H" for home) as a geocentric reference. Heading is measured with respect to the direction of first departure from the starting point, $\eta(0) = 0$, which is again a geocentric reference. Still, the procedure is idiothetic (independent of detectable landmarks) since the home location and the direction of first departure are not directly measured during the travel. Rather, they are updated by accumulating directional changes into η and by accumulating trigonometrically weighted forward speed into $\boldsymbol{x}$.

Other reference frames Equation 5.19 describes a geocentric and Cartesian scheme for path integration, but it is not the only way in which path integration can be carried out.

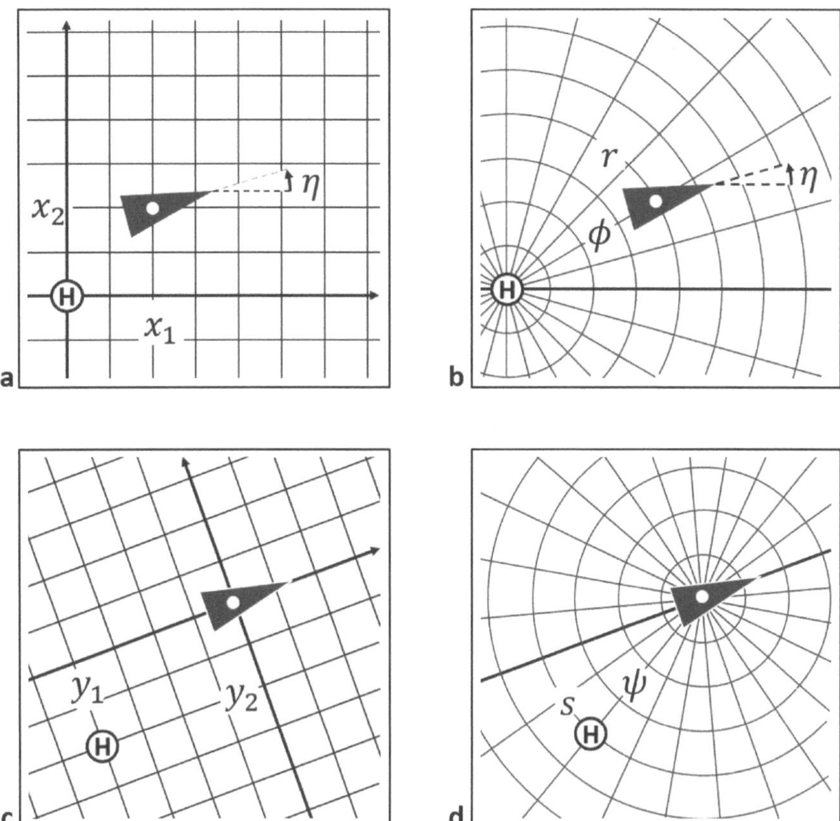

Figure 5.3
Coordinate systems used for path integration. H: home position (starting point). The black triangle marks the position and orientation of the agent with the white dot as a reference point. (a) Geocentric Cartesian. The horizontal axis is the axis of first departure from H. The coordinates of the agent are $(x_1, x_2) = (2, 2)$ (grid spacing = 1). η is the heading relative to the direction of first departure; in the picture, $\eta = 15°$. (b) Geocentric polar. Axis of first departure and heading angle as in (a). The agent's coordinates are $(r, \phi) = (4, 30°)$ (spacing of radii: 15°). (c) Egocentric Cartesian. The coordinate axes are the forward and left/right axes of the agent. The coordinates of home are $(y_1, y_2) =$ (–4, –2). Heading is not independently represented. (d) Egocentric polar. The coordinates of H are $(s, \psi) = (4, -150°)$.

In fact, other schemes have been proposed, based on behavioral experiments with various animal species. The most important distinctions are those between geo- and egocentric systems and between Cartesian and polar coordinate frames (Maurer and Séguinot 1995; Merkle, Rost, and Alt 2006; Cheung and Vickerstaff 2010).

Table 5.1
Systems of ordinary differential equations for idiothetic path integration. All schemes work with the measurements of forward speed v and rotation ω. Notations as in figure 5.3.

	Cartesian	Polar
Geocentric (represents agent position and heading)	$\dot{\eta} = \omega$ $\dot{x}_1 = v \cos \eta$ $\dot{x}_2 = v \sin \eta$	$\dot{\eta} = \omega$ $\dot{r} = v \cos(\eta - \phi)$ $\dot{\phi} = \frac{v}{r} \sin(\eta - \phi)$
Egocentric (represents home)	$\dot{y}_1 = -v + \omega y_2$ $\dot{y}_2 = -\omega y_1$	$\dot{s} = -v \cos \psi$ $\dot{\psi} = -\omega + \frac{v}{s} \sin \psi$

Geocentric systems such as the one discussed above use the starting point of the journey as a reference (figure 5.3a,c). We denoted this point as "home," but it should be clear that in principle, the starting point can be anything (for example, a large food item) that requires repeated visits or a landmark to which the section of a longer route can be anchored. In addition to a coordinate origin, the geocentric system requires a reference direction, which could be provided by a compass or distant landmark. In the idiothetic case discussed here, where no such cues are used, we chose the direction of first departure from home as a reference. The entire scheme then consists of three variables, two for the agent's position relative to home and one for its orientation.

In egocentric schemes (figure 5.3b,d), the coordinate origin is always the current position of the agent, and the reference direction is its heading. The schemes therefore work with just two variables; namely, the home position in the current egocentric frame. The two-dimensional vector specifying the home position in this system therefore also contains implicit information about the agent's heading. As a consequence, if the agent turns on the spot, the represented home position will move on a circular arc about the agent.

The equations for the four resulting cases appear in table 5.1. In the polar schemes, the distance variable (r or s, respectively) appears in the denominator of a fraction, which leads to problems if it approaches zero: that is, if the agent is close to home. This problem can easily be avoided if the agent uses some landmark cues for the final approach and if it homes and starts a new journey whenever home is close.

Mathematically, the difference between the various formulations of the path-integration equations is not obvious; in the absence of noise, they are equivalent. Figure 5.4a shows a simple trajectory as a heavy black line; its transformation into egocentric coordinates appears in figure 5.4b. The small ripples of this trajectory are reflected in the egocentric

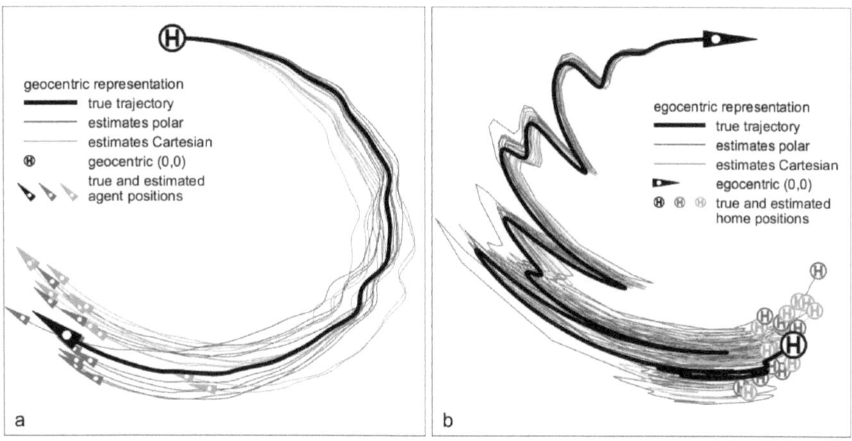

Figure 5.4
Numerical solution of the path-integration equations for a sample trajectory shown as a heavy black line. (a) Geocentric representation (home at coordinate origin, agent moves). (b) Egocentric representation (agent at coordinate origin, home representation moves). The thin lines show simulations with different noise representatives. All simulations use the same forward Euler algorithm with 100 equidistant steps and equal amounts of noise.

view by large circular oscillations resulting from body turns of the agent. These oscillations are larger the further the agent is away from home.

Taking this trajectory as the ground truth, the values of turn ω and forward speed v were calculated at 100 equidistant steps and disturbed by additive Gaussian noise. From these noise-affected "measurements," estimated trajectories were calculated based on each of the four formulations of the path-integration equations using a simple forward Euler procedure. The endpoints show the distribution of path-integration error. All simulations agree that return to home should start with a right turn of about 90 degrees, and walking distance should be about half the length of the outbound travel. No clear preference for one of the schemes is apparent. Note, however, that more sophisticated simulations of measurement error may lead to different results. For example, Merkle, Rost, and Alt (2006) found superior performance for the egocentric Cartesian scheme, while Cheung and Vickerstaff (2010) claim that all schemes but the geocentric Cartesian one are subject to catastrophic error in the long run. We will see in Section 5.4 that the mammalian grid cell system indeed seems to use a version of the geocentric Cartesian reference frame.

5.2.3 Allothetic Path Integration

Compasses and distant landmarks Path integration from the velocities of forward movement and turning can be improved by the use of additional allothetic information. One cue providing such information is a compass: that is, the measurement of a fixed geocentric direction, for example, the magnetic north in ship navigation. In biological navigation, several sources of compass information have been shown to play a role, most notably the celestial bodies (sun, moon, stars), the earth's magnetic field, and distant landmarks.

Of the celestial bodies, the sun as compass has already been mentioned above. As it takes its path across the sky, the solar azimuth obviously changes and has to be interpreted with respect to the time of day. This has been shown to be the case for ants and honey bees, the latter ones continuously adapting the direction of their waggle-dance, which indicates the changing angle between the sun and a food source when they dance for hours in the hive. In humans, Souman et al. (2009) show that a desired constant direction in extended outdoor hikes takes the solar azimuth into account. Walking direction turns at least slightly with the sun as long as it is visible and breaks down if the sun gets occluded by clouds. The same was demonstrated for the moon during night walks. Visibility of the sun also improves pointing performance to out-of-sight targets in human foragers in rainforests (Jang et al. 2019). Navigation by the starry sky was demonstrated in night migrating indigo buntings (*Passerina cyanea*) by Emlen (1970). These birds' experience with the starry sky was controlled by presenting them with a planetarium sky in which the normal star pattern was rotating about Betelgeuze (in the constellation Orion) rather than about Polaris in Ursa minor. When released and tested for migration under the real sky, these birds navigated as if Betelgeuze was the fixed compass direction: that is, they turned with the celestial rotation. This indicates that the direction toward the center of celestial rotation is a learned compass cue in these night-migrating birds.

Magnetic senses are found in many organisms, including bacteria and many animals, most notably, of course, in migrating birds (Wiltschko and Wiltschko 2005). Surprisingly, the magnetic compass of migrating birds is not based on the horizontal (earth-tangential) component of the magnetic field (declination) but on its inclination: that is, the angle between the magnetic field lines and the local horizontal plane. As a result, the bird's magnetic compass is not oriented north versus south but "pole-ward" versus "equator-ward" (i.e., the direction inverts when the equator is crossed). The physiological mechanisms of magnetosensation in birds are localized in the retina and dependent on simultaneous stimulation with light (Ritz, Adem, and Schulten 2000). Evidence for magnetic orientation in humans is weak, but some influences of magnetic field inclination on brain activity could be demonstrated (Wang et al. 2019).

Magnetic and celestial compass cues can also be combined. For example, Cochran, Mouritsen, and Wikelski (2004) studied northward migration of New World thrushes of the genus *Catharus* in the North American Great Lakes area, where geographic and magnetic north directions substantially deviate. These birds calibrate their magnetic compass using the direction of the setting sun as the true west. During the following night, the thrushes migrate with their magnetic compass, taking their calibrated west as a reference.

Celestial bodies are the classical case of "distant" landmarks whose absolute direction will not change with the movement of the observer. The same is true, to a lesser extent, also for distant mountain peaks or beacons, with the approximation becoming more and more error prone as the distance decreases. Still, the direction to distant landmarks can be used as a compass cue. For example, in the rodent head direction system (to be discussed below), it has been shown that if two landmarks supporting the same visual angle move in deviating directions, the neurons tend to anchor their preferred firing direction to the more distant landmark (Zugaro et al. 2004). Evidence for the use of distant ("global") landmarks for orientation in human navigators has been presented by Steck and Mallot (2000). For indoor environments, compass-like reference axes based on allothetic cues have been discussed in section 3.4.2.

Path integration with a compass In the geocentric path-integration mechanisms described in equation 5.19 and table 5.1, reliable compass information can be used to replace the first equation, which describes the updating of heading from turns. Heading η can be directly measured and does not need to be inferred from integration. This will, of course, reduce the overall path-integration error.

Müller and Wehner (1988) studied ant path integration in the presence of compass information in a triangle-completion task. Ants were running in a channel composed of two open tubes allowing a view of the sky. The angle α between the segments was varied across experimental conditions. When leaving the second segment, the ants would turn back to the nest but with systematic overshoot in the sense that they would intersect the initial path segment. Since the path-integration models summarized in table 5.1 do not produce such systematic errors, Müller and Wehner (1988) argued that the ant's path integrator, which is of course a product of evolution, is but an approximation of the computationally correct scheme. More specifically, they suggest a version of the geocentric polar integrator in which η is directly measured from the compass and the distance r is replaced by the path length l. The direction from home to the ant, ϕ, is then updated as

$$\dot{l} = v \tag{5.21}$$

$$\dot{\phi} = \frac{p(\eta - \phi)}{l}, \quad \text{with } p(\delta) = k\delta(\pi^2 - \delta^2). \tag{5.22}$$

The polynomial p is an approximation of the sin-function appearing in table 5.1, k is a constant that was determined experimentally, and l is the total previous path length. The idea that the angular change is weighted by previous path length rather than by the distance to the goal was introduced ad hoc but, together with the polynomial approximation of the sin-function, produces good fits of the experimental data.

In the egocentric schemes, compass information cannot be directly used to replace one of the equations but only to produce better measurements of ω. Remaining errors in the estimation of ω, however, will still accumulate, since heading is not explicitly represented.

Compass information would therefore only be of limited help to reduce error accumulation. In fact, this seems to be the major argument for preferring geocentric over egocentric schemes.

So far, we have considered path integration as the updating of a simple vector defined by two points, home and the agent's current position. If the compass direction is free of error, it can be used to replace the direction of first departure $\eta(0)$, leaving the structure of the underlying representation unchanged. The situation changes, however, as soon as multiple sources of compass information are available, which may be independently disturbed when landmark distance is large but not infinite or when compasses are only occasionally available while idiothetic path integration must take over in the meantime. Path integration then becomes an optimization problem in which directions to multiple beacons or distances to multiple landmarks need to be continuously updated. The resulting representation would be more like a map than a home vector and will therefore be discussed in sections 7.3 and 8.3.2.

Walking straight A problem related to path integration is to keep a straight course over long distances. This paradigm is hard to use in animals (but see Dacke et al. 2021), but human subjects can easily be instructed to do this task. In one experiment of the study by Souman et al. (2009), blindfolded subjects were instructed to walk along a straight line on a wide-open beach. Most of them ended up walking in circles and, in doing so, occasionally switched between different senses of rotation. Although there was a steady wind in a fixed direction, subjects did not use this as a compass cue but actually reported frequent changes of wind direction, even though these were only happening in their own, egocentric reference frames. The reason for the circling behavior is not clear; it seems that the perception of turn in the blindfolded condition is biased and that this bias occasionally switches to a new value, which is then again kept for a while.

Although animals cannot easily be instructed to walk straight, the ecological relevance of this performance seems quite clear, for example, in escaping a danger as fast as possible or in trying to find a new foraging area. Cheung et al. (2007) and Cheung and Vickerstaff (2010) have therefore argued that the generation of straight courses might be an evolutionary precursor of path integration and presented a thorough computational analysis of this behavior. The idea is to treat it as a "directed random walk": that is, a stochastic process in which the positions $\boldsymbol{x}_i$ of the agent form a series of random vectors. At each time step, the agent takes a step into a direction determined either from the previous direction (idiothetic case) or from a compass measurement (allothetic case) and disturbed by some noise process. The result is the distribution of endpoints after a long series of steps. In the idiothetic case, the expected value of this distribution is a fixed point at some distance from the start, which depends on the variance of the noise process. This is to say that on average, the agent cannot escape from the starting point by more than this expected value. In the allothetic case, a compass can be used, and arbitrary distances from the starting point can be reached even on average.

5.3 Path Integration in Humans

5.3.1 Triangle Completion

The standard paradigm for the study of path integration in humans is called "triangle completion" (see figure 5.5a,b). In this experiment, subjects are guided along two legs of a triangle that may vary in length and included angle. At the end of the second leg, subjects are asked to point to the start point (figure 5.5a), walk back to the start point (figure 5.5b), or both. Depending on which procedure is used, the turning angle from the end of the second leg, the trajectory, or the coordinates of the endpoint can be measured as dependent variables. It should be noted that the exact procedure for taking pointing measurements can be tricky. In figure 5.5a, we assumed pointing with the index finger, in which case the line from the nose (or the dominant eye) of the observer to their fingertip defines the pointing direction. When pointing with a device, such as a cane or laser pointer, the orientation of this device is a more natural definition. Particularly in the case of a laser pointer or a VR pointing device, one can also measure the position where the laser beam hits the ground (e.g., Wolbers et al. 2008). Finally, in desktop virtual reality setups, pointing is often realized by allowing the subject to rotate the scene until the intended direction is in the center of the screen (e.g., Restat et al. 2004). This procedure involves an imagined body rotation and may therefore be compromised by the failure of such imagined rotations to be incorporated in the representation of peripersonal space (see figure 3.5). For a discussion of pointing procedures, see Montello et al. (1999).

Loomis et al. (1993) used triangle completion with return (figure 5.5b) to study path integration performance in three groups of subjects: sighted but blindfolded, adventitiously blind (i.e., subjects had been born sighted but acquired blindness later in life), and congenitally blind people. Egomotion perception was thus restricted in all subjects to nonvisual cues. The study uses a variety of triangle configurations that lead to a range of required values for return length and return angle. Both parameters show a clear correlation between required and produced value but also an effect known as "compression to average": relatively long returns are underestimated while relatively short return distances are overestimated. Most interestingly, however, no significant difference was found between the three groups of subjects.

This latter result is surprising. If space perception without vision could be improved by training, one would expect that the blindfolded sighted group should perform poorer than the other ones, of which especially the congenitally blind subjects had a lifetime of training on using nonvisual egomotion cues. Contrary to this expectation, the results seem to indicate that nonvisual path integration is a mechanism that is not subject to more intensive learning in the blind but is available for sighted and blind people alike.

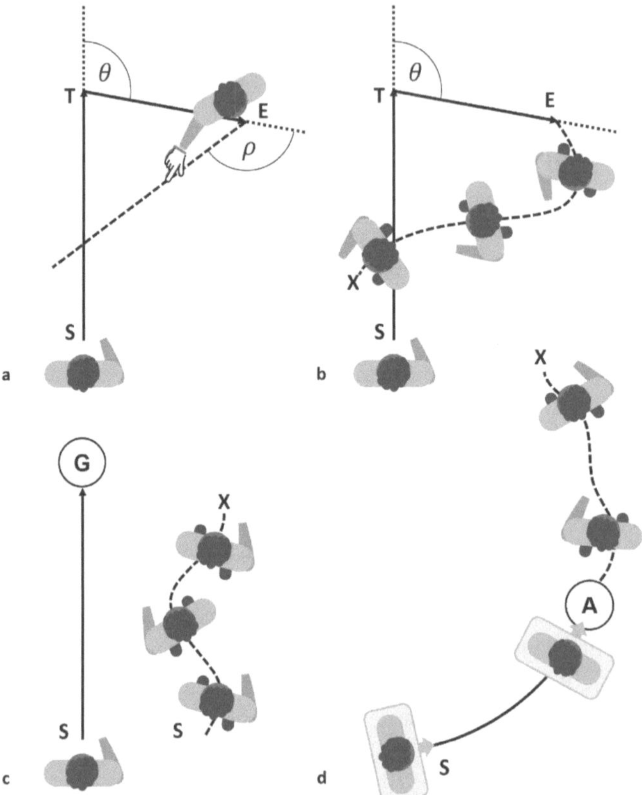

Figure 5.5
Experimental protocols for path integration in humans. (a) Triangle completion with pointing. S, start; T, turning point with turning angle θ; E, end point of guided section. The dependent variable is the pointing angle ρ. (b) Triangle completion with return. Dependent variables are the reported end point X of the return and other parameters of the return walk. (c) Walking up to a previously viewed goal at G. During the walk, the goal is removed or the subjects are blindfolded ("walking without vision"). The dependent variable is the endpoint of the walk, X, at the remembered goal position. (d) Reproduction of a passive transportation. Subjects are passively transported to a point A and asked to reproduce the movement. Dependent variables are the adjusted endpoint X as well as other path parameters.

In the study by Loomis et al. (1993), egomotion information is provided by nonvisual cues only. In the presence of vision, egomotion perception would be improved by optic flow while the general layout of the scene and contained landmarks provide additional reference information. Péruch, May, and Wartenberg (1997) tested triangle completion in a desktop virtual reality setup. Homing performance was comparable to the Loomis et al. study. Varying the amount of visual information by adjusting the field of view did not influence homing performance. Visual information about egomotion only (i.e., not including landmark information) can be provided in several ways. For example, Riecke, Veen, and Bülthoff (2002) used a virtual environment in which the landmarks present during the guided part of triangle completion task vanished and were replaced by completely different objects as soon as the return part started. Again, triangle completion performance is comparable to the results from blindfolded walking. Another way to exclude visual landmarks while still providing visual egomotion cues is to cover the ground floor with random dots, some of which are deleted after a limited "lifetime" and replaced by others. This pattern cannot be used as a landmark upon return to the starting point, but it does provide optic flow information (e.g., Wiener and Mallot 2006; Wolbers et al. 2007). In a recent study by Stangl et al. (2020), this ground texturing with limited lifetime random dots was augmented by a panorama of distant landmarks at infinity that provides an orienting reference direction (compass).

5.3.2 Continuous Updating vs. Configurational Memory

In triangle completion, the forward travel consists of two straight-line segments and an included angle. Similarly, in humanmade environments with streets or corridors, routes are often composed of straight sections connected by right-angle turns. This has led researchers to assume that subjects remember these discrete elements (leg lengths and included angle) and compute the return path only when reaching the end of the second leg ("encoding-error model"; Fujita et al. 1993; Harootonian et al. 2020). An alternative is the idea of the home vector discussed above, where path integration is a process of continuous updating and the segmentation of the travel into straight segments is irrelevant.

If a subject in a triangular completion task takes a step to the left and then back to the original course, continuous updating of the home vector predicts that these irrelevant movements are tracked and increase the overall performance error. In the encoding-error approach, however, the sidestep might already be ignored during encoding and should therefore not affect homing accuracy. May and Klatzky (2000) compared the effect of such irrelevant movements to that of verbal distractions happening at the same location along the path and found that performance is affected stronger by the irrelevant movements. This supports continuous models of path integration.

Another prediction of the discrete encoding-error scheme is that the number of linear segments should affect the precision of path integration. Sadalla and Magel (1980) asked subjects to walk unfamiliar indoor routes composed of straight sections and right-angle turns in both directions. Test routes were 61 m long and contained either two or seven right-angle turns. After completing the test routes and an additional shorter route, subjects judged the traversed distance of each route relative to the final, shorter route or relative to each other. Sadalla and Magel (1980) report an overestimation of the length of the seven-turns path and an underestimation of the two-turns path, both on the order of some 20 percent.

They suggest that the perception of traversed distance depends on the number of right-angle turns, which is one less than the number of straight segments in their experiment. Note, however, that a dependence on the total absolute turning angle, a quantity that may also be defined for less regular routes, might suffice to explain the results.

Wiener and Mallot (2006) tested the dependence of path integration error on path complexity by using forward travel paths with constant length and total turning angle but a variable number of straight-line segments. Thus, routes with more segments would have smaller turning angles between the segments. The experiment was carried out in virtual reality with unrestricted vision. The results clearly indicate that error is independent of the number of straight segments but does depend on total turning angle. Path integration thus seems to be a continuous updating process that does not decompose the travel path into straight segments.

Still, the idea that the overall configuration of the scene and the path taken in it affects path-integration performance cannot be completely abandoned. Wiener, Berthoz, and Wolbers (2011) explicitly asked subjects to solve a path-integration task either by continuously considering the start position or by remembering the configuration of the path segments. Subjects following the continuous strategy gave quicker responses, but the results in the configurational condition were more accurate. This suggests that the two mechanisms may be operating simultaneously. The continuous mechanism is nicely described by the theory of the home vector discussed in the previous section while the configurational approach assumes a richer representation, similar to the peripersonal space discussed in chapter 3. In this latter view, path integration is a kind of spatial updating in which the starting point of the current journey is just one of many other environmental locations whose egocentric position is continuously updated due to egomotion and landmark information.

Additional evidence for the involvement of richer representations in path integration comes from a study by Israël et al. (1997). Subjects were blindfolded and passively transported on a remotely controlled vehicle. They were then given the controller of the vehicle and asked to reproduce the distance previously traveled (cf. figure 5.5d). In this setup, vestibular cues were the only cues available for the judgment of egomotion. During the passive section of the travel, the vehicle was driving with one of three velocity profiles, a "triangular" one consisting of a constant acceleration followed by a constant deceleration, a "trapezoidal" one with a plateau of constant medium velocity in between, and a "rectangular" one with steep acceleration and deceleration and a long plateau in the middle. All profiles were scaled so that the total distance traveled was the same. If path integration is solely based on the home vector using one of the continuous update rules given in table 5.1, the velocity profiles should not matter, except maybe for some error variation. However, the results of Israël et al. (1997) clearly show that subjects reproduce not only the traveled distance but also the velocity profiles. This indicates that performance in this distance reproduction task is not based on simple vector accumulation alone but involves a richer, spatiotemporal representation of the passive part of the travel.

Distance reproduction in three dimensions was studied with subjects seated in the cabin of a cable robot by Hinterecker et al. (2018). Subjects showed a clear ability for vertical path integration, albeit with larger error and reduced gain as compared to planar tasks.

5.3.3 Sources of Path Integration Error

Without landmark usage, errors in path integration can have two major sources: errors in egomotion perception and errors in the integration mechanism, be it discrete or continuous. Lappe, Jenkins, and Harris (2007) used the same visual motion simulation in two different tasks performed by seated subjects. In one task, subjects were watching the simulated egomotion stimulus and were then required to adjust a depth marker to the distance traveled. In a second task, they saw the depth marker at some distance and were then asked to control the visual motion to stop at the position of the now invisible marker (figure 5.5c). Errors occurred in both tasks but with different biases: in the first task, the marker was placed too close, indicating an underestimation of traveled distance. In the second task, egomotion was stopped short of the marker position, indicating an overestimation of the distance traveled in the exact same stimulus. This difference cannot be caused by the visual motion sequence, which was the same in both cases, but must be due to the integration step or different strategies employed by the subjects in the two tasks.

In a more direct approach to path integration error, Stangl et al. (2020) used curved paths with actual walking and visual information provided by VR goggles. Subjects were guided along a smoothly curved path; at four points along this path, they were asked to verbally report the estimated distance from the start and to indicate the direction to the starting point by turning their body. From the error distributions obtained from a large number of measurements, a statistical model was fitted that is based on the geocentric Cartesian update (equation 5.19) augmented by noise. With this approach, the authors were able to identify noise in the estimation of velocity as the main source of error and showed that this error accumulated with traveled distance, not elapsed time. Noise in the perception of egomotion velocity was measured separately for each subject and was shown to increase with the subjects' age.

In the continuous update equations of table 5.1, time is treated as a physical variable, not as an additional variable that has to be measured by the agent. It is well known, however, that the perceived duration of stimuli may deviate from the veridical physical time in systematic ways (see, e.g., Ulrich, Nitschke, and Rammsayer 2006). If this were also true for path integration, systematic errors can be expected. Glasauer et al. (2007) used a dual-task paradigm to manipulate cognitive time in various path-integration paradigms. As a dual task, subjects were required to count backward in steps of seven from a number given by the experimenter and to speak out the results aloud. When this is done during the passive (encoding) phase of a path-integration experiment, perceived distance and duration are underestimated while these percepts are overestimated if the dual task is added during the reproduction phase. Additional judgments of egomotion speed are not affected. These results are consistent with the idea that the path integrator uses a veridical measurement of velocity and combines it with a cognitive representation of duration. Error is mostly due to the temporal component.

5.3.4 Neural Substrate

In rodents, it is generally accepted that the hippocampal and entorhinal systems are crucial for the performance of spatial tasks (see next section). In humans, the situation is more complex as more brain areas seem to be involved, probably related to the different mechanisms discussed above. Wolbers et al. (2007) investigated the brain areas recruited in a triangle completion task with pointing. Egomotion stimuli were simulated flights over a virtual plane, presented in the fMRI scanner. Behavioral results show overall good performance, again with the compression-to-average effect found already by Loomis et al. (1993). Considering all trials of all subjects together (within-subject analysis), the activation of the right, but not the left, hippocampus shows a significant positive correlation with performance: that is, subjects perform better if their right hippocampus is active. A more detailed analysis of all subjects individually showed that statistical pointing error is also affected by activity in the medio-prefrontal cortex, an area associated with working memory functions. The occurrence of systematic pointing biases can be predicted from activity in the human MT+ complex, a part of visual cortex known to be involved in motion perception. This may indicate that such biases are generated by erroneous estimates of egomotion, which cannot be compensated by the path-integration system proper.

Chrastil et al. (2017) used a modified task in which subjects were presented with a simulated movement along a circular arc that was shorter than, equal to, or longer than a complete circle. After watching the video, participants decided whether or not they ended up at the starting point. In order to do this, they had to track the goal location by path integration. Performance differences between subjects were positively correlated with morphometric measurements of the size[3] of three brain areas: hippocampus, medio-prefrontal cortex, and retrosplenial cortex. This latter area is part of the medial parietal lobe and is thought to be involved in heading representation and spatial imagery (Baumann and Mattingley 2010).

While these and a number of similar studies suggest that the hippocampus does play an important role in human path integration, Shrager, Kirwan, and Squire (2008) report that patients with bilateral lesions of both the hippocampus and the entorhinal cortex are still able to perform normally (as compared to age-matched controls) in a path-integration task. At the same time, long-term memory in these patients is severely impaired. For example,

3. The size of a brain structure is measured as the number of voxels, where a voxel is the minimal resolvable volume of the brain-scanning technique. In standard MRI, a voxel can be thought of as a cube with a 0.5-mm edge length.

only minutes after the experiment, they were not able to recall what they had been doing just before.

This finding is hard to interpret and raises questions about the function of the hippocampus in general. The discrepancies between mostly spatial processing in rodents and general memory functions in humans are not restricted to the case of path integration but found similarly also for other spatial performances such as route following and wayfinding. One might object that the assumed discrepancy is simply due to the fact that the brains of rodents and humans are not identical, but this immediately raises the question of the evolution of hippocampal function. Indeed, Buzsáki and Moser (2013) as well as Eichenbaum (2017) discuss an evolutionary view in which the hippocampus did evolve for dealing with space in basal mammals, but this function developed further into more general cognitive mechanisms in primates. Route knowledge might thus have been a precursor of event perception while the abilities for wayfinding and cognitive mapping might have been the origin of advanced performances in problem solving and decision-making.

5.3.5 Individual Differences

Path integration is a simple but ecologically highly relevant task that can be used as a marker of cognitive abilities. This was already the case in the study by Loomis et al. (1993), which compared normal and blind participants and found no substantial differences. One dimension along which differences have been found is aging, with a significantly poorer performance of the aged subjects (Stangl et al. 2020; Lester et al. 2019). Similarly, path integration has been used as a diagnostic tool in Alzheimer's disease, for example, by Mokrisova et al. (2016).

Sex differences in path integration have been reported, for example, by Kearns et al. (2002) who investigated the interaction of path integration with the landmark information provided in the environment. These landmarks would enable the subjects to home to the starting point of a triangle completion task by cues other than pure egomotion. In the male participants of the experiment, path-integration performance was largely unaffected by the additional cues. In the female participants, however, the estimated home location varied strongly between environments providing different types and amounts of landmark information. This indicates that the females relied more on the landmarks while the males based their judgments predominantly on the integration of egomotion cues. For a recent review of sex differences in navigation, see Nazareth et al. (2019).

Cultural differences in path integration related to spatial reference systems used in language have been reported by Levinson (2003). In one example, speakers of the Australian Guugu Yimithirr language were taken from their hometown in Queensland to places hundreds of kilometers away and then asked to point back to home. The results are stunningly accurate and far better than those of comparable experiments carried out with speakers of European languages such as Dutch or English. Levinson relates this performance to the usage of "absolute" reference schemes in verbal communication such as "north" rather than "left" when looking east, or "uphill" and "downhill" rather than forward and backward. Indeed, such absolute languages have also been described in American and African cultures, and the speakers of these languages show the same surprising accuracy in pointing.

5.4 The Computational Neuroscience of Path Integration

We now turn to the neural basis of path integration, which has been investigated with single-cell recordings in rodents and, more recently, also with brain-scanning techniques in human subjects. Neural models of path integration have to deal with two major problems, the encoding of egomotion and position parameters in neural activities, and the neural dynamics realizing the updating process. We will first discuss these two problems separately and then proceed to the overall model.

5.4.1 Population Coding

Tuning curves and estimators In the brain, information is not encoded in the spiking activity of neurons alone but needs to be interpreted in the light of each neuron's "specificity" for one stimulus or another. This fundamental fact was first noted by Johannes Müller in 1837 for the various sensory modalities: all neurons are activated electrically, but they give rise to perceptions of light, sound, and smell as well as any other sensory quality ascribed to each input by the cognitive apparatus. This principle of "specific nerve energies," as it came to be named, is not limited to the sensory pathways but prevails also in the representational and motor parts of the brain. Today, it is described by concepts such as tuning curves, receptive fields, spatial firing fields, or representational similarity, all of which relate a represented parameter to the neural activity of single neurons or neuron populations.

Let x denote a parameter to be encoded in neural activity: for example, heading direction or (as a vector) current position. The specificity of a neuron i would then be defined as the probability for i to fire, given the parameter value:

$$f_i(x) = \mathrm{P}(a_i = 1 \mid x) \tag{5.23}$$

where a_i is the activity of neuron i. This definition does not take into account effects of time: that is, the neuron may fire only some time after its preferred stimulus occurred, but we will not go into this problem here. The encoding of a parameter by the specificity of a population of "channels" is also known as labeled-line coding or ensemble coding; for review, see Pouget, Dayan, and Zemel (2000).

The function $f_i(x)$ is usually called a tuning curve if x is a scalar quantity. If it has a well-defined peak, we will say that the neuron is tuned to parameter x and define the value $\hat{x}_i$, where $f_i(x)$ takes its maximum as the preferred stimulus of the neuron i. As an example, consider the head direction cells mentioned already in chapter 1, figure 1.5c, where the encoded parameter is the geocentric heading direction. Another example is the encoding of the distance to a wall, as illustrated in figure 1.5d. If the parameter is a vector of spatial coordinates, f_i is called a firing field, which may be a localized peak as in the classical hippocampal place cells (figure 1.5a) or a multipeak landscape as in the entorhinal grid cells depicted in figure 1.5b. In general, a neuron may be tuned to multiple parameters, in which case x becomes a multidimensional vector, as is well known for neurons in the visual cortex, which may be simultaneously tuned for position in the visual field, edge orientation, spatial frequency, contrast, motion direction, and color (i.e., some seven independent stimulus dimensions).

Tuning curves are usually quite broad and have substantial overlap with the tuning curves of other neurons. The encoded parameter can therefore not be recovered from recordings of just one neuron but only from the activity of a population or ensemble of neurons. Each neuron can then be thought to "cast a vote" for its preferred stimulus value (the one it is most strongly reacting to), and the votes are weighted by each neuron's current firing rate. More formally, the encoded parameter value can be defined as

$$x^* = \frac{\sum_i a_i \hat{x}_i}{\sum_i a_i} \tag{5.24}$$

where $\hat{x}_i$ is the preferred parameter value of neuron i, and a_i is this neuron's activity (Georgopoulos et al. 1982). The activity is normally averaged over some time window, so that a_i becomes a continuous variable. x^* is also known as the center-of-gravity estimator or "population vector" even if x happens to be a scalar quantity. If the entire shape of the tuning curve is to be taken into account, a maximum likelihood estimator is more appropriate. Since $f_i(x)$ is the probability for $a_i = 1$ if the parameter takes the value x, the probability of a subset of neurons firing simultaneously equals the product of the values of their tuning curves for the same parameter, at least if we assume independence. We thus obtain the maximum likelihood estimator as

$$x^* = \arg\max_x \prod_{\{i \mid a_i = 1\}} f_i(x) \tag{5.25}$$

where a_i is assumed to be either 0 or 1.

Note that the tuning curves and the estimated x^* are only relevant for the experimenter; the neural dynamics itself is exclusively based on the corresponding pattern of activities $a_i, i = 1, \ldots, n$ alone. These signal the presence or absence of certain situations and are the basis of further actions taken by the organism while the tuning curves are not explicitly known to the system (see Barlow 1972). Tuning curves are a tool for the experimenter to determine the information implicit in the neural activity.

The concept of specificity can also be generalized to larger parts of the brain, in particular to the voxels of a fMRI scan. In this case, the stimulus parameter is often varied over time, and voxels whose activity vary in the same rhythm are then considered relevant for its encoding. The analysis amounts to a multivariate regression with the time pattern of stimulus presentation as regressors. As an example, we discuss a study by Nau et al. (2020)

on head direction cells below. A related approach uses algorithms from machine learning that are trained with brain scan data indexed with the parameter in question. If the algorithm manages to infer the parameter from the activity pattern of a given area, this area is again said to represent the parameter (representational similarity analysis). Examples will be discussed in sections 7.2.4 and 8.4.2.

Head direction cells Head direction cells are found in various areas of the brain, including the limbic system (lateral mammillary bodies), anterodorsal thalamus, and the presubiculum in the hippocampal formation (Taube 2007; Cullen and Taube 2017). They respond to the absolute (geocentric) direction of the head, both in forward running or during passive turning. This response is independent of the spatial position, as would be expected from their name. Tuning curves are rather sharp (figure 5.6) and look similar in the various brain areas; absolute firing rate, however, may vary substantially. In figure 5.6, tuning curves are shown with respect to the cardinal directions, which does not mean, however, that head direction response be based on a magnetic compass sense. In fact, the firing activity itself implicitly defines a reference direction, which can be used much as a compass even if no magnetic sense is involved.

If the animal turns and the heading direction changes, the activity of the previously active neurons will cease while neurons tuned to the new heading direction will increase firing. This is driven by cues from various sensory modalities, including the vestibular system, proprioceptive inputs from active movements, and visual inputs such as optic flow and landmarks. Except for the landmark cues, all these types of information relate to turning velocity, which means that the head direction signal must be the result of an integration over time.

The dependence of the head direction signal on landmarks was studied, for example, by Zugaro et al. (2004). A rat was placed on a small platform in the center of a cylindrical wall or drum. Cue cards (i.e., black pieces of cardboard) were placed in the drum and could be rotated about the central axis. If this happens, the preferred head direction of recorded head direction cells turns with the cue card, indicating that the card was used as a landmark defining the reference direction. If two cue cards are used and moved in different directions, most cells follow the larger one, even if the visual angle subtended by both cue cards is the same. This performance requires an estimate of distance, which seems to be based on peering movements and motion parallax: if the scene is illuminated stroboscopically, the perception of motion and motion parallax are disrupted, and the preference for the more distant landmark goes away. This behavior seems reasonable since larger landmarks are generally less likely to move and can be assumed to provide more stable reference directions than smaller ones.

Head direction cells may also be specific for pitch: that is, the vertical variation of heading. This is particularly relevant for flying animals. Finkelstein et al. (2015) recorded from head direction cells in flying bats and found that cells responding to movements in a given

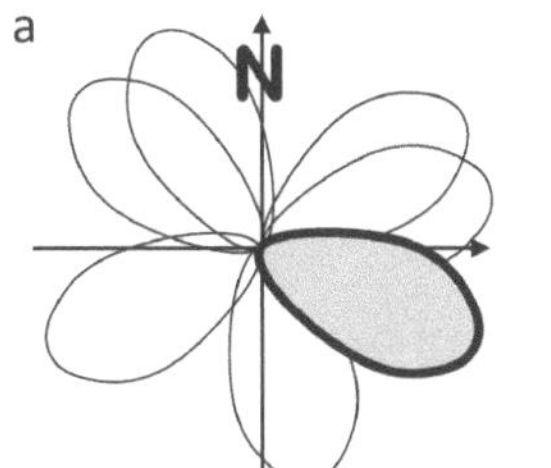

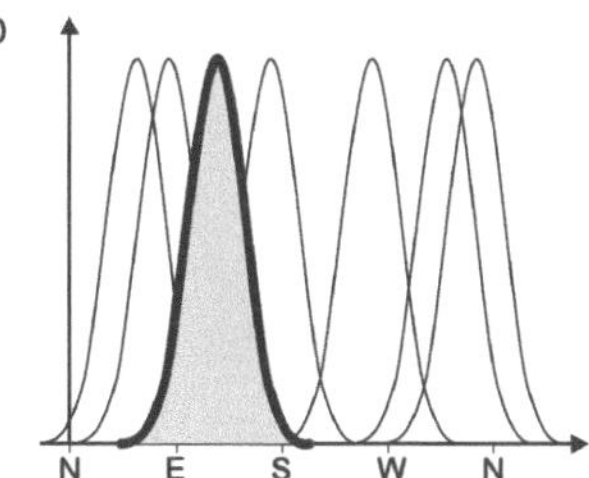

Figure 5.6
Population coding of heading direction. The figure shows schematic tuning curves of head direction cells for the forward direction of the animal's head. (a) Polar plot. (b) Cartesian plot of the same curves. Neural activity is shown relative to the maximal firing rate. The encoded parameter is geocentric heading, which is symbolized here by labeling with the cardinal directions N, E, S, and W. Note that real cell firing is often anchored to visual landmarks; if these are moved, tuning will therefore turn accordingly.

direction also respond to all headings on a circle passing through the preferred azimuth (horizontal direction) and the vertical up and down directions. In a bat flying inverted (back side down), a cell that was previously responding to northward movement will therefore respond to southward directions. Indeed, in a three-dimensional flight trajectory, path integration requires the encoding of azimuth, pitch, and roll (or torsion), and specificities for all three parameters or combinations thereof are reported by the authors.

Head direction cells are not confined to rodents and bats but have also been demonstrated in primates. In humans, direct recordings of single cells have been obtained from patients undergoing neurological surgery (Tsitsiklis et al. 2020). While the data basis available in this situation is of course limited, it clearly supports the existence of head direction tuning in the human medial temporal lobe. More extensive but indirect evidence comes from fMRI studies such as the one by Nau et al. (2020), who recorded brain activity during navigation in a virtual environment. Using an elaborated regression technique, brain regions could be identified, in which the activity varied with the virtual heading. With this technique, a wide variety of cortical areas was identified, including parts of the medial temporal lobe but also other areas, such as the occipital lobe. Of course, these studies have to be taken with caution because the subjects are stationary either in the scanner or under surgery, and no vestibular or proprioceptive heading cues are therefore present. Still, they support the idea that the head direction system does exist in humans.

Grid cells Population codes sample a parameter space by a finite set of channels. This works nicely for continuous variables defined on a limited domain such as the heading directions, which are confined to a circle: that is, the interval $[0, 2\pi]$. Position variables in the plane are principally unlimited and, at least for the purpose of path integration, should be represented with equal resolution everywhere. On the other hand, path integration is notoriously error prone, which compromises the usefulness of representations of large distances from the starting point. What is needed is a representation that is not limited in range, reflects route progress with high resolution, and needs only a limited number of neurons or channels.

These apparently incompatible requirements are met by the response characteristics of the entorhinal grid cells discovered by Hafting et al. (2005); for review, see Rowland et al. (2016). Figure 5.7a summarizes the experimental result: each grid cell has a multi-peaked firing field that extends over the entire experimental arena and probably beyond. Peaks are localized at the nodes of a highly regular hexagonal grid with a lattice constant or "wavelength" λ, depending on the individual cell (Stensola et al. 2015). Cells recorded from the same location in the brain show the same hexagonal pattern of activity peaks with the same grid spacing and orientation but shifted with respect to the other cells' grids. In different parts of the entorhinal cortex, grid cell spacing varies roughly between 40 and 120 cm. The figure also shows the symmetry axes of the hexagonal grid that occur in six directions, three of them connecting first-order neighbors (spacing λ) and three others in between, connecting second-order neighbors (spacing $\sqrt{3}\lambda$). These axes appear to be the same for all grid cells.

Figure 5.7b shows the so-called primitive cell of the hexagonal grid from which the entire grid can be reconstructed by repetition. It is a rhombus with four equal sides whose length is the lattice constant and angles of 60 and 120 degrees, respectively. The rhombus is composed of two equilateral triangles, but these cannot be the primitive cell, since rotations or mirrorings are required to construct a hexagonal grid from any of these triangles. The vectors $\boldsymbol{a}$ and $\boldsymbol{b}$ shown in the figure align with two of the nearest-neighbor symmetry axes shown in figure 5.7a; the third one is given by $\pm(\boldsymbol{a}-\boldsymbol{b})$. In Cartesian coordinates, we have $\boldsymbol{a} = \frac{1}{2}(-1, \sqrt{3})^\top$ and $\boldsymbol{b} = \frac{1}{2}(1, \sqrt{3})^\top$.

We can now use the axes $\boldsymbol{a}$ and $\boldsymbol{b}$ as an affine coordinate system to express each point (x, y) in the ground plane as

$$(x, y) = \lambda(a\boldsymbol{a} + b\boldsymbol{b}). \tag{5.26}$$

The parameter that the grid cell is tuned to can then be expressed as the vector composed of the remainders of a and b with respect to the division by 1. The remainder, or modulo, operation realizes the above requirement of combining high resolution for small progressions with low (in fact, zero) resolution for larger scale position. The fact that the possible values fill the interval $[0, 1) \times [0, 1)$ links the hexagonal grid to the geometry of a square, which will become important when modeling grid cell firing fields with an attractor neural network.

In order to regain some resolution also for larger scales, grid cells with different grid spacings are needed. Let λ_j denote the grid spacing of the jth grid cell and $\boldsymbol{s}_j$ the shift or

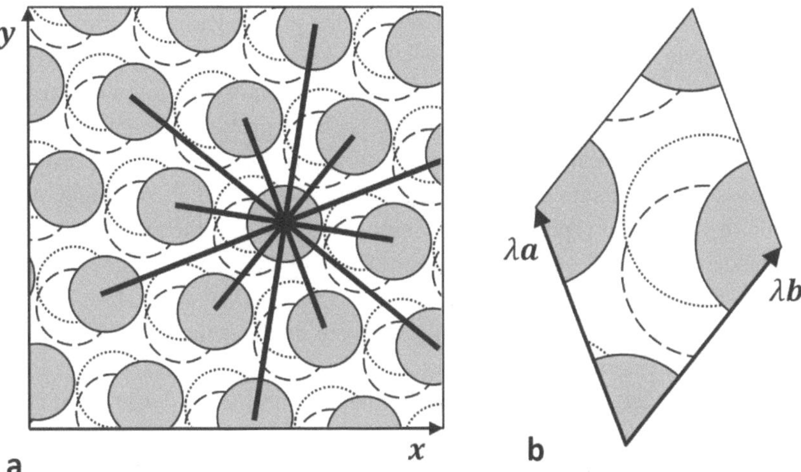

Figure 5.7
Population coding of position by entorhinal grid cells. The figure shows firing fields of three cells, one by filled gray circles, the other two by dotted and dashed lines. (a) Arrangement of firing fields in a plane. The pattern is thought to continue over large areas. The heavy black lines mark the directions of the six families of symmetry axes of the grid. Grid spacing and symmetry directions are the same for all three cells, which differ only by their offset from each other. (b) Primitive cell of the hexagonal grid from which the complete grid can be constructed by repetition. Each point in the rhomboidal cell can be defined as $a\boldsymbol{a} + b\boldsymbol{b}$: that is, by two coordinate values (a, b), which are confined to the square interval $[0, 1) \times [0, 1)$. $\boldsymbol{a}$ and $\boldsymbol{b}$ (printed in boldface) are unit vectors in the direction of two of the primary symmetry axes appearing in part (a) of the figure, and λ is the lattice constant controlling the spacing of the grid.

offset of its grid. This cell will be active in all places $\boldsymbol{x}$ satisfying

$$\boldsymbol{x} = \kappa\ \lambda_j \boldsymbol{a} + \nu\ \lambda_j \boldsymbol{b} + \boldsymbol{s}_j \quad \text{for some} \quad \kappa, \nu \in \mathbb{Z}, \tag{5.27}$$

where $\mathbb{Z}$ denotes the set of signed integers. A subsequent cell, such as a hippocampal place cell, connected to all grid cells active at a given position $\boldsymbol{x}$ will then be tuned to this position. If the λ_j would come in integer ratios, the range of uniquely represented space would be extended to the least common multiple of the grid spacings of the involved grid cell populations.

Another way to think about grid cell firing specificity is based on the theory of two-dimensional Fourier transforms (Solstad, Moser, and Einevoll 2006; Rodriguez-Dominguez and Caplan 2019). The Fourier transform of a hexagonal grid of activity peaks is a set of six peaks in the two-dimensional Fourier plane placed on the corners of a regular hexagon. In addition, a central peak at frequency $(0, 0)$ describes average firing activity.

This central peak can be removed by introducing some sort of unspecific inhibition. The distance of the six peaks from the center reflects the grid spacing λ_j. Grids with different offsets all generate the same six peaks, only that the Fourier phase at each peak will be different in a way specified by the Fourier shift theorem (multiplication with $\exp\{-i(\omega s_j)\}$). The activity of a set of grid cells with the same spacing λ provides a population code for the Fourier phase of the six peaks for this spacing. Other grid cell populations provide the same information for larger or smaller peak hexagons, depending on their spacing λ. All grid cell populations together thus represent a distribution of phase over the two-dimensional plane, which encodes an overall shift, or position. This position can be recovered by Fourier backward transformation.

Grid cells have been reported originally from the rat entorhinal cortex but are not limited to rodents. For example, Yartsev, Witter, and Ulanovsky (2011) found grid cells in the entorhinal cortex of the Egyptian fruit bat. In humans, indirect evidence for grid cell activity was presented by Doeller, Barry, and Burgess (2010), who had subjects perform virtual walks in different directions while recording brain activity in a scanner. The hexagonal structure of the grid cell system suggests that the number of active grid cells should vary with the direction of walking. When walking along one of the nearest-neighbor symmetry axes (short lines in figure 5.7a), a given grid cell will be activated more often than when walking along one of the symmetry axes connecting only second-order neighbors (long lines in figure 5.7). Since all grid cells share the same orientation of their grids, this might be visible in the fMRI scanner. Indeed, the authors report a sixfold rotational symmetry of activity with running direction, most strongly in the right entorhinal cortex but also in other areas.

Grid cell activity is driven by egomotion signals, as will be discussed in the next section. In addition, some variations of the grid pattern with environmental variables have been recorded. For example, grids recorded from rats moving in rectangular boxes tend to be compressed in the direction of the shorter side of the box (Barry et al. 2007). Also, in square confinements, one of the nearest-neighbor symmetry axes tends to be aligned with one side of the box (Stensola et al. 2015).

5.4.2 Neural Dynamics of the Updating Process

Stable attractors Besides the neural encoding of spatial parameters, the second problem that has to be solved by a neural theory of path integration is the integration or updating process itself. If the whole process is based on population coding, the update means that activity has to move from some neurons to others, whose preferred stimuli correspond to the new parameter value. Since path integration is a working memory process, no synaptic weight changes should be involved.

In modeling, this problem is generally addressed by attractor neural networks in which packets or peaks of sustained activity occur on layers of neurons connected in a lateral

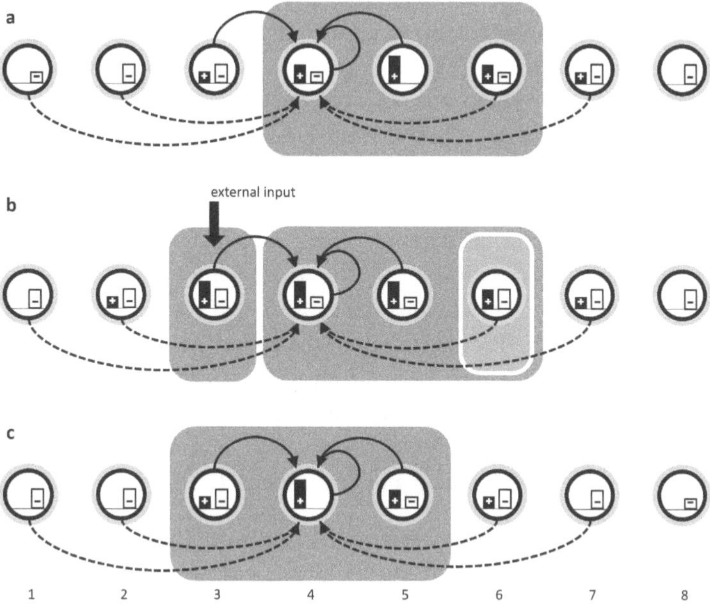

Figure 5.8
Attractor neural networks. (a) Eight neurons of a lateral inhibition network. The inputs of neuron 4 are shown as solid lines and arrows for excitation and dashed lines for inhibition; for the other neurons, the same pattern is assumed. The histograms inside each unit indicate total received excitation and inhibition. Assume that neurons 4, 5, and 6 are currently active and that a neuron will fire in the next time step if excitation exceeds inhibition. In this case, the three neurons will form a stable activity packet, marked by the gray oval. In (b), neuron 3 receives an additional external input that makes it fire. Neuron 6 will therefore receive additional inhibition (from neuron 3) and will cease to fire. The external input left of the packet will therefore cause a shift to the left. (c) Eventually, the activity packet will stabilize at its new position even after the external input stops.

inhibition scheme, while each neuron is tuned to a different parameter value. Based on earlier work by Wilson and Cowan (1973), Amari (1977) showed that such activity peaks can exist and stabilize on layers of neurons. This is mainly due to an interplay of short-range activation and long-range inhibition. If a group of neighboring neurons is active, this activation will be stabilized by the short-range activation, while the formation of other activity peaks is suppressed by long-range inhibition. Mathematically, the mechanism is reminiscent of the well-studied reaction-diffusion theory of morphogenetic "inducers" in early development (Turing 1952). For an intuitive explanation of peak formation, see figure 5.8.

The bifurcation analysis presented by Amari (1977) proves the existence of a number of different attractors,[4] most notably a localized activity peak whose width is determined by a range and strength of the excitatory and inhibitory coupling. Other attractors are regular grids of multiple peaks and periodic stripe pattern, which can arise if the range of inhibition is small. These latter attractors are, however, not considered in path-integration models.

Activity peaks can be moved on the neural layer by a simple mechanism also illustrated in figure 5.8. If an external input is delivered to neurons in the vicinity of an existing activity packet, they will be included in the packet. Since the area covered by the packet is now larger, inhibition will spread also inside to packet, particularly to the side opposite to the newly added neurons. As a result, the total activation of neurons at the margin opposite to the external stimulus will fall below firing threshold, and the activity packet will stabilize in the new position: that is, it will move in the direction of the external stimulus.

Attractors of this type are discussed in neural network models also of other types of working memory, such as the representation of future targets for eye fixation in frontal cortex. For an overview, see Zylberberg and Strowbridge (2017).

Ring attractor for the head direction system The idea of a ring attractor for the heading component of path integration was published independently in the same year by Hartmann and Wehner (1995) and Skaggs et al. (1995). While these models were designed to explain the situation in insects and in rodents, respectively, the basic idea is the same; it is summarized in figure 5.9.

The network consists of a set of head direction cells sampling the geocentric heading directions. These are assumed to be connected in a ring topology and a lateral inhibition coupling scheme. Each active head direction cell will therefore activate head direction cells tuned to similar directions and inhibit ones with more different preferred directions. Connectivity is not determined by spatial nearness of the actual neurons but by the similarity of their receptive fields. In figure 5.9, each cell's position symbolizes its preferred heading direction. Lateral inhibition is not shown explicitly but symbolized by the large gray ring connecting the head direction cells. On the ring of head direction cells, a stable activity packet is thought to exist, just as shown in figure 5.8a.

The second component is a set of two types of "shifter cells," one for leftward turns and one for rightward turns. Each shifter cell will be active if it receives simultaneous input from a head direction cell and an external signal for egomotion turn. If an activity peak is stabilized on the head direction ring, the connected shifter cells will receive input from the participating head direction cells. Since this does not suffice to exceed their threshold, they will stay silent. If an egomotion to the left, say, occurs, the subpopulation of leftward

4. In the theory of dynamical systems, an attractor is a subset of the system's state space toward which solutions are "pulled." These can be fixed points but also limit cycles or chaotic attractors. The attractors discussed in path-integration networks are usually simple fixed points in a high-dimensional state space, at least as long as activity is modeled as continuous spiking rate.

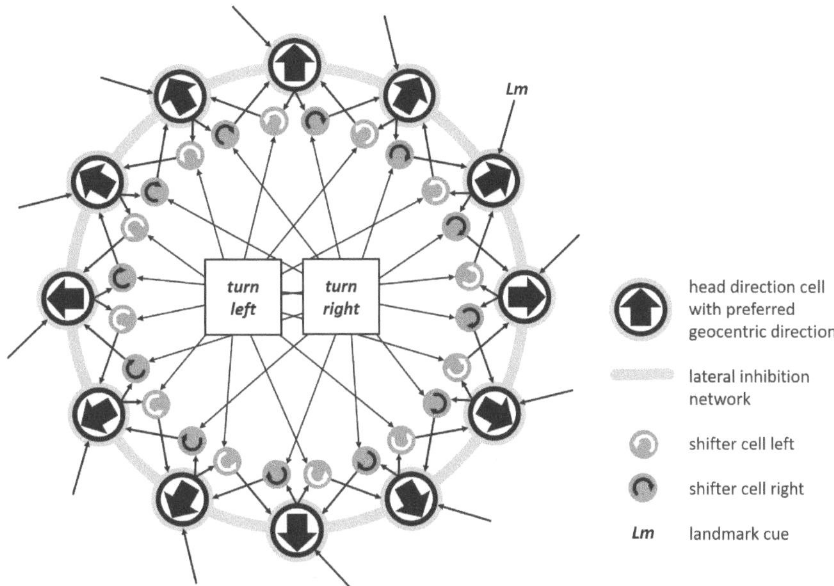

Figure 5.9
Ring attractor model for the heading component of path integration. The "turn left" and "turn right" signals encode heading turns sensed through vestibular or optic flow cues. Additional landmark cues act directly on the head direction cells. For further explanation, see text.

shifter cells will receive additional input, and those receiving both inputs will fire. This generates an additional input at the left margin of the head direction activity packet, which will shift the activity packet in a counterclockwise sense, just as illustrated in figure 5.8b,c. If the activity packet was previously focused on the north direction, it will now assume a westward component.

Egomotion signals may come from the vestibular system or optic flow cues. In addition, landmark or compass cues may activate individual head direction cells directly. The ring attractor nicely implements the first component of the path-integration equation (equation 5.19) or its polar version in the upper right part of table 5.1, where ω would be the egomotion turn signal and heading (η) the location of the activity packet on the ring. If additional landmark or compass information is considered, the simple path-integration equations have to be modified accordingly. Note that the mere existence of the head direction system is a compelling argument against the egocentric path-integration schemes in the lower row of table 5.1, which have no need for representing the heading direction η.

The model does not require that the head direction cells are actually arranged on a circle. It suffices that their connectivity follows the ring topology, which may be the case even if the cells themselves are placed irregularly. For visualizations of neural networks where the

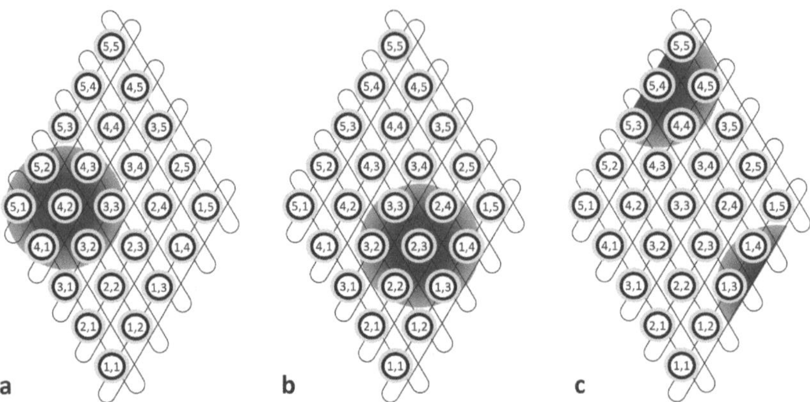

Figure 5.10
Grid cell attractor with periodic boundary conditions. The neurons sample the rhomboidal "primitive cell" of the hexagonal grid shown in figure 5.7b in a 5×5 lattice. With the notations introduced there, hexagonal coordinates are given as $(5a, 5b)$. Thin lines mark connectivity in the lateral inhibition network with periodic boundary conditions; for example, cell $(1,5)$ is connected not only to cells $(1,4)$ and $(2,5)$ but also to $(1,1)$ and $(5,5)$. The topology is therefore that of a torus. An activity packet centered around cell $(4,2)$ in (a) is shifted in the direction of the animal's movement (b). (c) When reaching the lattice boundary at the lower right, the activity packet will reenter from the upper left. If the rat runs straight for an extended distance, each cell will repeatedly become part of the activity packet, which generates the periodic firing field pattern of the grid cells.

cells are plotted at the position of their preferred response parameter in a parameter map, Samsonovich and McNaughton (1997) have coined the term "chart." In this sense, the ring in figure 5.9 is a chart, not an anatomical map. Indeed, the ring topology has not been anatomically demonstrated in mammals (for review, see Angelaki and Laurens 2020). In the fruit fly *Drosophila*, however, moving activity peaks encoding heading direction have been directly observed in the protocerebral bridge area (Kim et al. 2017).

Periodic attractor for positional advancements The second part of path integration is the advancement of the currently represented position in the direction provided by the head direction system. For the repetitive position variables (a, b) encoded by the grid cells (equation 5.26), this can be achieved by another attractor network, this time in two dimensions (Samsonovich and McNaughton 1997; McNaughton et al. 2006). Figure 5.10 shows a five-by-five lattice of grid cells sampling the hexagonal (a, b) coordinates of the rhomboidal primitive cell shown in figure 5.7b. As in the case of head direction, the arrangement is a chart of the encoded parameter: that is, preferred position in the hexagonal grid, not the anatomical position of the actual neurons. Connectivity is that of a two-dimensional lateral inhibition network allowing for activity packets to stabilize on the network.

A peculiarity of the model is the assumption of periodic boundary conditions: that is, neurons at the margin of the lattice are connected to neurons at the opposite margin. This is indicated in the figure by the connections closing each row and column into a loop. The overall topology resulting from periodic boundary conditions is a torus, as can be seen if the rhomboidal lattice is sheared into a square geometry. The connection of the top and bottom rows of this square results in a cylinder or tube whose ends are then connected to form a torus. The torus is an infinite surface on which the activity packet can be shifted endlessly in the same direction. If this happens (i.e., if the rat is running straight for a long distance), the packet will continue to move and will hit a given neuron in regular intervals. This would then generate the repetitive firing pattern found in the entorhinal grid cells.

For computational models of the toroidal attractor, a distance function is needed that takes into account the periodic boundary conditions of the sheet and allows the modeling of lateral interactions between neurons, as shown by the dark area in figure 5.10. Equations for such distance function have been presented, for example, by Guanella, Kiper, and Verschure (2007). With the notations introduced in equation 5.26, the toroidally wrapped distance between two points $\boldsymbol{x}, \boldsymbol{x}'$ in the rhomboidal primitive cell can be simply expressed as

$$d(\boldsymbol{x}, \boldsymbol{x}') = \lambda \min_{\kappa, \nu \in \{-1,0,1\}} \|\boldsymbol{x} - (\boldsymbol{x}' + \kappa \boldsymbol{a} + \nu \boldsymbol{b})\| \tag{5.28}$$

where λ is the lattice constant and $\|\cdot\|$ is the ordinary Euclidean norm. The distance is measured along the shortest connection possible in the periodic boundary conditions, as shown in figure 5.10. For example, the distance between the neurons (2, 5) and (2, 1) would be $\lambda/5$, not $4\lambda/5$, because it is measured along the marginal connection. In the equation, this is taken care of by padding a ring of rhomboidal cells around the central one, calculating the distances to the corresponding points in all cells, and taking the minimum.

The system of the ring attractor for head direction and the periodic attractor for positional advancement implements a slightly modified version of the geocentric Cartesian path-integration system (equation 5.19), which thus seems to be the scheme used by the brain.

Details of the periodic attractor model are still a matter of debate. This concerns, for example, the representation of egomotion speed and the role of grid cell populations with different wavelengths (Burak and Fiete 2009). Also, the interplay with the hippocampal place cell system and the integration of landmark information are active fields of research. For recent reviews, see Moser, Moser, and McNaughton (2017); Bicanski and Burgess (2020); Li, Arleo, and Sheynikhovich (2020); and Baumann and Mallot (2023).

No equivalent of the grid cell system has been reported in insects. A model that extends the ring attractor for heading to also encode distance information in insects has been suggested by Stone et al. (2017).

Key Points of Chapter 5

- Path integration is the construction of a metric representation of the agent's current position from egomotion cues, with or without additional compass information.
- The simplest possible representation is the home vector from the current position to the start point of an excursion. It can be calculated from egomotion data by integration of the according Frenet–Serret equations.
- Compass information is required for path integration over extended distances or for keeping a straight course over extended periods of time.
- Path integration in humans is studied with paradigms such as triangle completion and walking without vision.
- In mammals including humans, path integration recruits parts of the entorhinal and hippocampal cortices.
- Neural network models of path integration consist of two components, a network of head direction cells representing the current forward direction and a network of entorhinal grid cells representing position on a local two-dimensional chart.
- The integration of egomotion into position in both components is achieved by shifting a bump of neural activity (an "attractor") on a layer of neurons with spatial specificities.
- In insects, a similar attractor representation of head direction has been demonstrated directly.

References

Amari, S.-I. 1977. "Dynamics of pattern formation in lateral-inhibition type neural fields." *Biological Cybernetics* 27:77–87.

Angelaki, D. A., and J. Laurens. 2020. "The head direction cell network: Attractor dynamics, integration within the navigation system, and three-dimensional properties." *Current Opinion in Neurobiology* 60:136–144.

Barlow, H. B. 1972. "Single units and sensation: A neuron doctrine for perceptual psychology?" *Perception* 1:371–394.

Barry, C., R. Hayman, N. Burgess, and K. J. Jeffery. 2007. "Experience-dependent rescaling of entorhinal grids." *Nature Neuroscience* 10:682–684.

Baumann, O., and J. B. Mattingley. 2010. "Medial parietal cortex encodes perceived heading direction in humans." *Journal of Neuroscience* 30:12897–12901.

Baumann, T., and H. A. Mallot. 2023. "Gateway identity and spatial remapping in a combined grid and place cell attractor." *Neural Networks* 157:226–339.

Bicanski, A., and N. Burgess. 2020. "Neuronal vector coding in spatial cognition." *Nature Reviews Neuroscience* 21:453–470.

Brun, R. 1914. *Die Raumorientierung der Ameisen und das Orientierungsproblem im allgemeinen.* Jena: G. Fischer.

Burak, Y., and I. R. Fiete. 2009. "Accurate path integration in continuous attractor network models of grid cells." *PLoS Computational Biology* 5 (2): e1000291.

Buzsáki, G., and E. I. Moser. 2013. "Memory, navigation and theta rhythm in the hippocampal-entorhinal system." *Nature Neuroscience* 16:130–138.

Cheung, A., and R. Vickerstaff. 2010. "Finding the way with a noisy brain." *PLoS Computational Biology* 9 (11): e112544.

Cheung, A., S. Zhang, C. Stricker, and M. V. Srinivasan. 2007. "Animal navigation: The difficulty of moving in a straight line." *Biological Cybernetics* 97:47–61.

Chrastil, E. R., K. R. Sheerill, I. Aselcioglu, M. E. Hasselmo, and C. E. Stern. 2017. "Individual differences in human path integration abilities correlate with gray matter volume in retrosplenial cortex, hippocampus, and medial prefrontal cortex." *eNeuro* 4:e0346–16.2017.

Cochran, W. W., H. Mouritsen, and M. Wikelski. 2004. "Migrating songbirds recalibrate their magnetic compass daily from twilight cues." *Science* 304:405–408.

Cullen, K. E., and J. S. Taube. 2017. "Our sense of direction: Progress, controversies and challenges." *Nature Neuroscience* 20:1465–1473.

Dacke, M., E. Baird, B. El Jundi, E. J. Warrant, and M. Byrne. 2021. "How dung beetles steer straight." *Annual Review of Entomology* 66:243–256.

do Carmo, M. P. 2014. *Differential geometry of curves and surfaces.* 2nd ed. Mineola, NY: Dover.

Doeller, C. F., C. Barry, and N. Burgess. 2010. "Evidence for grid cells in a human memory network." *Nature* 463:657–661.

Eichenbaum, H. 2017. "The role of the hippocampus in navigation is memory." *Journal of Neurophysiology* 117:1785–1796.

Emlen, S. T. 1970. "Celestial rotation: Its importance in the development of migratory orientation." *Science* 170:1198–1201.

Finkelstein, A., D. Derdikman, A. Rubin, J. N. Foerster, L. Las, and N. Ulanovsky. 2015. "Three-dimensional head-direction coding in the bat brain." *Nature* 517:159–164.

Frisch, K. von. 1950. "Die Sonne als Kompaß im Leben der Bienen." *Experientia* 6:210–221.

Fujita, N., R. L. Klatzky, J. M. Loomis, and R. G. Golledge. 1993. "The encoding-error model of path-way completion without vision." *Geographical Analysis* 25:295–314.

Georgopoulos, A. P., J. F. Kalasak, R. Caminiti, and J. T. Massey. 1982. "On the relation of the direction of two-dimensional arm movements and cell discharge in primate motor cortex." *The Journal of Neuroscience* 2:1527–1537.

Glasauer, S., E. Schneider, R. Grasso, and Y. P. Ivanenko. 2007. "Space-time relativity in self-motion reproduction." *Journal of Neurophysiology* 97:451–461.

Guanella, A., D. Kiper, and P. Verschure. 2007. "A model of grid cells based on a twisted torus topology." *International Journal of Neural Systems* 17:231–240.

Hafting, T., M. Fyhn, S. Molden, M.-B. Moser, and E. I. Moser. 2005. "Microstructure of a spatial map in the entorhinal cortex." *Nature* 436:801–806.

Harootonian, S. K., R. C. Wilson, L. Hejtmánek, E. M. Ziskin, and A. D. Ekstrom. 2020. "Path integration in large-scale space and with novel geometries: Comparing vector addition and encoding-error models." *PLoS Computational Biology* 16 (5): e1007489.

Hartmann, G., and R. Wehner. 1995. "The ant's path integration system: a neural architecture." *Biological Cybernetics* 73:483–497.

Hinterecker, T., P. Pretto, K. N. de Winkel, H.-O. Karnath, H. H. Bülthoff, and T. Meilinger. 2018. "Body-relative horizontal-vertical anisotropy in human representations of traveled distances." *Experimental Brain Reseach* 236:2811–2827.

Israël, I., R. Grasso, P. Georges-François, T. Tsuzuku, and A. Berthoz. 1997. "Spatial memory and path integration studied by self-driven linear displacement. I. Basic properties." *Journal of Neurophysiology* 77:3180–3192.

Jang, H., C. Boesch, R. Mundry, V. Kandza, and K. R. L. Janmaat. 2019. "Sun, age and test location affect spatial orientation in human foragers in rainforests." *Proceedings of the Royal Society B* 286:20190934.

Kearns, M. J., W. H. Warren, A. P. Duchon, and M. J. Tarr. 2002. "Path integration from optic flow and body senses in a homing task." *Perception* 31:349–374.

Kim, S. S., H. Rounault, S. Druckmann, and V. Jayaraman. 2017. "Ring attractor dynamics in the *Drosophila* central brain." *Science* 356:849–853.

Labhart, T., and E. P. Meyer. 1999. "Detectors for polarized skylight in insects: A survey of ommatidial specializations in the dorsal rim area of the compound eye." *Microscopy Research and Technique* 47:368–379.

Land, M. F., and D.-E. Nilsson. 2012. *Animal eyes.* 2nd ed. Oxford: Oxford University Press.

Lappe, M., M. Jenkins, and L. R. Harris. 2007. "Travel distance estimation from visual motion by leaky path integration." *Experimental Brain Research* 180:35–48.

Lester, A. W., S. D. Moffat, J. M. Wiener, C. A. Barnes, and T. Wolbers. 2019. "The aging navigational system." *Neuron* 95:1019–1035.

Levinson, S. C. 2003. *Space in language and cognition.* Cambridge: Cambridge University Press.

Li, T., A. Arleo, and D. Sheynikhovich. 2020. "Modeling place cells and grid cells in multi-compartment environments: Entorhinal-hippocampal loop as a multisensor integration circuit." *Neural Networks* 121:37–51.

Loomis, J. M., R. L. Klatzky, R. G. Golledge, J. G. Cicinelli, J. W. Pellegrino, and P. A. Fry. 1993. "Nonvisual navigation by blind and sighted: Assessment of path integration ability." *Journal of Experimental Psychology: General* 122:73–91.

Maurer, R., and V. Séguinot. 1995. "What is modelling for? A critical review of the models of path integration." *Journal of Theoretical Biology* 175:457–475.

May, M., and R. L. Klatzky. 2000. "Path integration while ignoring irrelevant movement." *Journal of Experimental Psychology: Learning, Memory and Cognition* 26:169–186.

McNaughton, B. L., F. P. Battaglia, O. Jensen, E. I. Moser, and M.-B. Moser. 2006. "Path integration and the neural basis of the 'cognitive map'." *Nature Reviews Neuroscience* 7:663–678.

Merkle, T., M. Rost, and W. Alt. 2006. "Egocentric path integration models and their application to desert arthropods." *Journal of Theoretical Biology* 240:385–399.

Mittelstaedt, H., and M.-L. Mittelstaedt. 1982. "Homing by path Iitegration." In *Avian navigation,* edited by F. Papi and H. G. Wallraff, 290–297. Berlin: Springer Verlag.

Mokrisova, I., J. Laczo, R. Andel, I. Gazova, M. Vyhnalek, Z. N. Nedelska, D. Levik, J. Cerman, K. Vlcek, and J. Hort. 2016. "Real-space path integration is impaired in Alzheimer's disease and mild cognitive impairment." *Behavioral Brain Research* 307:150–158.

Montello, D. R., A. E. Richardson, M. Hegarty, and M. Provenza. 1999. "A comparison of methods for estimating directions in egocentric space." *Perception* 28:981–1000.

Moser, E. I., M.-B. Moser, and B. L. McNaughton. 2017. "Spatial representation in the hippocampal formation: A history." *Nature Neuroscience* 20:1448–1464.

Müller, J. P. 1837–1840. *Handbuch der Physiologie des Menschen für Vorlesungen.* Coblenz: J. Hölscher.

Müller, M., and R. Wehner. 1988. "Path integration in desert ants, *Cataglyphis fortis*." *PNAS* 85:5287–5290.

Müller, M., and R. Wehner. 1995. "The hidden spiral: Systematic search and path integration in desert ants, *Cataglyphis fortis*." *Journal of Comparative Physiology A* 175:525–530.

Nau, M., T. N. Schröder, M. Frey, and C. F. Doeller. 2020. "Behavior-dependent directional tuning in the human visual-navigation network." *Nature Communications* 11:3247.

Nazareth, A., X. Huang, D. Voyer, and N. Newcombe. 2019. "A meta-analysis of sex differences in human navigation skill." *Psychonomic Bulletin & Review* 26:1503–1528.

Papi, F., ed. 1992. *Animal homing.* London: Chapman & Hall.

Péruch, P., M. May, and F. Wartenberg. 1997. "Homing in virtual environments: Effects of field of view and path layout." *Perception* 26:301–311.

Pouget, A., P. Dayan, and R. Zemel. 2000. "Information processing with population codes." *Nature Reviews Neuroscience* 1:125–132.

Restat, J., S. D. Steck, H. F. Mochnatzki, and H. A. Mallot. 2004. "Geographical slant facilitates navigation and orientation in virtual environments." *Perception* 33:667–687.

Riecke, B. E., H. A. H. C. van Veen, and H. H. Bülthoff. 2002. "Visual homing is possible without landmarks—A path integration study in virtual reality." *Presence: Teleoperators and Virtual Environments* 11:443–473.

Ritz, T., S. Adem, and K. Schulten. 2000. "A model for photoreceptor-based magnetoreception in birds." *Biophysical Journal* 78:707–718.

Rodriguez-Dominguez, U., and J. B. Caplan. 2019. "A hexagonal Fourier model of grid cells." *Hippocampus* 29:37–45.

Rossel, S., and R. Wehner. 1986. "Polarization vision in bees." *Nature* 323:128–131.

Rowland, D. C., Y. Roudi, M.-B. Moser, and E. I. Moser. 2016. "Ten years of grid cells." *Annual Review of Neuroscience* 39:19–40.

Sadalla, E. K., and S. G. Magel. 1980. "The perception of traversed distance." *Environment & Behavior* 12:65–79.

Samsonovich, A., and B. L. McNaughton. 1997. "Path integration and cognitive mapping in a continuous attractor neural network model." *Journal of Neuroscience* 17:5900–5920.

Shrager, Y., C. B. Kirwan, and L. R. Squire. 2008. "Neural basis of the cognitive map: Path integration does not require hippocampus or entorhinal cortex." *PNAS* 105:12034–12038.

Skaggs, W. E., J. J. Knierim, H. S. Kudrimoti, and B. L. McNaughton. 1995. "A model of the neural basis of the rat's sense of direction." In *Advances in neural information processing systems 7,* edited by G. Tesauro, D. S. Touretzky, and T. K. Leen, 173–180. Cambridge, MA: MIT Press.

Solstad, T., E. I. Moser, and G. T. Einevoll. 2006. "From grid cells to place cells: A mathematical model." *Hippocampus* 16:1026–1031.

Souman, J. L., I. Frissen, M. N. Sreenivasa, and M. O. Ernst. 2009. "Walking straight into circles." *Current Biology* 19:1538–1542.

Stangl, M., I. Kanitscheider, M. Riemer, I. Fiete, and T. Wolbers. 2020. "Sources of path integration error in young and aging humans." *Nature Communications* 11:2626.

Steck, S. D., and H. A. Mallot. 2000. "The role of global and local landmarks in virtual environment navigation." *Presence: Teleoperators and Virtual Environments* 9:69–83.

Stensola, T., H. Stensola, M.-B. Moser, and E. I. Moser. 2015. "Shearing-induced asymmetry in entorhinal grid cells." *Nature* 518:207–212.

Stone, T., B. Webb, A. Adden, N. B. Weddig, A. Honkanen, R. Templin, W. Wcislo, L. Scimeca, E. Warrant, and S. Heinze. 2017. "An anatomically constrained model for path integration in the bee brain." *Current Biology* 27:3069–3085.

Taube, J. S. 2007. "The head direction signal: Origins and sensory-motor integration." *Annual Reviews in Neuroscience* 30:181–207.

Tsitsiklis, M., J. Miller, S. E. Qasim, C. S. Imman, R. E. Gross, J. T. Willie, E. H. Smith, et al. 2020. "Single-neuron representations of spatial targets in humans." *Current Biology* 30:245–253.

Turing, A. 1952. "The chemical basis of morphogenesis." *Proceedings of the Royal Society (London) B* 237:37–72.

Ulrich, R., J. Nitschke, and T. Rammsayer. 2006. "Perceived duration of expected and unexpected stimuli." *Psychological Research* 70:77–87.

Wang, C. X., I. A. Hilburn, D.-A. Wu, Y. Mizuhara, C. P. Cousté, J. N. H. Abrahams, S. E. Bernstein, A. Matani, S. Shimojo, and J. L. Kirschvink. 2019. "Transduction of the geomagnetic field as evidenced from alpha-band activity in the human brain." *eNeuro* 6:e0483–18.2019.

Wehner, R. 2020. *Desert navigator: The journey of an ant.* Cambridge, MA: Harvard University Press.

Wiener, J. M., A. Berthoz, and T. Wolbers. 2011. "Dissociable cognitive mechanisms underlying human path integration." *Experimental Brain Research* 208:61–71.

Wiener, J. M., and H. A. Mallot. 2006. "Path complexity does not impair visual path integration." *Spatial Cognition and Computation* 6:333–346.

Wilson, H. R., and J. D. Cowan. 1973. "A mathematical theory of functional dynamics of cortical and thalamic nervous tissue." *Kybernetik* 13:55–80.

Wiltschko, W., and R. Wiltschko. 2005. "Magnetic orientation and magnetoreception in birds and other animals." *Journal of Comparative Physiology A* 191:675–693.

Wohlgemuth, S., B. Ronacher, and R. Wehner. 2002. "Distance estimation in the third dimension in desert ants." *Journal of Comparative Physiology A* 188:273–281.

Wolbers, T., M. Hegarty, C. Büchel, and J. M. Loomis. 2008. "Spatial updating: How the brain keeps track of changing object locations during observer motion." *Nature Neuroscience* 11:1223–1230.

Wolbers, T., J. M. Wiener, H. A. Mallot, and C. Büchel. 2007. "Differential recruitment of the hippocampus, medial prefrontal cortex and the human motion complex during path integration in humans." *Journal of Neuroscience* 27:9408–9416.

Yartsev, M. M., M. P. Witter, and N. Ulanovsky. 2011. "Grid cells without theta oscillations in the entorhinal cortex of bats." *Nature* 479:103–107.

Zugaro, M. B., A. Arleo, C. Déjean, E. Burguière, M. Khamassi, and S. I. Wiener. 2004. "Rat anterodorsal thalamic head direction neurons depend upon dynamic visual signals to select anchoring landmark cues." *European Journal of Neurocience* 20:530–536.

Zylberberg, J., and B. W. Strowbridge. 2017. "Mechanisms of persistent activity in cortical circuits: Possible neural substrates for working memory." *Annual Review of Neuroscience* 40:603–627.

6 Places and Landmarks

Next to egomotion and its integration into metric position information, the second major source of information about navigational space is provided by local cues and landmarks and organized as knowledge of places. In the simplest case, this information is a "snapshot": that is, a memory of the retinal image whose vantage point is the represented place. After a theoretical discussion of the concept of "place," this chapter describes the snapshot mechanism in insects and explains how it extends to the idea of local position information and more elaborate mechanisms of place recognition. One dimension of this elaboration is the amount of image processing needed for the definition of a landmark. We will discuss the role of depth information, including the "geometric module" idea, and the notion of identifiable landmark objects that links place recognition to the well-studied field of object recognition in visual cognition. Finally, we give an overview of the neural encoding of place in the hippocampal system of place and vector cells as found in rodents and primates.

6.1 Here and There

Place as a cognitive category The idea of "place" is at the core of spatial cognition and spatial representation. The timeline of ever-changing sensory and motor signals that an agent perceives or produces can perceptually be segmented into more or less discrete events limited by event boundaries such as the beginning and ending of a rain shower or the entering and leaving of a park. Zacks and Tversky (2001) suggest that such event boundaries are perceived in instants where perception changes substantially or, more correctly, when predictions of upcoming perceptions turn out to be wrong. Events have an inner structure that can be described by event "schemata" in the observer's mind. These schemata provide a list of variables such as actors with different roles (agent, patient), a description of what is happening or being done (in linguistic theories a verb phrase), the outcome, and causal reasons why the event happened. Many event schemata can be summarized by answering the question "Who did what to whom?" The variables (question words) are bound to specific values in the process of event recognition or in later memory processes. In addition to the variables named above (i.e., who, what, and to whom), events also have temporal and spatial variables. We might therefore want to augment our schema by asking when and where the event happened. Temporal relations on the timeline of events such as precedence and partial or complete overlap have been used by Bertrand Russell (1915) to explain the perception of time (see also Lambalgen and Hamm 2005).

This approach to the definition of experienced time can be extended to the experience of space by considering the "where" variable of the event schema. It is bound to the location where the event "takes place" and will be called the "place code" in the sequel. While each event is unique in an agent's lifetime, its place is not reserved to a given event but may be the site also of other events happening at different times. A place may therefore be defined as a property shared by a certain set of events from the agent's lifeline or, technically speaking, as an equivalence relation over the set of events. Using a term from Gärdenfors (2020), we could also say that events are "fungible" with respect to place: that is, the place does not change if it becomes the stage of other events.

Other equivalence relations or shared properties of events can be considered: for example, all events happening to the same patient or all events subserving the intake of food. Among these, the spatial property stands out as the one that the agent can actively control by locomotion behavior (i.e., by moving to the site in question). Spatial cognition therefore means first and foremost the ability of an agent to discover spatial ordering as a type of locomotion-related invariance or perceptual constancy in the timeline of events.

The definition of place as a variable or "type" in an event schema is related to the idea of peripersonal space introduced in chapter 3. Like the current "here" discussed in that chapter, remembered places (that we might collectively refer to as "there") are not geometric points but describe environmental situations in which we may have effected, encountered, or suffered certain events. In linguistics, both answers, "here" and "there," would be valid tokens to fill the "where?" question provided in an event schema. The place codes of remote places (i.e., their memories) are built on elements of the peripersonal space experienced and constructed during earlier visits. They may be consolidated to various degrees in order to allow efficient spatial planning and behavior in the future.

As may be expected from this discussion, spatial memory is closely related to the memory of the events that took place at each location and is therefore sometimes considered to be part of episodic memory. In this view, space and events are not fully separated, primary categories of cognition since spatial knowledge is derived from and dependent on the understanding of events. Using again the theory of primary categories by Gärdenfors (2020), we could say that, while events are "fungible" with respect to place, the reverse is not true. Events do change if they are replaced to other locations. The two categories are therefore not mutually fungible and cannot be strictly distinguished. We may thus end up recharting the cognitive terrain that Kant (1781) described by the categories of space, time, and causality in a somewhat different way as a primary category of events together with subcategories for location, duration, and sequence.

Still, partial distinctions between the spatial domain and the domain of events do exist in two important respects: first, we can go back to places where something happened, even if we cannot go back to the event itself, and second, the relation of various places is characterized by a symmetric and quantifiable notion of distance, which is quite different from the directed relation of events given by time and causality. Put differently, spatial memory is distinguished from episodic memory by the representation of at least some aspects of geometry. The role of geometry in spatial memory will be discussed in detail in section 8.3.

Experimental paradigms In the experimental study of places as a cognitive entity, we need to operationalize the concept in terms of a behavioral paradigm. A classical approach to this problem is the plus-maze: that is, an arrangement of two corridors with a central crossing. A rat or another subject is released in one arm (say the southern one) and trained to receive a reward in the eastern arm. This can be achieved by learning to always turn right at the junction, a strategy known as response learning. Alternatively, if the rewarded arm possesses distinguishing cues, the agent might learn to use these as a guidance to enter the correct arm. This strategy is known as place learning. In the test phase, the agent is released from the northern arm, which looks exactly like the southern one from which training took place. If the response "turn right" was learned, the agent will enter the western, unrewarded arm. If, however, place learning occurred, the rewarded arm will also be entered from the new start position.

This paradigm was introduced by Tolman, Ritchie, and Kalish (1946), who suggested that place learning would be easier than response learning. However, both mechanisms may be operating together and the usual rules of cue integration will likely apply. For a modern review, see Goodman (2021).

A much-used operationalization of "place" in which the available landmark cues can be clearly defined is the Morris water maze (Morris 1981) depicted in figure 6.1a. It consists of a water tank filled with cloudy water that prevents vision of submersed objects. At one point, a platform is placed slightly below the surface so that no direct visual cues of the platform position are available. A rat swimming in the tank will occasionally hit the platform and mount it to take a rest. The next time the rat is put into the water, it will try to find the platform right away. This is supported by distant landmark cues such as structures of the room (windows, lamps) or additional cue cards placed in the environment. The rat thus seems to know how the configuration of landmarks should look from the goal location and which movements are needed to make the current perception match the remembered configuration. Indeed, the experiment also works when the rat is initially placed in different starting positions in the tank. This distinguishes the performance from response learning or vector navigation based on path integration: if the rat would store a "goal vector" from the start to the goal, rotation of the starting point by some angle should lead to corresponding errors in goal finding.

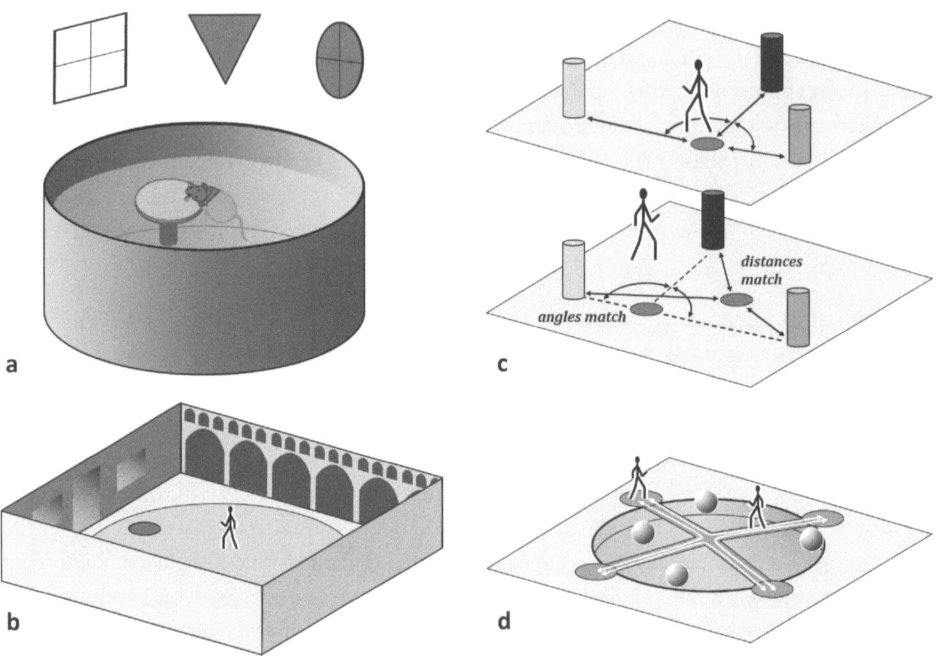

Figure 6.1
Experimental paradigms for the study of place learning and place recognition. (a) Water maze (Morris 1981). A rat is swimming in cloudy water until it finds a submersed platform on which it may rest. In later trials, the rat will be able to swim directly to the platform, even if released from different starting points. Landmarks are placed outside the tank. (b) Virtual reality version of the water maze used by Jacobs et al. (1998). Subjects are initially trained to find a place marked on the floor. In the test phase, they have to return to this place without the marker. (c) Landmark replacement experiment by Waller et al. (2000). Subjects are trained to a marked place (upper panel). When returning, one landmark has been replaced such that two possible targets can be hypothesized: a place where the inter-landmark angles match and a place where the landmark distances match. (d) Incidental learning paradigm used by Mallot and Lancier (2018). In the learning phase, subjects are walking across a plus-shaped bridge, performing 90-degree turns at the crossing. Landmarks are four balls hovering in mid-air above a pond. In the test phase, all structures except the landmarks are covered by ground fog. Subjects are asked to walk to the midpoint of the bridge, which has not been an explicit goal during training.

Similar experimental protocols have been developed for other animals as well as humans. For example, Mizunami, Weibbrecht, and Strausfeld (1998) had cockroaches run in a circular arena of hot metal that was cooled down at one spot, which was found and remembered by the animals. Hort et al. (2007) designed a circular chamber for human experimentation. One of the first versions of a "water maze" in virtual reality is the setup of Jacobs et al. (1998), depicted in figure 6.1b. In this version, the goal location is initially marked on the floor. In the test phase of the experiment, the marker is removed and the subjects are asked to find their way back to it from memory. Another procedure for the learning phase is "teleportation" of the subject to the goal location: that is, presenting the view from the goal location upon button hit. The subjects may then study the goal and the view from it. They are then "teleported" back to a starting location from where they are required to walk to the goal in closed loop (Gillner, Weiß, and Mallot 2008).

Another paradigm for the study of place recognition pioneered by Tinbergen and Kruyt (1938; see next section) is landmark replacement. A version used by Waller et al. (2000) in a study with humans is shown in figure 6.1c. A subject is trained to a place in an environment with three landmarks. In the test phase, one landmark is replaced such that different hypotheses about the cues used for return can be tested. In one place, the distances to the landmarks in the changed configuration equaled the distances during training, while in another location, the interlandmark angles remained unchanged. Waller et al. (2000) show that landmark distances are the most relevant cue in this condition. Note, however, that a similar experiment on honey bees conducted by Cartwright and Collett (1983) found the reverse preference, as will be discussed in the next section. Note also that landmark replacement experiments always involve cue conflict, which may cause the agent to use different strategies at least if the conflict is recognized. This is a general problem in cue integration studies that has to be carefully considered.

A final place learning paradigm shown in figure 6.1d is incidental learning. In this case, the goal is not explicitly marked but has to be discovered as a place of special significance. Mallot and Lancier (2018) designed a virtual environment consisting of four places connected by a plus-shaped bridge across a pond. During training, subjects walked across the bridge several times, each time taking a 90-degree turn at the bridge center. Landmark information was provided by four colored spheres hovering in midair above the pond. In the test phase, bridge and pond were rendered invisible by adding ground fog to the VR simulation. The subjects were then asked to walk to the center of the bridge, using the spheres as the only available landmark information. We will come back to this experiment below. For further place-learning paradigms, see also figures 6.2, 6.3, and 6.9.

We conclude this brief overview of place-learning paradigms by a discussion of the dependent measures used in the experiments. The most important such measure is of course the error of the positional judgment, measured either as a scalar distance or as a two-dimensional vector. The vectorial measure allows one to distinguish a bias (systematic error, accuracy) as the average deviation from the goal and a confusion area (error ellipse, precision) defined by the variance of the positional judgments. Another much-used measure is the catchment area: that is, the set of all starting points from which the goal can be found. Homing to a goal is often modeled as minimizing some measure of the difference between the current perception and the remembered place representation. The catchment area is then the "basin of attraction" of this optimization, the bias is the offset of

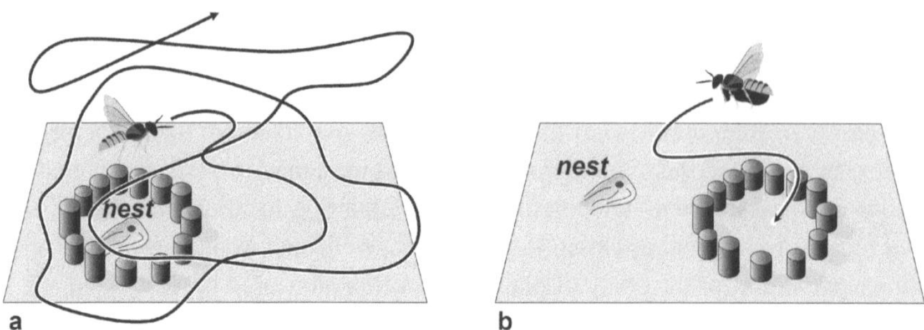

Figure 6.2
Landmark usage in the beewolf *Philanthus triangulum*. (a) A beewolf leaving the nest flies a series of loops around the nest entrance. These are thought to support the memorization of landmark information. In the experiments by Tinbergen and Kruyt (1938), the nest entrance has been marked with a ring of objects such as clay cylinders or pine cones. (b) The beewolf returns carrying a prey item (a honey bee). In the meantime, the ring of landmarks has been displaced. The beewolf will then search in the center of the ring of objects. This indicates that the objects are used as landmarks. After Tinbergen (1958).

the similarity peak from the true goal location, and the confusion area is the width of this peak.

6.2 Snapshot Homing

6.2.1 Landmark Usage in the Beewolf

The simplest form of a long-term memory of a place is the snapshot memory studied in various species of hymenopteran insects such as bees, wasps, and ants. Charles Darwin noted already in 1877 that honey bees have a memory of places as they "know the position of each clump of flowers in a garden" (Darwin 1877, 423). The issue was studied systematically by Tinbergen and Kruyt (1938) in the European beewolf *Philanthus triangulum*, a species of digger wasp preying on honey bees. Beewolves dig holes into sandy ground that reach a length of several tens of centimeters. When a wasp has caught a bee, the prey is paralyzed and placed together with an egg in a brood cell deep in the burrow. The brood cell is sealed, and when the larva hatches from the egg, it will feed on the prey. As more prey becomes available, more brood cells are added to the burrow. Every time the wasp goes out hunting, it hides the nest entrance with sand. On its way back, it has to find the covered entrance and to distinguish it from many other nest entrances built by other digger wasps living in the same area. For more details on *Philanthus* lifestyle, see Tinbergen (1958) and Evans and O'Neil (1991).

Tinbergen and Kruyt (1938) placed a ring of pine cones around one nest while the wasp was inside, as shown in figure 6.2a. When the wasp came out, it performed a "location study" by circling around the nest entrance in loops of increasing radius. Similar behaviors are also known as orientation flights or walks from other insect species (see Collett and Zeil 2018). In the common wasp (*Vespula vulgaris*), as well as in other species, the orientation flights are carried out in sideways flight with the viewing direction remaining toward the nest. After completing its location study, the beewolf flies away looking for more food. In the meantime, the experimenters picked up the pine cones and placed them in an identical ring a short distance away. When the wasp came back for the next time, it searched for the nest in the center of the ring of cones but in vain (figure 6.2b).

This result is clear evidence for some sort of landmark-based memory in insects. Tinbergen and Kruyt (1938) performed a number of additional experiments to identify more specifically the type of landmark information used. In one example, they used rings of twelve landmarks, six elevated clay cones and six hollow ones stuck in the ground, which were placed alternately around the nest. When the wasp was away, the authors replaced the elevated and the hollow cones into two separate rings of six objects, centered at different locations. When the wasp came back, it preferred the place marked by the elevated cones. While this is an interesting result, it does not allow clear conclusions on the detailed type of information extracted from the visual image and the mechanism underlying this performance. It raises, however, an important question: what exactly is the information picked up and stored by the wasp as a code or cue for the remembered place?

6.2.2 Honey Bees

Homing experiments A model of landmark-based homing has been presented by Cartwright and Collett (1982) based on measurements in honey bees (*Apis mellifera*). Bees were trained to receive food from a feeder placed close to landmark objects such as upright cylinders. The bees collect some food, bring it back to the hive, and then come back to the feeder. This work cycle is repeated at a high pace, allowing for good quantitative measurements. In the test phase, the feeder is removed and the area cleaned to avoid olfactory cues. The authors then record the time spent by the searching bees in a set of measurement squares laid out around the feeder. They find a clear peak centered at the former location of the feeder, which is sharper if three landmarks are used instead of one.

The fact that the experiment works even with one landmark is remarkable. Since the landmark is a homogeneous cylinder, it does not provide in itself any information about the side of the landmark on which the bee is currently foraging. One might therefore expect that the search distribution should be a ring around the landmark, which is, however, not the case. The reason for this may be the use of other cues present in the environment but not intentionally placed as landmarks by the experimenter or the solar compass that bees and many other insects are able to use (see section 5.2.3). Landmark distance and its bearing relative to a compass direction would indeed suffice to specify the feeder location with just one landmark.

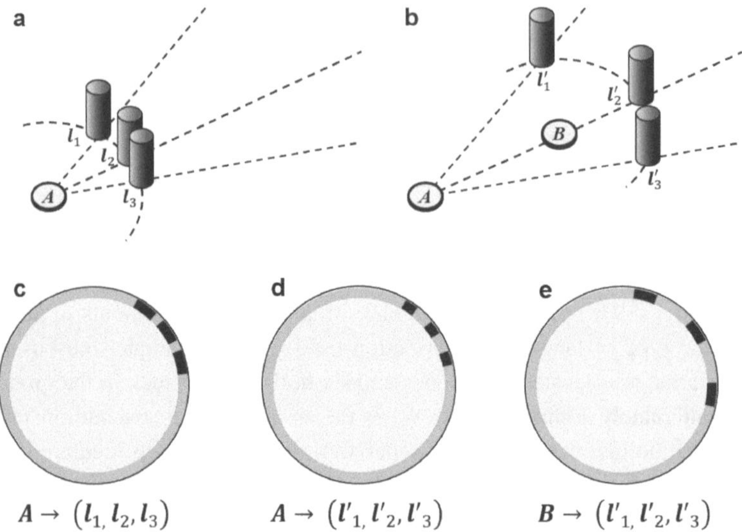

Figure 6.3
The landmark displacement experiment by Cartwright and Collett (1982). (a) Training phase. A honey bee is trained to a feeder at A marked by three landmarks placed on a circle about A. (b) Test phase. The landmarks are displaced to the new locations l'_1, l'_2, l'_3 placed on the view lines from A to the former landmark positions and on a circle about a new center B such that the landmark distances from B equal the previous landmark distances from A. The honey bees search preferably in location A, where the landmark bearings are the same as during training, not at B where the distances match. (c-e) Schematic of the panoramic images or snapshots involved in the experiment. (c) If the bee looks from A, the original configuration (l_1, l_2, l_3) will appear as a group of three spots close by. (d) After landmark displacement, the view from A will have the landmark images at the original image locations, but reduced in visual angle. (e) When looking from B, the visual angles subtended by each landmark are the same as during training, but the positions are changed. The preference for feeding site A implies that for the bee, the views (c) and (d) look more similar to each other than views (c) and (e).

Cartwright and Collett (1982) proceeded by performing a number of landmark replacement experiments that were designed to elucidate the exact type of stored information used in the approach of the goal. The most important one uses three landmarks positioned on a circle with a radius of about 1 m around the feeder and an interlandmark angle of about 60 degrees (figure 6.3). In the test phase, the landmarks are placed on a circle centered at a new location (B in figure 6.3b) at the intersection points with straight lines drawn from the feeder position A through the original landmark positions. Two hypotheses about the bee's behavior can then be formulated: first, if the bee remembers the distances to the landmarks, it should search at the center of the new landmark circle B. If, however, it uses the angles between the landmarks, it should search at the correct location A. The results show that the latter hypothesis is true: that is, search patterns in the test phase center at *A*, albeit with a larger variation than the search pattern performed without landmark displacement. It therefore appears that distance information is neglected in favor of visual angle. Since the visual angles between the landmarks correspond to their position in a picture taken from the feeder location, the authors suggest that the honey bees use a visual "snapshot" recorded at the goal location to remember the place. Clearly, the orientation flight behavior described above would be helpful in acquiring this snapshot information.

The Cartwright–Collett model Based on this and further experimental work, Cartwright and Collett (1982, 1983) presented an influential model of visual homing by moving so as to reduce the differences between the current retinal image and a stored snapshot. For just one landmark, the approach is sketched in figure 6.4. The snapshot is acquired at the goal position and stored as a one-dimensional panorama or ring image representing the viewing directions without vertical extension. This image is segmented into a part showing the landmark and the rest, (i.e., a segment showing a background). The snapshot is oriented relative to some geocentric reference as might be available from a compass or the rotational component of path integration. Segmentation is also applied to the current retinal images obtained when moving around in the environment.

At the goal, snapshot and current images are the same. At other positions, the landmark segment in snapshot and current images may differ with respect to two parameters, midpoint position on the ring (bearing) and size (subtended angle). In the case of pure bearing differences (position A in figure 6.4), the agent will need to move in a direction orthogonal to the view line directed at the landmark and away from the position of the landmark image in the stored snapshot. This is indicated in the figure by two arrows pointing eastward, one centered at the landmark and one centered at the background segment. If snapshot and current images appear at the same bearing but with different subtended angles (position B in figure 6.4), the agent can simply move toward the landmark if its image is too small and away from it otherwise. In the general case where both parameters differ (e.g., position C in figure 6.4), both types of movements are combined by vector summation. Note that these arguments can be made also for the background segment and lead to the same vectors.

In the case of multiple landmarks, the situation is more complicated. Each landmark image and gap (i.e., each segment of the image ring) may suggest a movement that is either tangentially or radially leading away from this segment, but the direction (left/right or forward/backward) of each movement can only be determined from the correspondence of image and memory. Cartwright and Collett (1982) solve this problem by assigning to each landmark and background segment in the snapshot the closest segment of the same type in the current image. This may lead to errors, for example, if many landmarks enclose the goal in their middle. For the experimental cases studied by Cartwright and Collett (1982), however, the model works very well.

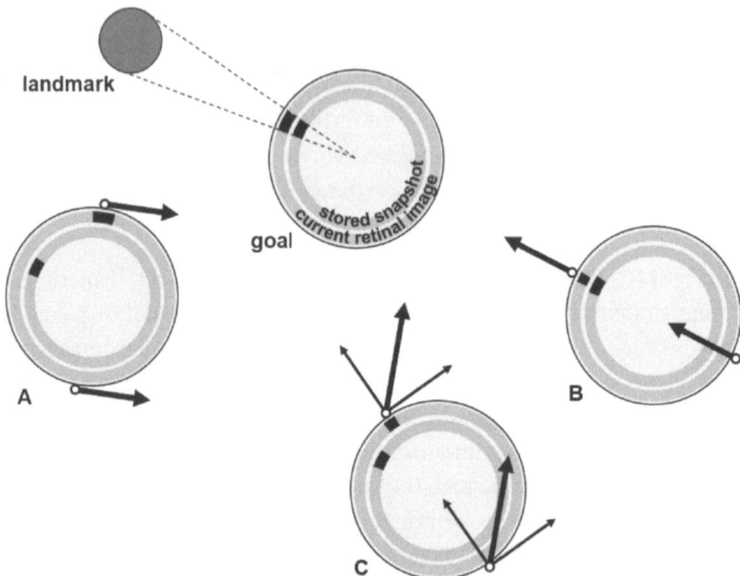

Figure 6.4
Schematic of the model for visual homing proposed by Cartwright and Collett (1982) shown for a single landmark. The current retinal image and the stored snapshot are shown as concentric circles for four positions. Both images consist of two segments, the landmark (black) and the "gap" or background (gray). Note that both snapshot and current image are oriented relative to a geocentric reference. If the agent turns on the spot, the figure would remain unchanged. At the **goal** position, the snapshot is stored and therefore matches the current image exactly. Position **A** has the correct distance from the landmark such that the current and stored images of the landmark have the same size. To also match their positions, the agent would have to move sideways relative to the feature position. At **B**, the positions match, but the landmark image is too small and the gap too large. The agent moves in the direction of the landmark image. In general positions such as **C**, the agent has to move sideways to match image positions and toward or away from the landmark to match image size.

Correspondence-based homing is computationally related to optic flow if we consider the current and the goal snapshot as two frames of a motion sequence. If they are similar enough to allow the computation of a motion field, the associated "egomotion" vector would be the sought homing direction. The algorithms for the computation of egomotion from optic flow as discussed in chapter 2 can thus be applied.

The snapshot model predicts that if a landmark is replaced by a larger one, the homing animal should search farther away from this landmark at a place where its image is smaller and thus better fits the snapshot. This was tested in an experiment with wood ants (*Formica rufa*) by Durier, Graham, and Collett (2003). After being trained to forage at a location halfway between two equal landmarks, the ants were tested in an environment

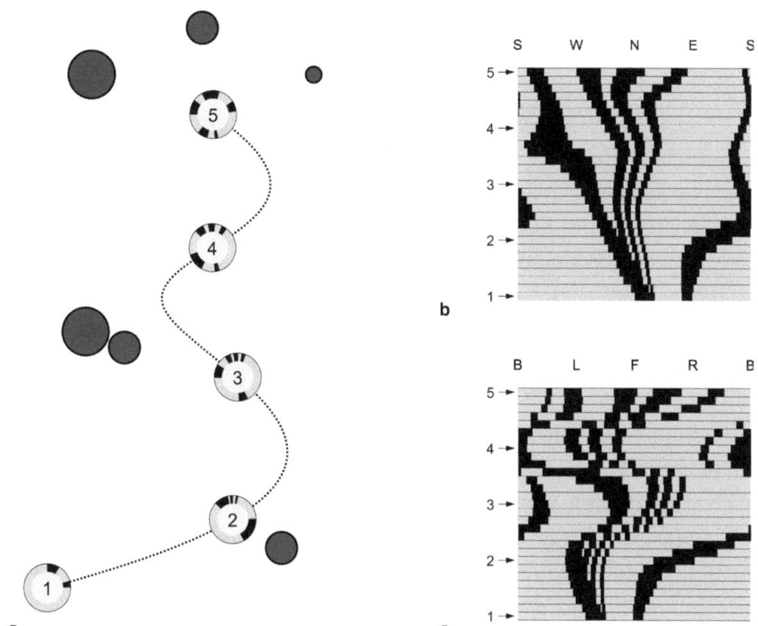

Figure 6.5
Snapshot homing. (a) As an agent is moving through an environment, different snapshots can be recorded. (b) Linearly laid-out snapshots for different positions along the trajectory. The numbers 1–5 correspond to the snapshot positions shown in (a). The viewing directions within each snapshot are shown in geocentric orientation; S–W–N–E–S is south–west–north–east–south. (c) The same for egocentric alignment; B–L–F–R–B is backward–left–forward–right–backward. Except for their alignment, the snapshots in (b) and (c) are identical. Note the smoother picture with continuous bands in (b), which results from the fact that geocentric snapshots do not change upon body turns.

with a large and a small landmark. As expected, the search location was shifted toward the smaller landmark. In a triangular landmark configuration with a feeder in the center, the simultaneous reduction of the sizes of all landmarks did not change the search location but led to higher variability in the approach direction to the goal. This supports the idea that the quality of snapshot matching increases search fidelity.

6.2.3 The Computational Theory of Visual Homing

Following the initial suggestions by Cartwright and Collett (1982), a large number of approaches to snapshot-based navigation have been published; for an overview, see Möller and Vardy (2006). The main elements of these algorithms can be delineated as follows.

Snapshot formation: What image information is used in the snapshot? For this question, the visual information available at a given position of the agent is called the "local position information" (Trullier et al. 1997). From this, the snapshot is obtained by a number of simple operations, including sampling, thresholding, and maybe edge detection. Higher processing such as depth detection or object recognition will be discussed later in the context of landmark usage. Completely unprocessed snapshots are also called "raw."

Goal recognition: For place recognition (i.e., if we want to decide that the snapshot position has been reached) we need a comparison operation such as the sum of squared differences between the pixels of the stored snapshot and current view. Since the local position information will vary continuously with the observer position, the area in which recognition obtains is a measure for the performance of the recognition step. It is sometimes called the "confusion area" of the snapshot.

Snapshot homing: If the snapshot differs from the current image, can we predict the approach direction? If we have such a mechanism, it can also be used for goal recognition, for example, by assuming that the goal is reached when the length of the suggested approach steps drops below some threshold. The set of starting points from which a goal can be reached by snapshot homing is called its "catchment area"; it will be large if long-distance landmarks are available but can get very small in cluttered environments. If we consider the field of homing directions suggested by the homing algorithm as a vector field, the catchment area is the basin of attraction containing the goal.

With these three questions in mind, we now discuss some examples for snapshot homing algorithms.

Minimal place codes: The average landmark vector The Cartwright and Collett (1982) model is certainly a very parsimonious approach to place recognition. Each place is represented by a circular snapshot, which is just the horizontal line of a panoramic image. Simulation with a resolution of less than 100 pixels along the horizontal line leads to reasonable results. Thus, a place memory would be given by the list of pixel intensities: that is, by a vector of length one hundred or even less.

An even more parsimonious place memory and homing algorithm has been suggested as the "average landmark vector" by Möller (2000). Like the Cartwright and Collett model, it uses circular snapshots with black and white landmark images. The directions of black pixels (landmark images) on the circle are expressed as unit vectors (the landmark vectors) and averaged as in a circular mean.[1] The result is a two-dimensional vector that will be large if all landmark points occur in similar directions but may be close to zero if the landmark points are distributed equally in the image. It is reminiscent of the center-of-gravity

1. In circular statistics, average angles are calculated by considering the unit vectors $(\cos\alpha, \sin\alpha)^\top$ in the direction of each angle α. These unit vectors are averaged by components, and the "resultant" vector is transformed back into an angle by the two-argument arctangent function. By this procedure, the average of +179° and –179° is correctly obtained as $\pm 180°$, whereas standard averaging would yield $0°$.

descriptor of a place suggested by O'Keefe (1991). If the agent moves, the individual landmark vectors will change in a way that can be described by subtracting a multiple of the movement vector from each of the landmark vectors. As a result, the average landmark vector will change approximately in the same direction. Thus, the movement vector (i.e., the motion direction required for reducing the distance to the goal) can be calculated by simply subtracting the average landmark vectors of the current and goal locations.

The algorithm has been tested in simulations and on a mobile robot called "Sahabot 2" navigating the habitat of the desert ant *Cataglyphis* (Lambrinos et al. 2000). It is able to home in to a stored position in environments with just a few landmarks that stand out clearly from the background. However, if multiple places in cluttered environments are to be memorized, the representation by the average landmark vector will likely lead to confusions between different places, a problem known as spatial aliasing.

Matching without correspondence: The morphing algorithm One advantage of the average landmark algorithm is that it does not rely on feature correspondences that are computationally hard to obtain. An alternative is to use plain image comparison: that is, the sum of squared differences of the individual pixels. If no geocentric reference direction is available, this must be done for each possible rotation of the snapshot relative to the current image. This is equivalent to finding the peak of the cross-correlation of image and snapshot, again a computationally costly operation.

One such approach was proposed by Franz et al. (1998) and later dubbed the "morphing" algorithm by Möller and Vardy (2006). During homing, the agent probes a number of different directions for the next movement step. The retinal image is therefore transformed by enlarging the portion in the probed motion direction and compressing the other parts of the image: that is, by making a simple prediction of the image changes resulting from a movement into the probed direction. By comparing the transformed (morphed) retinal image with the snapshot, an expected improvement of the match resulting from the probed motion is computed. The procedure is repeated for a sample of possible movement directions, and the agent then proceeds in the direction with the largest expected improvement. The algorithm was tested with a miniature robot in an environment of toy houses by measuring the catchment areas for a number of goal locations and showed good performance.

Skyline cues An experiment allowing a more specific characterization of the image information used was carried out with the Australian ant *Melophorus bagoti* by Graham and Cheng (2009). Ants were trained to a permanent feeder and then allowed to return to the nest. In this case, the ants remembered the return direction by the landmark cues available at the feeder. Graham and Cheng (2009) took panoramic photos of the surround of the feeder from the ant's point of view. They then constructed a panoramic arena with walls of variable height and adjusted the height to the angular elevation of the highest object visible in each direction. The walls were black and traced the skyline of the original scene (i.e., the lowest visible points of the sky). The arena was placed around the feeder, and the ants went off in the correct direction. If, however, the arena was rotated by some angle, the ants would choose this angle for their attempt to return to the nest. This shows that the memorized snapshot may indeed be the skyline: that is, the elevation of the lowest sky pixel as a function of azimuth.

The detection of the skyline is probably based on spectral cues. Möller (2002) showed that sky pixels can be distinguished from non–sky pixels by their UV-versus-green spectral contrast. This would largely simplify feature extraction and lead to robust results invariant to illumination and time of day. Note that the usage of skyline cues is consistent with the finding by Tinbergen and Kruyt (1938) about the preference for elevated cones as landmarks. A computational algorithm for homing based on skyline cues was presented by Basten and Mallot (2010).

6.2.4 Local Position Information

Snapshots are an example of the more general concept of local position information (Trullier et al. 1997), which includes the complete set of sensory input available at the given position. Of course, this may be different at different visits at a given place, due to changes in the scene configuration, moving objects, changing illumination and weather conditions. The skyline information discussed in the previous section is an example of a local position information that can be expected to be stable over extended periods of time. Here we give a computational account of local position information based on the concept of the image manifold.

The image manifold The idea of representing different places by snapshots rather than by coordinate values taken with respect to some frame of reference may appear strange to a land surveyor but is natural for a cognitive agent lacking the devices for measuring such coordinates. Indeed, this is the natural situation encountered by animals, humans, or robots in the absence of aids such as the global positioning system (GPS), detailed topographic maps, or laser range finders. The space thus defined is not a two- or three-dimensional point space, but a manifold of images.[2] Examples for paths in this image manifold appear in figure 6.5b,c. Spatial coordinates may be inferred from this manifold as a parameterization but are not required at least for simple navigational tasks. In robotics, navigation in the image manifold is also studied as "appearance-based" navigation; see, for example, Lowry et al. (2016).

For a mathematical account of the situation, consider a static scene viewed from a viewpoint (x, y) with a fixed viewing direction. We consider black-and-white images with n pixels as n-dimensional vectors. As the observer moves, the image will change and the set of all images possible in a given scene becomes a manifold (n-dimensional function of two

2. It is interesting to note that this idea of the local position information is closely related to the plenoptic function of Adelson and Bergen (1991), which is motivated by problems of visual scene perception.

or three variables) $\boldsymbol{I}(x, y)$ or $\boldsymbol{I}(x, y, z)$: that is, a mapping from the two or three-dimensional environmental space to the n-dimensional image space.

Root mean square image distance Zeil, Hoffmann, and Chahl (2003) have measured the image manifold in a small volume of outdoor space using a three-dimensional gantry and a panoramic camera to scan the volume point by point. At every position in three-dimensional space, an image is obtained; we denote this image as $\boldsymbol{I}(x, y, z)$ and its $n \times m$ pixels as $I_{ij}(x, y, z)$. In order to estimate possible catchment areas, Zeil, Hoffmann, and Chahl (2003) defined one of these images as the reference image $\boldsymbol{I}_o = \boldsymbol{I}(x_o, y_o, z_o)$ and measured the root mean square (r.m.s.) difference between this and all other images as

$$d^2_{\text{r.m.s.}}(x, y, z) = \frac{1}{nm} \|\boldsymbol{I}(x, y, z) - \boldsymbol{I}_o\|^2 = \frac{1}{nm} \sum_{i=1}^{n} \sum_{j=1}^{m} (I_{ij}(x, y, z) - I_{o,ij})^2. \tag{6.29}$$

This difference is then plotted as a function of the viewpoint position (x, y, z). The image difference is of course zero for the position from which the reference image was taken but rises quickly for viewpoints offset from the reference point. Rise is sharp up to about a distance of a few centimeters, but an additional increase of image distance up to a distance of about 1 m is also observed. This distance marks an upper limit of the diameter of "basins of attraction" or catchment areas within which the goal (reference snapshot) can be reached by moving downhill in the image difference landscape. For larger distances, there is little hope that snapshot homing by minimization of the root mean square distance of raw images could work.

Local image variation The rise of image difference with the distance of the viewpoints depends also on the distance of the imaged objects. If these are far away, the basin of attraction in the image difference function will be shallow and possibly wider than in cluttered environments. Zeil, Hoffmann, and Chahl (2003) recorded snapshots in a narrow but uncovered corridor such that the main image variation occurred in the upper image region picturing relatively distant treetops. This view allowed much larger catchment areas than the sideways views to the nearby structures.

The amount of image change resulting from small changes of viewpoint is called the *local image variation*, l.i.v. (Hübner and Mallot 2007; see figure 6.6). For movements in the plane, it can be defined as the areal magnification in the image manifold of a small patch of x, y-space, which can be calculated from the first fundamental form of the manifold as

$$\text{l.i.v.} = \sqrt{\boldsymbol{I}_x^2 \boldsymbol{I}_y^2 - (\boldsymbol{I}_x \boldsymbol{I}_y)^2} = \|\boldsymbol{I}_x\| \, \|\boldsymbol{I}_y\| \sin \angle \boldsymbol{I}_x \boldsymbol{I}_y. \tag{6.30}$$

Here, $\boldsymbol{I}_x$ and $\boldsymbol{I}_y$ denote the image changes occurring upon small movements in the x and y directions (i.e., the partial derivatives of $\boldsymbol{I}$); they are vectors with the same dimension as the image itself.

For example, local image variation will be small when standing in the center of a large, empty room, since small movements of the observer position will not change the image

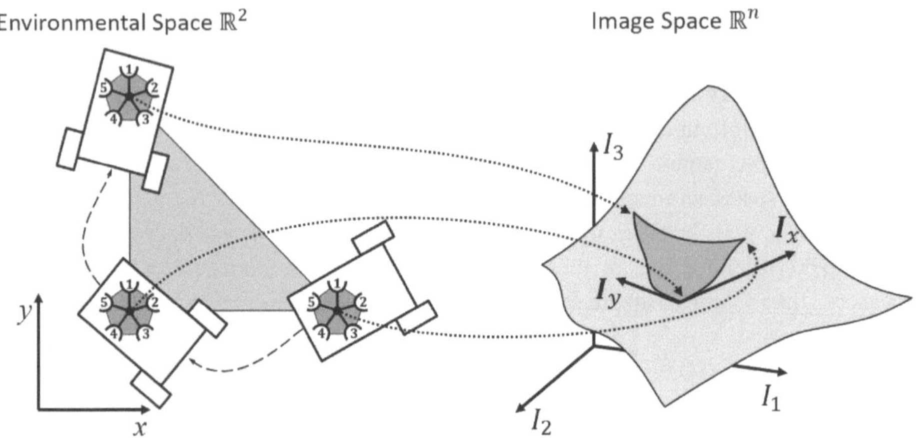

Figure 6.6
Image manifold and local image variation (l.i.v.). The left part shows an agent carrying a sensor head with $n = 5$ sensors (pixels) and fixed geocentric orientation (i.e., sensor 1 is always pointing north). The sensor image can be thought of as a n-dimensional vector $\boldsymbol{I} = (I_1, I_2, \ldots, I_n)$ changing with the position of the agent. The resulting image manifold is shown on the right side for three of the n pixels (vector dimensions). A triangle spanned by the agent's position in environmental space corresponds to a curvilinear triangle in the image manifold. The area of this triangle, divided by the area of the positional triangle in environmental space, is the local image variation defined in equation 6.30. In infinitesimal formulation, it can be calculated from the partial derivatives of the image manifold $\boldsymbol{I}_x$ and $\boldsymbol{I}_y$ also shown in the figure.

very much. However, when standing on the threshold of a door, small changes of position will open or occlude large fields of view, leading to large image variation. Related measures of how image variation depends on parameters of the scene include the fractal dimension (Shamsyeh Zahedi and Zeil 2018) or spatial frequency content (Stürzl and Mallot 2006). Indeed, the latter study demonstrates that the low spatial frequency component of panoramic snapshots can be efficiently used to determine homing directions.

Changing scenes If the panoramic scan of the environment is repeated at a different time of day with changed directions of illumination by the sun, the image similarities are generally much reduced even between images taken at the same position. Some general similarities remain that might allow an approach from about a meter distance, but exact approaches seem almost impossible.

The Zeil, Hoffmann, and Chahl (2003) study shows that the raw snapshot algorithm with pixel-based root mean square image difference is unlikely to work under real-world conditions. To make snapshot homing work, more elaborated schemes for image comparison are needed that provide some invariance for changes in illumination or small object motion (e.g., leaves moving in the wind). Note that the edge-based algorithm proposed by Cartwright and Collett (1982) and the skyline cue suggested by Graham and Cheng (2009) provide such invariance and should do better than suggested by the root mean square data.

6.2.5 Snapshot Memories in Humans

Humans are also able to use raw image information for navigation. Gillner, Weiß, and Mallot (2008) devised a virtual environment consisting of a circular room with the wall covered by a continuous cyclic color gradient with red appearing in the north direction, green in the southeast, and blue in the southwest. Colors changed continuously with viewing direction, such that no edges could be detected on the wall. Subjects were placed in this room and could walk around and turn their head to explore the scene. If they approached one part of the wall (e.g., in the section colored in red), the visual angle subtended by that wall segment increased while the color gradient visible at the opposite wall would be compressed due to perspective. In this sense, every position in the room was associated with one specific view, and these views differed by the angles subtended by each color. Floor and ceiling were colored uniformly white and black, respectively.

In a test trial, subjects were virtually relocated to one of five positions in the room and viewed the environment from this viewpoint by turning their head. After this cueing phase, their virtual position was switched back to the starting location, and they were now asked to walk to the position that had been presented before. Note that this is a working memory task, not a long-term memory task. We will come back to this differentiation in the next chapter. Subjects show good overall performance; walking trajectories are directed toward the goal from the beginning, and the accuracy of finding the goal is way above chance. Since no other cues are available, this indicates that raw snapshot information is used.

The role of raw snapshot information can be further investigated by considering the eventual localization error. The statistical variance, or "precision," of localization judgments should depend on the amount of image variability introduced by small shifts of the viewpoint (i.e., the local image variation defined above). In the experiment, it can be varied in two ways. First, if image contrast is reduced, the local image variation over a given range of viewpoints is reduced. Second, if the diameter of the circular room is increased, viewpoint changes will have a smaller effect on image variation. Similarly, the error should be larger for goals in the center of the room. The effect is clearly visible in the data: if image variation decreases, the homing error increases accordingly. In mechanisms based on feature matching, this effect is predicted only for the enlargement of the room but not for the reduction of contrast which should have no effect as long as the features remain at all visible.

The error distributions produced in many trials can be used to test more specific mechanisms. With the raw r.m.s. algorithm (i.e., by moving so as to reduce the root mean square

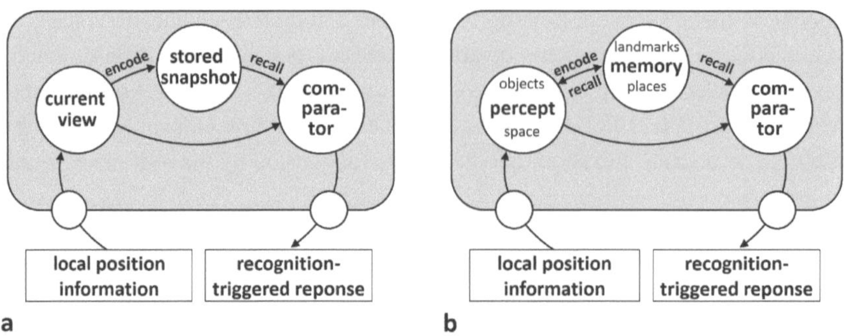

Figure 6.7
(a) Summary of the plain snapshot mechanism of homing and place recognition. The representation of the agent's current position is simply the current view. If a place is to be remembered, its view is transferred to a long-term memory as a stored snapshot. For homing, the goal view is loaded into the comparator, which calculates the required approach direction from the downloaded snapshot and the current view. (b) In elaborated place recognition models, the current snapshot is replaced by a richer representation of peripersonal space, which may also include the recognition of landmarks. For this, the "percept" stage needs to interact with a long-term memory of places, scenes, and landmarks. The result is a complex place description that can again be stored in long-term memory, if the current place is to be remembered. The comparator now acts on the richer place codes (i.e., descriptions of peripersonal space.)

difference between current and reference images), roughly circular error distributions are predicted that scale with local image variability. As an alternative model, Gillner, Weiß, and Mallot (2008) tested the boundary vector cell model by Barry et al. (2006). This model predicts that subjects remember the distance to the wall, which they might perceive by motion parallax, and the geocentric direction to the closest wall segment. This geocentric direction will rely on landmark usage and may be subject to error, which increases for reduced image contrast. Calculations based on this mechanism predict that the error distributions should differ in shape for different goal locations. Error distributions would be elongated tangentially along the wall for goal locations close to the wall and circular for central goal. No such dependence was found in the data. Thus, subjects appear to have used raw image difference.

Another experimental approach to this problem was presented by Gootjes-Dreesbach et al. (2017) and Pickup, Fitzgibbon, and Glennerster (2013). In these experiments, subjects were asked to locate themselves in relation to three thin, colored, vertical poles used as landmarks. In this case, view-based or snapshot models rely on the angles between the landmarks while depth-based models take into account also their distances from the target location. Expected error distributions were calculated for both models, and a better fit was obtained for the view-based one.

6.3 Including Depth Information

6.3.1 Encodings of "Place"

Snapshot-based place recognition consists of three elements: the retinal image as a sensory input (outer rings in the Cartwright–Collett model; figure 6.4), stored snapshots as memory traces of the remembered place (inner rings in figure 6.4), and a comparison mechanism for estimating the approach direction or acknowledging the arrival at the place associated with the current snapshot (figure 6.7). This basic scheme can be elaborated in various ways. For example, the sensory input can be processed to allow invariances for illumination, incidental object motion, weather conditions, and so on, as already mentioned above. Such inputs may still be thought of as snapshots as long as mechanisms such as edge detection, skyline detection, or depth detection are used. Computations would then still be based on pixelwise comparisons, not on the recognition of landmark objects. Put differently, we speak of snapshot mechanisms as long as the image processing involved does not exceed the level of "early" or "preattentive" vision: that is, neighborhood operations for the local analysis of contrast, texture, disparity, motion, or color (Marr 1976). The representations involved are "snapshots" in the sense that they form two-dimensional, image-like arrays.

Snapshots are but one way of encoding places for storage in spatial memory. Some other possibilities are shown in figure 6.8 ranging from the raw retinal image (a) to edge-enhanced images (b), depth maps (c), or local charts of recognized and labeled objects (f). Two possible intermediate representations are the "spatial layout" (d) and the isovist (e). The spatial layout is basically a bounding box formed by walls or other boundaries of an environment. Thus, the spatial layout of a room does not change if the furniture is removed or rearranged so as to form new configurations of possible landmarks. Neurons responsive to the spatial layout of a scene have been identified in the human parahippocampal place area and retrosplenial cortex (Epstein 2008; Bilalić, Lindig, and Turella 2019). The isovist (figure 6.8e) is a concept introduced in architecture (Benedikt 1979) that emphasizes possible usage of the viewed space. It is the set of all points on the floor that can be directly seen from a given viewpoint. In a multiroom environment, this also includes portions of adjacent rooms visible through open doorways. In the perception of indoor spaces, isovists are useful descriptors of navigability and experience of space (Wiener et al. 2007). Depth map, spatial layout, isovist, and chart all capture aspects of the peripersonal space as discussed in chapter 3.

The place codes shown in figure 6.8a–f differ by the depth of processing applied to compute each code. They also differ in their information content. For example, each pixel of the raw retinal image may change already upon small movements of the observer, whereas a scene description with identified objects will be recognized from a much larger catchment area.

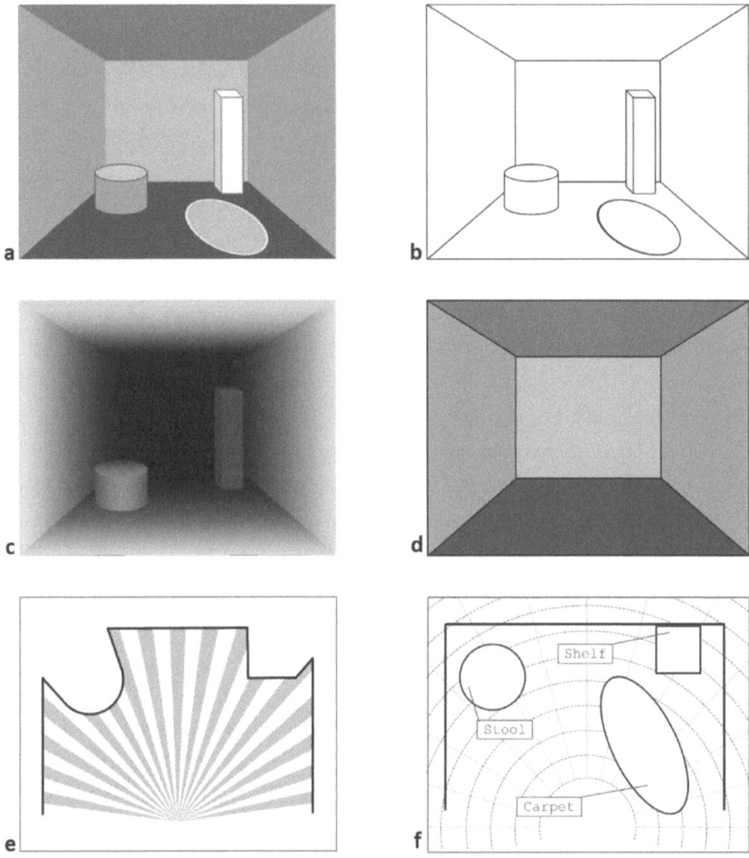

Figure 6.8
Representations of local position information obtained by various amounts of image processing. (a) Raw snapshot. (b) Contour image after edge detection. (c) Depth map shown by gray levels (darker means more distant). Note that the carpet does not show up, because it has the same depth as the floor. (d) Spatial layout (bounding box without furniture). (e) Isovist, or map of the visible portion of the floor. Note that this is not a retinal image, but a map of peripersonal space. The blue frame indicates a reference frame. (f) Local chart with egocentric coordinates of peripersonal space shown in blue and configuration of identified objects.

6.3.2 The "Geometric Module"

As an example for the experimental approaches into landmark definition, we consider the relation of snapshot-like image information and depth descriptors in the so-called geometric module literature (Cheng and Gallistel 1984; Cheng 1986; Cheng, Huttenlocher, and Newcombe 2013). Rats are placed in an open box sized 120 by 60 cm. The corners of the box hold cue cards that are easily distinguishable both by visual texture and by odors. In

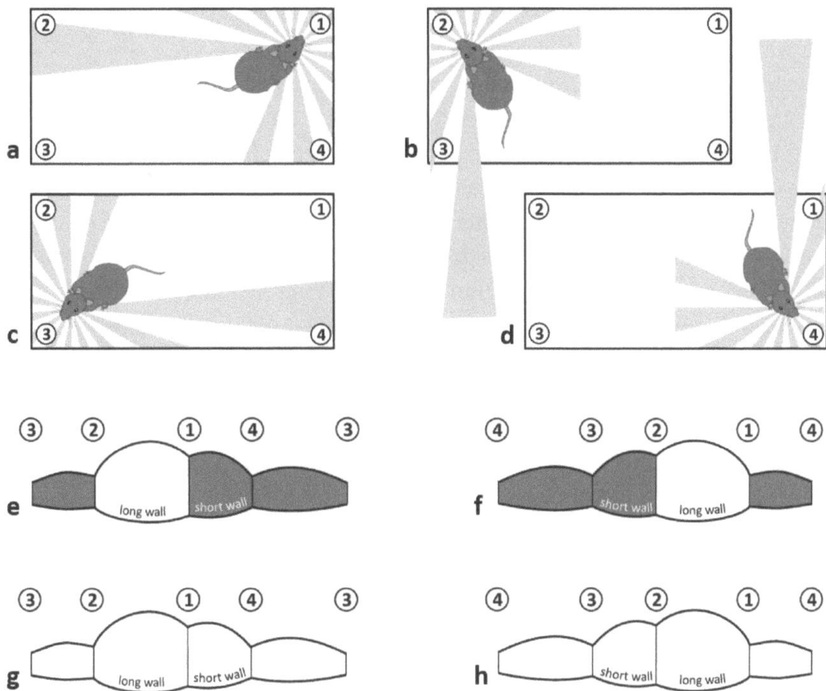

Figure 6.9
The "geometric module." (a–d) A rat is foraging in a rectangular box and finds some food in corner 1. The gray fan marks the local depth signature that might be used to remember the corner. After removal from the box, disorientation, and placement in a fresh, identical box, the rat mostly searches in corners 1 and 3 but less often in corner 2 or 4. This is in line with the observation that the egocentric depth signatures for corners 1 and 3 are identical but differ from those in corners 2 and 4. (e) Panoramic image (unrolled cylinder image) of the box when facing corner 1. Note the black and white coloring of the walls not visible in figures (a) to (d). (f) Same for corner 2. As an effect of coloring, the views of all four corners are different and should not be confused (corners 3 and 4 not shown). (g,h) Same as edge image. The images for corners 1 and 3 and for corners 2 and 4 are identical.

addition, the short walls and one of the long walls of the box are colored in black while the other long side is colored white. The floor is covered with pine chips, below which bait can be hidden. In one experiment, food was placed near to one of the corners of the box and a rat was allowed to search. After having found the bait, but before it had time to completely eat the reward, the rat was taken out of the box, disoriented by moving it around, and placed in an identical box without bait. The experimenter then observes where the rat will search for the remainder of the bait. Results show a preference for the correct location in about 44 percent of the trials. Of the remaining trials, about half (25 percent of all) lead to the diagonally opposing corner, while the other half (31 percent of all) were trials in which the rat searched elsewhere or did not search at all.

The searches in the diagonally opposing corner were called "rotational errors" by Cheng (1986). Their large number indicates that the rat largely ignores the feature cues and odors provided by the cue cards and also the black and white coloring of the walls. Instead, it seems to use the geometric layout of the reward situation: that is, the fact that the relevant corner has a long wall to its left and a short wall to its right (or vice versa). In perception and egocentric representation, this cue is indeed the same in a corner and its diagonal opposite but not for the other two corners (figure 6.9a–d). Cheng and Gallistel (1984) suggested that this result is evidence for a "purely geometric module" that contains geometric but no featural information and is separate from other modules or even "encapsulated" in the sense of Fodor (1983). The view has since been criticized, and we will come back to this issue below.

The task as it is described here is a working memory task, and it is not clear whether place representations used in working memory and peripersonal space are the same as long-term memory codes of place. Indeed, Cheng (1986) devised a long-term memory version of the experiment but found less clear results.

The same paradigm was used in an experiment with humans by Hermer and Spelke (1994). Subjects were tested in a small chamber of 1.9×1.2 m (6.25×4 ft) with four identical blinds placed in the corners. In two conditions, the walls were either of equal color or one of the walls was colored differently from the others. The experiment was carried out with adults and with small children aged 18 to 24 months, who were led into the chamber through a fabric door. Inside the chamber, the experimenter presented a toy object and placed it behind one of the occluders. The participants were led out of the room, disoriented, and led in again. They were then allowed to search behind one of the occluders, and the children were allowed to keep the toy if it was found in the searched place. The children produced rotational errors in both conditions, irrespective of the colors used to mark the walls. Adults, however, produced rotational errors only if the wall colors were equal, in which case no cue to the correct corner is available. If the wall colors differed, adults performed correctly.

Besides the age of the subjects, the effect also depends on the size of the room. Learmonth, Nadel, and Newcombe (2002) compared a small room (1.9×1.2 m) with a larger one where both sides were doubled. Children older than forty-six months as well as adults performed correctly (did not produce rotational errors) in this larger room. In the small room, rotational errors were found for children up to five years but not for six-year-old children.

The conclusions from the "geometric module" literature were challenged also by a series of papers by Stürzl et al. (2008), who repeated the original Cheng (1986) experiment with a simulated agent. A robot with a panoramic camera was placed in a rectangular box as

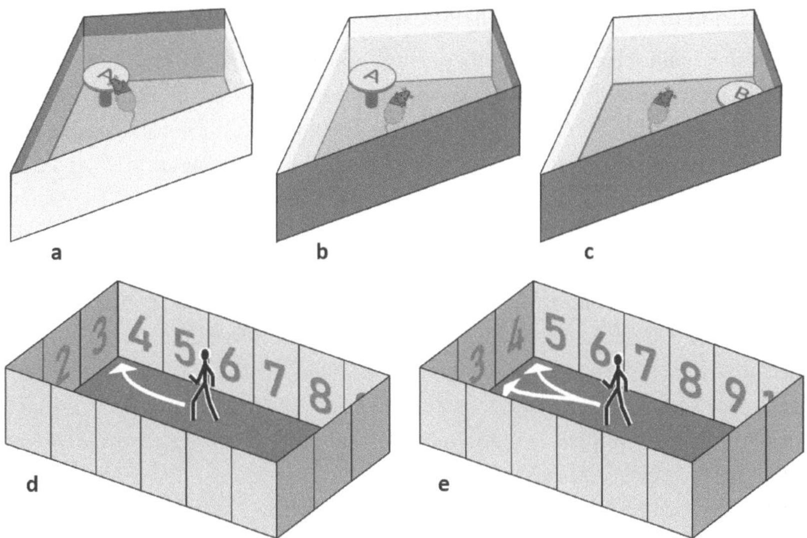

Figure 6.10
Experimental paradigms for the study of interactions between pictorial and geometric place cues. (a–c) Kite-shaped water maze used by Graham et al. (2006). A rat learns to swim to a submersed platform at position A defined by two cues, the geometry (long wall left) and the wall color (dark). This control condition is then presented alternatingly with one of the mazes shown in (b) and (c). Performance is better in the a–b alternations (position defined by geometric cue) and poorer in the a–c alternations (position defined by wall coloring). (d, e) Schematic of the "Streetlab" environment used by Bécu et al. (2020). The numbers symbolize pictures of the ground floor of houses as seen when walking in the street. They are mounted on panels and can be moved as indicated in the figure. (d) A subject learns a place in the corner between panels 3 and 4. (e) After moving the panels, subjects may search between panels 3 and 4 at the short side of the maze, or again in the corner, which now shows panels 4 and 5. Preference of pictorial and geometric cues was shown to depend on age.

was used by Cheng (1986). It took images in the corners of the box as snapshot memories and applied a simple feature-based homing algorithm to simulate the rat's behavior (figure 6.9e–h). The agent did perform rotational errors even in the presence of cue cards and a colored wall. This is due to the fact that the cue cards are rather small in a panoramic image and that the dark uniform coloring of the wall is removed by bandpassing, as is standard for feature extraction algorithms. What remains is the perspective image of the corners and margins of the walls, which is of course the same in one corner and its diagonal opposite. This example shows clearly that the distinction between pictorial and geometric image information is artificial and led to the rejection of the geometric module idea.

6.3.3 Interaction of Depth and Pictorial Cues

The claim of the "geometric module" theory was that geometric information (i.e., depth) is processed independently from (other) image cues such as color (the black or white wall) or features (the cue cards). Of course, this claim is problematic since both depth and image cues depend on the retinal image, and it is not clear why they should be separated. An alternative view of the role of pictorial and geometric (depth-related) cues to place recognition is that these and maybe other cues are not "encapsulated" each into its own module but that they all make their specific contributions to a joint description, or "representation," of a place. This view is similar to the idea of cue integration discussed in section 2.6; see also Cheng et al. (2007). It leaves open the possibility that information from various cues is weighted higher or lower, depending on the assumed reliability and accuracy of each cue. Indeed, such cue weighting was observed already by Cheng (1986), where pictorial and odor cues were not completely ignored but resulted in a slight preference of the correct over the diagonally opposed corner.

Clear evidence for this type of cue interaction in rodents was presented by Graham et al. (2006), who had rats swimming in a pool with a kite-shaped geometry (figure 6.10a–c). The pool had four corners, two with right angles and, along the long axis of the pool, one corner with a sharp and one with an obtuse angle. As in the Morris water maze paradigm, the pool was filled with water made opaque by adding some milk. Submersed under the surface of the water, a platform was put in on which the rat might rest from swimming. This platform could occur in either one of the 90-degree corners: that is, left or right of the obtuse angle when looking into the long axis of the kite (positions A and B in the figure). In addition to this "geometric" cue, a color cue was given by painting one of the walls in black. If this color cue was predictive of the platform position, the rats easily learned to use it and ignored the geometric cue. Vice versa, if the geometric cue was predicting the position of the platform, it could also be used but with a lower probability.

In human navigation, the issue of cue preference was addressed, for example, by Bécu et al. (2020), who crossed pictorial and geometric cues in their "Streetlab" environment (figure 6.10d,e). This is a rectangular arena with street-level views of houses placed on posterboards along the walls. The subject is trained to a location, for example, in one corner of the environment. This place may be encoded either by geometric cues (the ordering of long and short walls adjacent to the corner) or by pictorial cues (i.e., the particular house depicted in the corner). In the test phase, the posterboards are cycled either clockwise or counterclockwise such that the geometric and pictorial cues now hint to different places. While individual subjects showed different preferences, the group analysis revealed a clear effect of age: younger participants (aged nineteen to thirty-seven years) preferred the pictorial cues while older ones (aged sixty-one to eighty-one years) used the geometric cues for their judgment. This effect was independent of sex.

Evidence for a weighted integration of depth and pictorial cues was also found in a landmark replacement study by Waller et al. (2000); see also figure 6.1c. Subjects learned a place defined by three landmark objects (vertical cylinders) and were tested after one of the landmarks had changed position. Position changes were arranged such that two particular points could be defined: point D where the distances to the three landmarks were the same as during learning, and point A, where the interlandmark angles remained unchanged. Subjects searched the goal closer to point D than to point A: that is, they preferred the distance cues, but the effect depended on the details of the landmark configuration (presence or absence of right angles) and whether or not the goal was within or outside the triangle defined by the three landmarks.

6.3.4 Depth-Based vs. View-Based Models

Direct tests of depth-based and view-based models of place recognition can be attempted in two ways: (i) by the use of stimuli that exclusively provide only one or another type of cue and (ii) by the construction of explicit quantitative models for each cue that can be fitted to experimental data. The first approach requires stimuli that provide visual depth information without view-based or pictorial cues. While this may seem hard to achieve at first glance, such stimuli have been designed as "limited lifetime random dot" patterns for the study of the kinetic depth effect by Sperling et al. (1989). In such displays, each dot is presented only for a short period of time, but during this time, the image of the dot moves according to the observer's motion parallax and therefore clearly defines the depth of a surface containing the dot. When the dot's lifetime ends, a new dot is generated at another position randomly in the image, which will now move according to the depth and motion parallax of the surface appearing in this new image position. As long as the observer and the displayed surface stay in place, the stimulus is just a featureless field of random flicker. As soon as the observer moves, however, the surface becomes clearly visible from motion parallax cues.

Mallot, Lancier, and Halfmann (2017) created a virtual kite-shape room defined by limited-lifetime random dots and had subjects study goal locations by performing parallactic movements of their heads. In the test phase, subjects were asked to navigate to the goal from various start locations. In this situation, homing by snapshot matching is not possible since the image changes continuously even if the subject is not moving. Still, human subjects performed well in this task, with no substantial difference to similar experiments in rooms with ordinary textured walls. This result shows clearly that depth information can be used for the recognition of and the return to remembered places.

Quantitative models of homing performance that allow a comparison of view and depth based place recognition schemes can be constructed as Bayesian or maximum likelihood estimation. The place representation is modeled as a vector of measurements $\boldsymbol{m} = (m_1, ..., m_n)^\top$ based on either bearing or depth measurements obtained from the environment. The measurements obtained at the goal are stored in memory as a place code $\boldsymbol{c} = (c_1, ..., c_n)^\top$; see also figure 6.11c. The interesting variable is the probability of obtaining

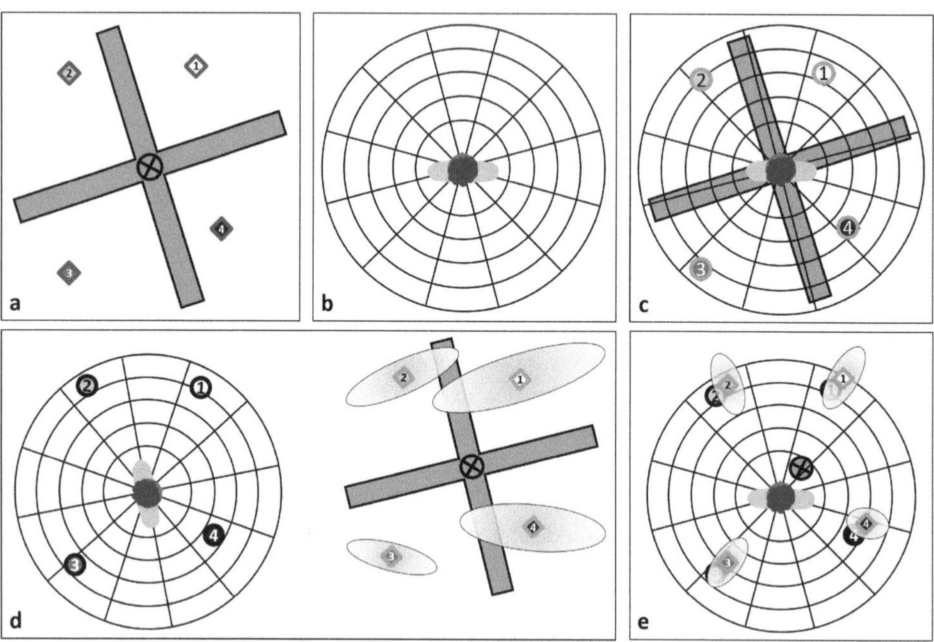

Figure 6.11
Maximum likelihood model of landmark recognition from depth and bearing data. (a) Environment used in the place recognition experiment by Mallot and Lancier (2018). Subjects walk on the plus-shaped bridge and see four landmarks (diamonds), which are numbered 1–4 to indicate landmark identity (see also figure 6.1d). In the test phase, subjects are requested to walk to the center of the bridge. (b) Agent with local chart in which landmark position can be stored. (c) During encoding, the agent establishes a place code made up of the four landmark positions; the landmark representations are marked by circles. (d) When away from the goal, the agent still has the place code (left part). At the same time, the real landmarks are perceived at an error-prone position, leading to measurement distributions $P(m_i|\boldsymbol{x})$. The distribution densities of position measurements are indicated by ellipses. Note that distance error is assumed to be larger than bearing error. (e) When returning to the goal, the bridge has been made invisible. The landmarks are closer and perception therefore becomes more veridical (smaller ellipses in the figure). The match between remembered and perceived landmark position can be described by the log-likelihood function $LL(\boldsymbol{x})$ from equation 6.32.

the same measurement from another position $\boldsymbol{x}$,

$$L(\boldsymbol{x}) = P(\boldsymbol{m} = \boldsymbol{c} \mid \boldsymbol{x}). \tag{6.31}$$

$L(\boldsymbol{x})$ is the probability of "recognizing" the goal when the agent is actually at an arbitrary position $\boldsymbol{x}$. When treated as a function of this position, it is called the likelihood function. Ideally, L will take its maximum at the goal location ($\boldsymbol{x} = \boldsymbol{x}_G$), in which case the position estimation will be "unbiased": that is, the goal location will be found, at least on average.

$L(\boldsymbol{x})$ can also be interpreted as the firing field of a "vector cell" (see section 6.5.2), for example, an object vector cell if just one landmark is considered.

If we assume that the various measured cues, $m_1, \ldots, m_n$, are statistically independent, we may write the right side of equation 6.31 in components. Taking the logarithm, we obtain the log-likelihood function

$$LL(\boldsymbol{x}) = \sum_{i=1}^{n} \ln P(m_i = c_i \mid \boldsymbol{x}). \tag{6.32}$$

The conditional probabilities $P(m_i = c_i \mid \boldsymbol{x})$ are usually modeled as Gaussians centered about the veridical value of each measurement; they may also take into account special properties of the sensor. For example, if m_i is the distance to the ith landmark located at position $\boldsymbol{l}_i$, $P(m_i \mid \boldsymbol{x})$ might be a Gaussian centered at $\|\boldsymbol{l}_i - \boldsymbol{x}\|$ (the true distance to the landmark) and variance increasing with this distance, since distance measurements get less reliable the farther the target is away. This is shown in the error ellipses in figure 6.11d,e, which are elongated toward the observer (distance error is larger than bearing error) and get smaller as the observer approaches the landmarks. In view-based models, the same landmark will be represented as a visual angle or retinal image position relative to some global north direction. The peak of the resulting likelihood function models the expected average of the place judgments. If it is acute, predicted error is low, while shallow peaks of the likelihood function predict larger errors. Technically speaking, the error ellipse of the estimate is given by the Hessian of L at peak position. The model thus makes predictions about both biases and the statistical scatter of positional judgments, which can both be used in experimental testing.

The results from such modeling studies are mixed. Pickup, Fitzgibbon, and Glennerster (2013) and Gootjes-Dreesbach et al. (2017) had subjects learn and return to a position in front of a row of poles serving as landmarks and found better fits by view-based models (see also section 6.2.5). In contrast, Mallot and Lancier (2018), using the incidental learning paradigm shown in figure 6.1d, found a clear advantage of depth-based models. Among other things, this difference may be related to the fact that the target in this latter study was enclosed by the landmarks, while in the former study, all landmarks appeared on one side of the goal. Enclosure of the goal by the landmarks also seemed to play a role in the study by Waller et al. (2000) discussed above.

6.3.5 The Role of Boundaries

An alternative view of the "geometric module" focuses on the special meaning of boundaries in the memory of places. For indoor environments, boundaries are usually the walls of a room and, in the classification of figure 6.8, would show up in the spatial layout, the isovist, and the local chart representations. In a virtual outdoor environment, Doeller and Burgess (2008) and Doeller, King, and Burgess (2008) used a circular arena surrounded by a cliff that rose to about eye height and allowed for viewing of distant landscapes beyond the cliff. Within the arena, subjects learned the position of an object (e.g., a vase) in relation to a landmark (a traffic cone), the boundary, or both. When trained and tested with one cue only, subjects performed about equally well in both cases: that is, when relying on the landmark or the boundary cue. In the main experimental conditions, subjects were trained with both cues in combination but tested with only the landmark or the boundary cue in isolation. Performance was unchanged when the boundary was provided but dropped significantly if the landmark was provided as the sole cue. In the terminology of learning theory, this means that the boundary cue "overshadows" the landmark cues but not vice versa: if the boundary cue is present during learning, it is always used (incidental learning), whereas the landmark cue is learned by reinforcement only. Ecologically, this behavior makes a lot of sense, since boundaries such as cliffs or brick-built walls tend to be more reliable cues than movable landmark objects.

The differences in learning mechanisms found for place learning relative to landmarks and boundaries are reflected also by a dissociation in the brain areas involved (Doeller, King, and Burgess 2008): learning locations relative to the boundary was found to activate the right hippocampus while learning locations relative to landmarks activated the right dorsal corpus striatum.

6.4 Identified Landmark Objects

6.4.1 Landmarks and Other Objects

A landmark is a piece of local position information stored in spatial memory for purposes of spatial behavior such as place recognition or route selection. Landmarks can be quite different things such as recognized buildings (e.g., the Eiffel Tower), smaller objects (e.g., a statue or a sign post), conspicuous street junctions (e.g., the acute angle between Broadway and 7th Avenue at Times Square, New York), or distinguishable landscape elements without clear boundaries (e.g., the Matterhorn or Monte Cervino in the Swiss and Italian Alps). An extreme case is given by image structures that depend on viewpoint and do not correspond to any real object in the world. An example of this would be a view axis in which two towers in a town align. Sometimes, places are even named after nonexisting objects, as is the case for the "Four Lakes View" (Vierseenblick) close to the German town of Boppard, where four "lakes" are visible, which are actually sections of the river Rhine, visually disconnected by intervening mountain ranges.

This list makes it quite clear that landmarks are not just objects but differ in a number of ways from the objects studied in visual object recognition and naming, in grasping, or in manipulation. Landmarks are usually large and, as in the Matterhorn example, may lack clear boundaries. They cannot be picked up and turned around in the hands to study their appearance from different viewing directions. Landmark recognition always happens on the individual level; that is, a rather low level of abstraction. Indeed, recognizing something as a "tree" or a "house" (i.e., as a member of a certain category; see Rosch 1975) is of little avail in spatial cognition. In order to be useful as landmarks, objects have to be recognized individually: that is, as "the oak tree I used to climb as a child" or "the home of my friend." Since landmarks are meant to characterize places, positional invariance will be counterproductive in landmark recognition. Indeed, the confusion of similar landmarks appearing at different places is a common problem in place recognition known as landmark aliasing. Even orientational invariance, another important issue in visual object recognition, is usually undesirable for landmark recognition, since different approach directions should not be confused. Landmarks may also be characterized by unnameable features as, for example, slope of the terrain (Nardi et al. 2021) or the angle included by a street junction, as was the case in the above Times Square example (Janzen et al. 2000).

In the literature on visual processing, a dissociation between the recognition of objects and landmarks is known as the bifurcation of the visual pathways from the primary visual cortex (occipital lobe) into a "dorsal" (parietal) and a "ventral" (temporal) processing stream (Kravitz et al. 2011). In these studies, "landmarks" are usually small markers in grasp space, placed next to a feeder from which reward may be obtained. Choices based on the placement of such landmarks recruit parietal areas (the "where" pathway), while choices based on object identity rely on processing in the temporal lobe. The landmark function is thus applied to grasp space, not to navigational space. Still, an involvement of parietal cortex in landmark processing is also well established in large-scale spatial cognition (see below).

Behavioral evidence for the different cognitive status of landmarks and other objects was provided also in a study by Biegler and Morris (1993). Rats were foraging in an arena with a clear spatial reference frame provided by a structured boundary. In the open center of the arena, two distinguishable landmarks marked the position of feeders placed in a fixed spatial relation to each landmark. Reward could only be obtained from one of the feeders while the bait in the second feeder was inaccessible. The rats learned the configuration and reliably found the rewarding feeder. In a second condition, the landmark-feeder pairs were moved around randomly between trials, keeping the fixed spatial relation between each landmark and its feeder. In this situation, rats did not learn to localize the feeder relative to its landmark. Still, they did distinguish the two landmarks and searched more in the vicinity of the landmark belonging to the rewarding feeder. Within this vicinity, they found the feeder by search. That is to say, the distinction of the landmark identities was learned, but the landmark was not used to guide further search. Biegler and Morris (1993) describe this as a dissociation between spatial learning (feeder position relative to landmark) and discrimination learning (which landmark indicates reward?). This latter performance relates to object recognition but not to the object's use as a landmark.

A behavioral study crossing object and geometrical cues used in route decisions while balancing the amount of information available in either type of cue has been presented by Stankiewicz and Kalia (2007). Subjects navigated a virtual maze made of straight hallways in a north–south or east–west direction. Hallways started, ended, or intersected at points taken from a regular grid. A number of grid points were selected for testing; at these points, the straight forward direction was always open, while joining hallways might or might not open to the sides. Thus, each test point was characterized by its geometric configuration as no junction (straight hallway only), junction left, junction right, or crossing. For the object-based place cueing, two pictures were used (showing a fish and an apple) and placed in combination before the junction. Again, four combinations are possible: fish/fish, fish/apple, apple/fish, and apple/apple. The repetition of the identical images all over the place might seem a bit surprising, but it has the advantage that the information provided by the structural and object-based cueing scheme is easily comparable.

After exploring and learning the maze, either the junctions or the pictures were covered by dark walls (distinguishable from the other wall sections) and the subjects were asked to recall the missing information from memory. Structural information was remembered better than object information. In an additional long-term study, the test was repeated after one year. Although some information had of course been forgotten, the difference between memories for structural and object knowledge remains.

These examples should make it clear that the recognition of landmarks has to be distinguished from the recognition of objects as studied in visual cognition. In order to discuss landmark recognition in its own right, three types of questions can be asked:

1. Landmark definition: How are landmarks extracted from the local position information?
2. Landmark selection: Which candidates for landmarks are selected and at which positions?
3. Landmark usage: How are landmarks used in navigation?

The first question, landmark definition, has already been discussed in section 6.3.1 and figure 6.8. We now turn to the remaining two questions.

6.4.2 Landmark Selection

Of the many visual cues impinging on the eye at every location, only very few are remembered and can thus be used as landmark information. This is particularly true if landmark information is recovered by processes of attentive (rather than preattentive) vision and requires costly computations. A number of criteria can be named that may play a role in landmark selection, including attention itself, the salience of a cue, and its relevance for navigation. The latter criterion involves also the uniqueness of the remembered cue in the environment (violations of this uniqueness are known as aliasing in the robotics literature) and the expected permanence: that is, the probability that the object will still be there when the place is visited for the next time. Thus, a fancy car might make a salient cue, it may be relevant (e.g., if it is parked close to a junction or other decision point of a navigational problem), but it does not make a good landmark since it may be gone next time the observer comes by. In the sequel, we will discuss examples for some of these criteria. For a review of landmark selection in urban environments, see Yesiltepe, Dalton, and Torun (2021).

Salience and eye movements Little is known about the role of eye movements in wayfinding. Wiener et al. (2012) studied gaze fixations in a search task presented as a sequence of static images in each of which a navigational decision to the left or right side had to be made. Decision points differed in the spatial layout, or the "isovist": that is, the set of visible floor points from a given viewer perspective (see figure 6.8e). Subjects were instructed to go through the various rooms and find a reward. Eye movements during search were distributed roughly equal between the left and right sides of the image and not predictive of the motion decision. After learning, however, when subjects tried to reproduce a path, the side of fixation was highly predictive of the motion decision, indicating that subjects had looked for cues relevant to the current decision. Of course, it would also be possible that a leftward decision would be associated with a cue appearing on the right side of the image, but this does not seem to be a frequent case.

Yesiltepe et al. (2021) had human observers watch movies in which a boat on a lake was apparently trying to navigate to various goals appearing along the shore. Subjects reported later which objects or features in the scene they used to describe the navigational behavior. These reports were compared to theoretical salience measures of the scenery produced by neural networks applied to the image sequence. The networks had previously been trained on human gaze data obtained from object recognition tasks. The authors found no correlation between the reported landmark features and the predicted salience. This may indicate that object salience and landmark salience are different things.

Attention and blocking Hamilton, Driscoll, and Sutherland (2002) used a virtual reality version of the Morris water maze (figure 6.1a) to study place learning in humans. In this experiment, a circular pool was placed in a square room with landmarks placed on all four sides of the room. Three groups of landmarks were used, called A, B, and C, where one object from each group would appear at each side of the room and outside the circular pool. Note that each group of landmarks provided sufficient information to solve the task.

In the simplest condition (control I), both landmark groups A and B were presented together during training. In the test phase, only group B was presented, and the subjects performed well. In the main experimental condition, an additional training phase was added at the beginning of the experiment in which only landmark group A was presented. The remainder of this experiment was identical to the control I condition: that is, a training phase with landmark groups A and B followed by a test with landmark group B only. In this condition, subjects performed significantly poorer than in the test I condition. The authors discuss this as an instance of "blocking," a well-known effect in associative learning, in which the initial learning of landmarks A blocks the later learning of landmarks B since no additional information would be obtained. In the test phase, when landmarks A are gone, the failure of learning landmarks B leads to the reduced performance. Note, however, that another reason for ignoring landmark group B might be their lack of permanence: their appearance only in the second training phase indicates that their presence cannot be relied on. In a second control condition, the initial learning phase is carried out with yet another set of landmarks, C, followed by the same phases as before. This time, A and B are learned alike, as in the control I condition.

Relevance The relevance of landmarks is often studied by comparing memory for landmarks at decision points with memories for equally salient objects occurring along the route at places where no decision is required. In one such study, Aginsky et al. (1997) used a change detection paradigm in which the façade color of buildings in a virtual environment was changed after exploration. Subjects were then asked to report which of the buildings had undergone this change. Correct change reports were more likely for buildings placed near decision points (street junctions) than for buildings along the straight street section where no navigational decisions were required.

During outdoor navigation, eye movements will be performed in support of multiple concurrent tasks, including, among others, course control (cf. figure 4.5), taking notice of interesting events, or the selection and encoding of landmark information. Wenczel, Hepperle, and Stülpnagel (2017) used a mobile eye tracker to compare fixation patterns during two walks, each guided by the experimenter. In the first walk, subjects thought that they would not be required to report the route, and learning was therefore assumed to be incidental. After finishing the first walk, however, they were indeed asked for a description and also instructed that the same would happen after the second walk, during which learning was thus intentional. Fixation patterns in the two walks were similar for landmarks at decision points or at straight sections of the path but differed at so-called potential decision points, where a junction occurred while the actually walked path continued straight. Landmarks at potential decision points were fixated during incidental learning but almost ignored during the intentional learning phase.

A somewhat different view on landmark relevance relates to the range of landmark visibility. Global landmarks, such as structures on a mountain ridge, towers, or even celestial bodies, can be used as compasses in path integration (see section 5.2.3) but also to guide approaches from afar. For behavioral studies of the role of local and global landmarks in human wayfinding, see, for example, Credé et al. (2020) and the study by Steck and Mallot (2000) discussed in the next chapter (see figure 7.13).

Landmark relevance is also an issue in child development. Comparisons of the wayfinding performance of adults with that of schoolchildren beginning to explore their neighborhood reveal interesting differences in the landmark usage between the two age groups. For example, Allen et al. (1979) report that children often used displays in shop windows that are locally distinctive but are hard to tell apart from a distance. As a consequence, wayfinding is more likely to fail. Adults also performed better than children in a recognition task with photographs taken from salient landmarks in the explored neighborhood. It

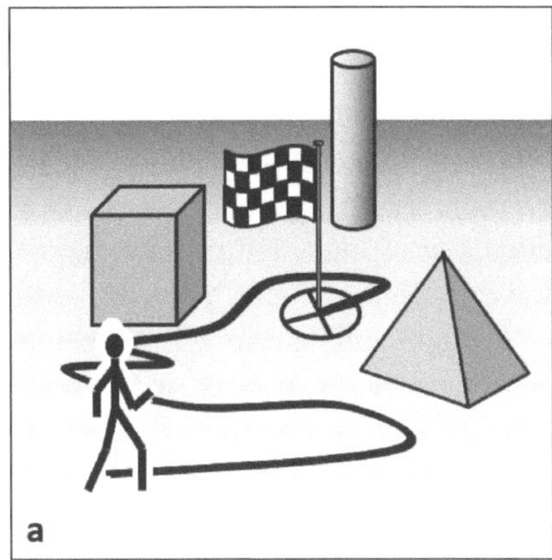

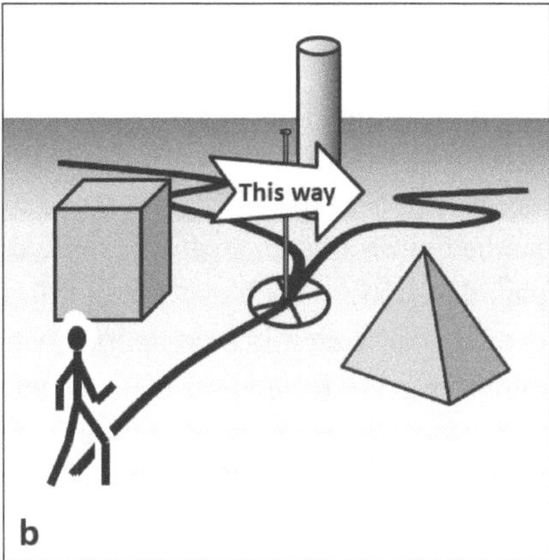

Figure 6.12
Landmark usage. (a) Guidance. The agent reaches a goal by matching local position information to a stored place code. (b) Direction. The memory contains not only the place code but also a description of the action that has to be taken once the goal is reached. This can be a motor command (e.g., "turn right") but also a new guidance to the next (sub)goal.

thus seems that children need to learn what makes a good landmark and that the criterion of landmark relevance is not obvious from the beginning. For an overview of landmark usage in child development, see Cornell and Heth (2006).

6.4.3 Landmark Usage

Once a landmark is defined and stored in memory, different ways of usage are possible. Two basic cases are usually distinguished (see O'Keefe and Nadel 1978, chap. 2).

In "guidance" (figure 6.12a), the agent moves so as to obtain a particular sensory input matching the stored information. A typical example is snapshot homing, in which case the current retinal image is compared to the stored snapshot. Other examples include the approach of a beacon or the following of a wall: that is, moving such that the retinal image of the wall stays stable. Guidance may also be provided by a guide who is walking in front and can be followed by keeping the guide's image stable in the follower's eye. The mechanism is reminiscent of the "menotaxis" discussed in section 4.1 where the agent moves such as to stabilize the bearing angle or the retinal image position of a stimulus. However, it differs from menotaxis by the addition of a remembered landmark or snapshot to which the current retinal image has to be compared. In guidance, this landmark knowledge is the only memory component, which distinguishes it from the second type of landmark usage to be discussed below. Guidance is also known as piloting or beacon navigation. These terms are borrowed from nautical navigation in which a ship is kept on course by a pilot monitoring the bearing to one or several beacons.

The second type of landmark usage is "direction" (figure 6.12b). In this case, memory contains not just information about a place but also a representation of the action that is to be taken when the place is reached. These two components, landmark knowledge and action, form an association pair in which the recognition of a place or landmark triggers a behavioral response ("recognition-triggered response"; see Trullier et al. 1997). The action part can be quite variable, such as simple directional turns to the left or right, more complex path descriptions such as walk uphill, or the activation of a remembered snapshot of the next subgoal, which is then approached by snapshot homing. We will see below that recognition-triggered responses can be concatenated into long sequences, which are called "routes." A typical experiment for studying route behavior is wayfinding in mazes with many compartments. We will come back to route memories in section 7.4.

Landmarks may also be used in contexts not immediately relating to wayfinding, most notably in communication about space. This problem is part of the larger issue of spatial language, which is beyond the scope of this book; for treatments of this problem, see, for example, Bloom et al. (1996), Levinson (2003), and Gyselinck and Pazzaglia (2012). In the case of verbal direction giving, Michon and Denis (2001) showed that the issues of landmark salience and relevance do play a role. However, additional language-specific issues arise, such as landmark nameability or the interaction of place names with the presence or absence of the named structures (e.g., the name "Park Road" may apply even in the absence of a park).

6.5 Neurophysiology of Place Recognition

6.5.1 Population Coding in Place Cells

Of the neuron types with spatial specificities listed in figure 1.5, the one most directly related to the representation of place is of course the place cell. As has already been discussed in section 1.4, the firing probability of each place cell depends on the current position of the rat and can be modeled as a more or less bell-shaped function of space centered at a position specific for each place cell (figure 6.13a–c). The area where firing probability exceeds some threshold is called the firing field; for the three examples from figure 6.13a–c, firing fields are outlined by the dashed lines in part (d) of this figure. From the simultaneous activity of a population of place cells, the current position of the rat can be nicely estimated as long as the firing fields are known: that is, if they have been determined in advance.

If many place cells with known firing fields are recorded simultaneously, the position represented in the population activity of these cells can be calculated and compared to the true position of the rat. Wilson and McNaughton (1993) recorded from eighty hippocampal neurons, about half of which showed specificities for place. During maze exploration, firing rates $\bar{\boldsymbol{r}}(\boldsymbol{x}) = (\bar{r}_1(\boldsymbol{x}), \ldots, \bar{r}_{80}(\boldsymbol{x}))$ were recorded where the bar stands for averaging over all visits of location $\boldsymbol{x}$ during the exploration phase and the index numbers the cells. The

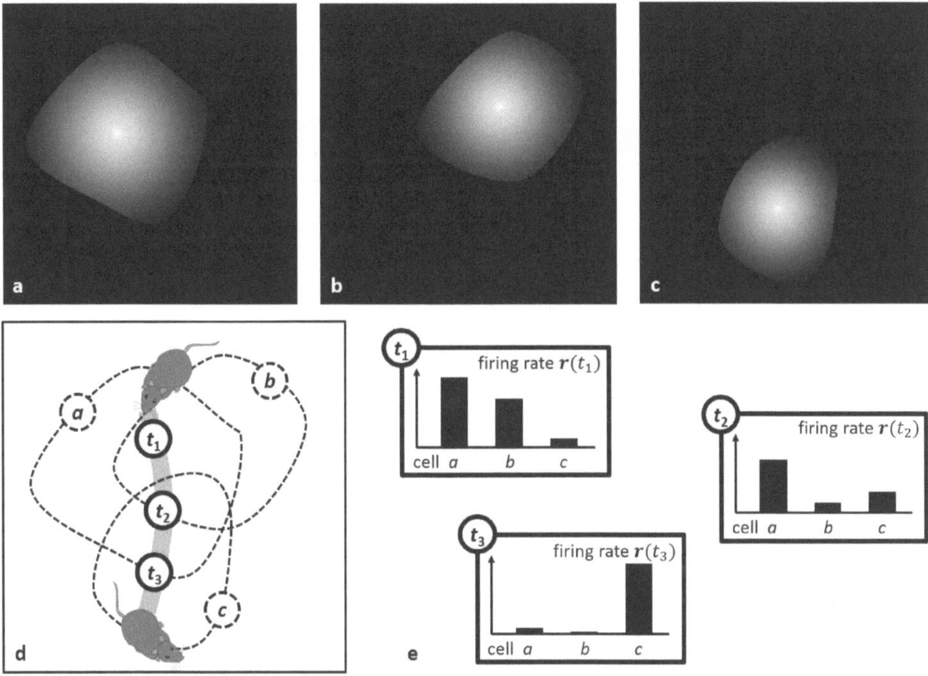

Figure 6.13
Population coding of place by hippocampal place cells. (a–c) Firing fields of three place cells shown as heatmaps (intensity functions $\bar{r}_a(\boldsymbol{x}), \ldots, \bar{r}_c(\boldsymbol{x})$). (d) Same firing fields shown as outlines (dashed lines). When a rat is running along the gray path, the firing rates of these cells will change. (e) Firing rates of the three cells for three instants in time ($t_1, \ldots, t_3$). From these firing rates, an estimate of the position occupied at each instant can be calculated.

function $\bar{r}_i(\boldsymbol{x})$ is the firing field profile of cell i, as depicted in figure 6.13a–c. The instantaneous firing activities (within one-second windows) are denoted by $\boldsymbol{r}(t) = (r_1(t), \ldots, r_{80}(t))$; they appear in figure 6.13e for three instants numbered t_1, t_2, and t_3 and three cells named a, b, and c. The represented place is then reconstructed as

$$\boldsymbol{x}^*(t) = \arg\min_{\boldsymbol{x}} (\boldsymbol{r}(t) - \bar{\boldsymbol{r}}(\boldsymbol{x}))^2, \tag{6.33}$$

that is, by finding the position $\boldsymbol{x}^*$ in the recorded activity profiles where they most closely resemble the current activity pattern. Wilson and McNaughton (1993) found good overall agreement between the actual position of the rat and the estimate calculated from the place cell activities. Place encoding by place cell populations remains stable over days or even weeks but gets increasingly blurred as place cells change their firing fields (Ziv et al. 2013).

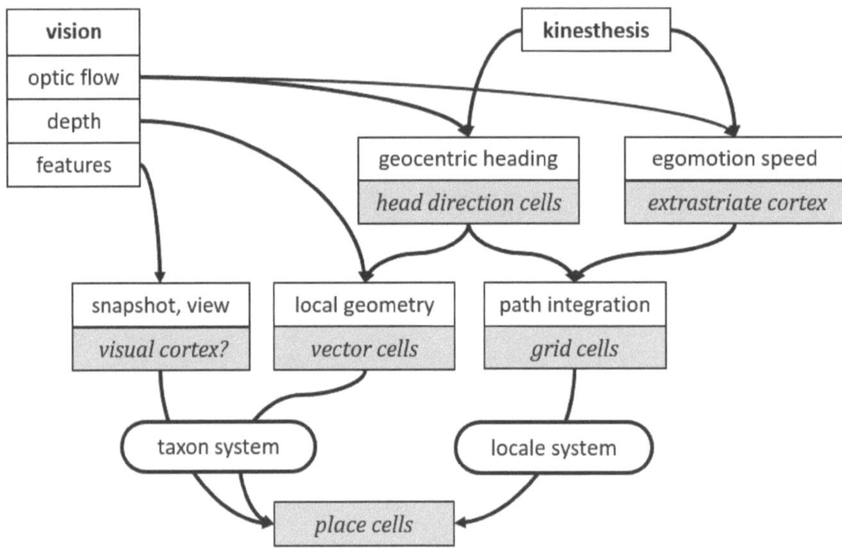

Figure 6.14
The hippocampal network for place. Summary of the sensory inputs (top) and the most important cell types. For explanation, see text.

Note that the reconstructed position $\boldsymbol{x}^*$ is an interpretation by the experimenter. The brain itself works with the activity vector $\boldsymbol{r}(t)$, which encodes the position in an implicit way.

6.5.2 The Network of Space Selective Cells

How do the place cells know the rat's current position? The answer to this question has several parts. First of all, there is an input from path integration, as is clearly demonstrated by the fact that place cell activity keeps its specificity even in complete darkness (i.e., in the absence of landmark information) for an extended period of time. This stability is thought to depend on the grid cell system discussed in section 5.4, which is able to run in the absence of visual input purely by proprioceptive egomotion information.

The second type of place cell input provides various forms of landmark information. In the simplest case, this may be the retinal image (or snapshot) preprocessed by the usual feature detectors as in the model suggested by Sheynikhovich et al. (2009). In figure 6.14, this input is attributed to the visual cortex, but other structures in the parietal and medial temporal lobes may be also involved (see Vann, Aggleton, and Maguire 2009). Additional landmark information is provided by so-called vector cells, whose firing specificity can be described both by a spatial firing field (as in place cells) and by a receptive field in which a cue must occur to trigger the neuron (Bicanski and Burgess 2020). Boundary vector cells fire at a fixed geocentric distance from a boundary, irrespective of the rat's head or body orientation (see Solstad et al. 2008 and figure 1.5d). This independence is probably provided by the head direction cells and their (putative) ring attractor that acts much like a compass (figure 5.9). Firing of a boundary vector cell thus indicates that a boundary occurs at a certain distance and geocentric direction, which are characteristic of each cell. For example, in a square room, a boundary vector cell signaling a boundary to the west and 1 m away will fire in all locations about 1 m east of the western wall. The receptive fields of the boundary vector cells (i.e., the boundary locations to which the cell reacts) thus chart the environment of the rat in the form of a local map. This map is centered at the current position of the rat as indicated by place cell firing (i.e., egocentric), but oriented along the current "north" of the head direction system (Bicanski and Burgess 2020). The resulting map may thus be said to be geo-centered and "ego-oriented," as is discussed in more detail in section 7.3.

In addition to the boundary vector cell, a second type of vector cell has been described that responds to the position of discrete objects (Grieves and Jeffery 2017; Høydal et al. 2019). These "object vector cells" encode object position and orientation in much the same way as discussed for the boundary vector cells. Specificity is mostly for object location, while object identity is largely ignored.

A simplified summary diagram of the hippocampal system for place appears in figure 6.14. Sensory inputs from the visual system provide cues to egomotion as well as local position information (depth and features). By "kinesthesis," we summarize proprioceptive and vestibular information, as discussed in section 2.2.2. Egomotion information is represented by head direction cells (geocentric heading) and as the speed in the egocentric forward direction. In primates, egomotion-sensitive neurons have been found in mediotemporal cortex (area MT, or hMT+ in humans), named as extrastriate visual cortex in the figure. Egomotion information feeds into the path-integration system but also into the geocentric positional information represented by boundary and object vector cells. The final input to place cells is formed by three branches: egocentric image information, the two types of vector cells, and path integration. In the original nomenclature of O'Keefe and Nadel (1978), image, boundary, and object cells constitute the "taxon" system while path integration makes up the "locale" system. For further details of the hippocampal place network, see Grieves and Jeffery (2017); Moser, Moser, and McNaughton (2017); and Bicanski and Burgess (2020).

6.5.3 Encoding of Place in the Hippocampal System

Place cell activity generally reflects the current position of a rat in a two-dimensional environment. However, this simple rule does not completely describe place cell behavior. We discuss three important cases that also relate to the concept of "place" itself: multiple reference frames, multiple repetitive rooms, and extensions to the third dimension.

Multiple frames of reference When we travel in the backseat of a car, we may say that we have been sitting in the same place during the entire journey, while the car was of course

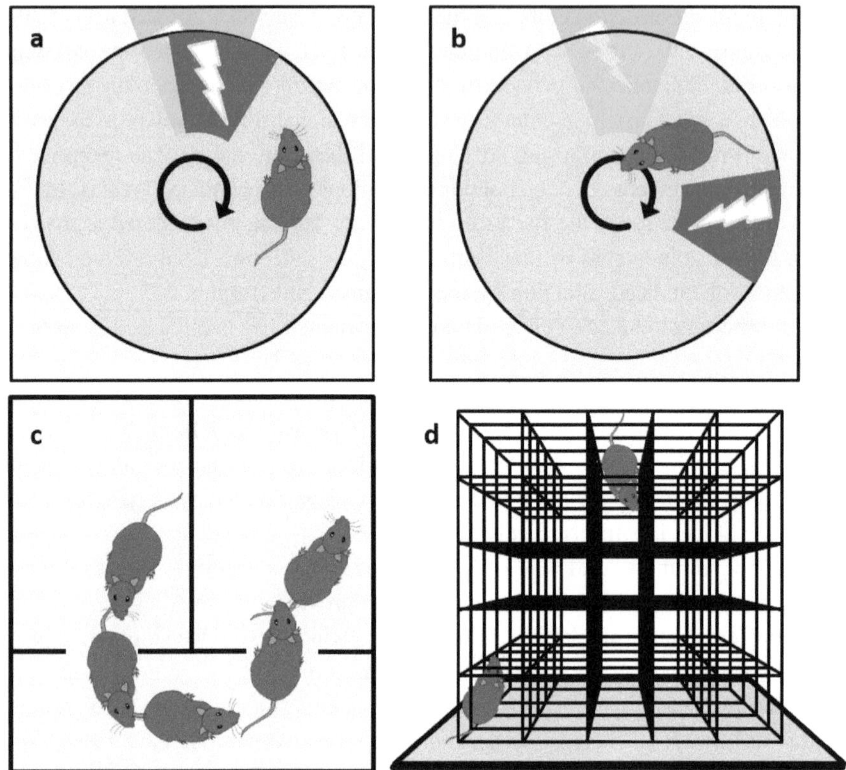

Figure 6.15
Place cell coding in complex situations. (a, b) A rat running on a rotating disk learns to avoid two regions in which electric foot shocks are delivered. One area is defined relative to the surrounding room (cyan) while the other moves with the rotating disk (red). Place cells defined for either reference frame exist (Kelemen and Fenton 2010). (c) Two-room maze with connecting corridor used by Skaggs and McNaughton (1998) in which rats show place-field repetition. After prolonged training, the place cells "remap" and become specific for one of the two rooms. (d) In rats climbing a three-dimensional wire lattice, volumetric place fields are found (Grieves et al. 2020).

moving around. In dynamic environments, such as a car cruising in a static world, places therefore need to be defined with respect to some frame of reference: in the example, that of the car or that of the outside world. Kelemen and Fenton (2010) investigated this situation with rats running on rotating disks on which they received electric foot shocks when entering any one of two areas (figure 6.15a, b). One area was defined on the rotating disk while the other one remained fixed in the reference frame of the surrounding room. Thus, foot shocks would also be received when the rat was sitting on the disk but was rotated into the room-fixed area by the rotation of the disk. Rats learn to avoid both areas simultaneously. Firing activities for individual place cells are localized better in either the rotating or the static reference frame. Multiunit recordings show that when approaching one of the avoidance regions, most active neurons were from the group using the appropriate reference frame, which allows the rat to plan the required motions. Reference frame preference was not stable, however: that is, a given place cell might use one or the other frame during the course of the experiment.

Multiple repetitive spaces In environments made of two or more identical boxes that are connected by a corridor, place cells tend to fire at the "same" location in each box relative to the local reference frame provided by the box itself (see Skaggs and McNaughton 1998 and figure 6.15c). For example, a cell might fire in the right back corner of each box. This behavior may be expected if only locally available landmark information was used but not if place cell firing would be based on metric information provided by path integration (i.e., if firing would mark the position on a global map). In any case, the place cell changes the position of its firing field, from its initial location in box 1 to a different, albeit visually similar, location in box 2. This effect is called place cell reorganization or place cell remapping. It shows that place cell firing fields are not static; indeed, in large environments, a given place cell may be used in different regions with different meaning.[3]

Place cell remapping is accompanied by a simultaneous remapping of the grid cell system and may in fact be caused by this (Fyhn et al. 2007). As discussed in chapter 5, path integration is prone to error accumulation and will be most useful in the vicinity of anchor points at which errors can be compensated by resetting the system. The nature of such anchor points is unknown; in multicompartment environments, they may be related to the entry gates of each compartment. Remapping would then be the reset of the path-integration system at anchor points that are confused due to their visual similarity.

Interestingly, place field repetition does not occur if the rooms are rotated with respect to each other, as in a multiarm maze. This is thought to result from the head direction cell input into the boundary vector cells, which makes misaligned rooms look different for the rat and would thus prevent the confusion of the respective anchor points. For reviews and theoretical models of place cell remapping, see Grieves, Duvelle, and Dudchenko (2018); Li, Arleo, and Sheynikhovich (2020); and Baumann and Mallot (2023).

In the two-room apparatus, Carpenter et al. (2015) demonstrated grid cell remapping as a discontinuity or break in the hexagonal firing patterns occurring at the room boundary. After prolonged exploration, however, the discontinuity goes away, and a continuous hexagonal pattern is established that extends seamlessly across both rooms. This is an improved representation of space in which the rat would now be able to distinguish visually similar places in the two rooms. It is interesting to consider areas in which grid cell firing forms a continuous pattern as a higher-level "region": that is, a super-ordinate entity

3. Note that the representation of space by place cells is not continuous: that is, neighboring place cells do not generally have neighboring firing fields. In fact, if such continuity existed, it would be destroyed by remapping.

in a hierarchical representation of space. For general discussion of spatial hierarchy, see section 8.4.

The third dimension Place cells have originally been studied in ground-dwelling animals where the firing fields are largely two-dimensional areas. Experiments with bridges or multiple floor levels indicate that these firing fields extend in the upward direction much like vertical columns. Jeffery et al. (2013) describe this anisotropy of mental space as "bicoding," in which the horizontal and vertical dimensions are represented differently. In flying animals such as flying foxes or bats, where place cells have also been recorded during free flight (Yartsev and Ulanowsky 2013; Wohlgemuth, Yu, and Moss 2018), volumetric firing fields have been found that chart three-dimensional space roughly isotropic in all dimensions. This is likely related to the higher relevance of the vertical dimension for these animals. However, even rats are not completely confined to the two-dimensional plane. In rats exploring fully three-dimensional lattices as shown in figure 6.15d, place cell firing fields have been found that show some specificity also for the third dimension. The firing fields are intermediate between vertical columns and isotropic spheres, with less vertical resolution in directions rarely traveled by the rat (Grieves et al. 2020).

6.5.4 Neural Place Representations in Primates, Including Humans

Place cells were first discovered in the rat hippocampus (O'Keefe and Dostrovsky 1971), and most of the subsequent research has been done in rodents. Some examples from other mammalian groups exist, most notably from bats, where place cells have been studied in free flight (see above). In the primate hippocampus, spatially selective cells have been found that simultaneously encode viewpoint and viewing direction (see Rolls 2021). Such "spatial view cells" are active when a monkey occupies a certain position in space and looks at a certain object or landmark. This difference between hippocampal encoding in rodents and monkeys may be related to the fact that primates generally have small visual field (e.g., about ± 90 degrees in humans), while rodents have almost panoramic vision. The restricted visual field is due to the roughly parallel orientation of the view axes of the two eyes and the large overlap of the monocular visual fields. The local position information available to primates is thus derived from restricted views but at the same time does contain fine-grained stereoscopic depth information, another consequence of visual field overlap. In any case, the dependence of view cell firing on viewing direction adds an egocentric element to the representation that may play a role in wayfinding and planning.

Single-cell recordings from human subjects can be obtained only as part of a medical treatment and are therefore rare. Ekstrom et al. (2003) were able to record from patients who had received intracranial electrodes for medical reasons. When these patients navigated a virtual environment, three types of neuronal activities could be demonstrated. Neurons responsive to places were found mostly in the hippocampus, while recordings from parahippocampal cortex indicated a specificity for views. Cell responses correlating with navigational goals were also found outside the hippocampal formation, in the frontal and temporal lobes.

These results are largely in line with findings from fMRI brain scanning. Epstein and Kanwisher (1998) identified an area in the posterior parahippocampal gyrus that is significantly more active during place recognition tasks than during comparable object recognition tasks. This study uses a wide variety of place descriptions, including views of furnished rooms, the spatial layout of the same rooms (cf. figure 6.8d), landmarks, landscapes, or urban outdoor scenes. The parahippocampal place area (PPA) showed a strong activity for all of these stimulus types but not for nonspatial stimuli such as faces, objects, or even the furniture removed from the furnished rooms and shown in isolation. Response is larger for novel than for familiar stimuli and is therefore thought to be involved more in the encoding than in the recall of spatial information.

Spatial functions of brain areas can also be studied by the encoding of landmarks placed at decision points or elsewhere (see section on landmark relevance above). Janzen and Turennout (2004) used a rectangular grid maze in which some connections were closed; thus, when walking down one block, way choices may or may not be possible. Grid points were marked with objects, some of which were children's toys and others were nontoys. Subjects were instructed to learn a route through the maze for later guiding a children's tour. In the test phase, subjects were shown objects and had to decide whether the object had been encountered during the exploration of the maze. fMRI scans show increased activity in the PPA for objects marking decision points as compared to objects from straight sections of the route. This effect occurred for both toy and nontoy objects: that is, no contrast for toy vs. nontoy objects was found in the PPA. This contrast, however, was present in a region of the fusiform gyrus, which thus seems to be involved in object recognition. This result also supports the distinction between objects and landmarks discussed above. Similar results are also reported in a study by Kanwisher and Yovel (2006) comparing the recognition of places and faces.

Spatial specificities have also been found in additional brain areas, including the medial parietal and retrosplenial cortex as well as the occipital place area (OPA). These areas, however, seem to be related to issues of perception and spatial working memory, which will be discussed in the next chapter. For review, see Epstein and Baker (2019), Ramanoël et al. (2020), and Baumann and Mattingley (2021).

Key Points of Chapter 6

- Places are representations or memory items abstracted from the timeline of events. They are recognized by local cues from the environment and the adjacency to other places.
- The simplest example of place knowledge is the snapshot memory studied in insects.
- The Cartwright–Collett model of snapshot homing is an example of a larger class of view-based models describing the approach toward remembered places.

- Local position information in the visual domain is completely described by the image manifold. Feature extraction algorithms from early vision applied to each image lead to more robust place codes.
- Place recognition in humans also relies on depth information. Feature and depth information are not separated in encapsulated modules but interact in the recognition of places.
- Identified landmarks constitute a special kind or class of objects that can be partially dissociated from objects recognized in other contexts.
- Landmarks can be used in two ways: guidance allows one to orient with respect to landmarks while direction associates landmark recognition with a navigational action.
- The firing activity of hippocampal place cells represents spatial position in a population code. Place cells integrate local position information from boundary and object vector cells with metric information provided by the entorhinal grid cell system.
- Place cell firing depends on the currently employed frame of reference. It also shows spatial specificities in the third dimension.

References

Adelson, E. H., and J. R. Bergen. 1991. "The plenoptic function and the elements of early vision." In *Computational models of visual processing,* edited by M. S. Landy and J. A. Movshon, 3–20. Cambridge, MA: MIT Press.

Aginsky, V., C. Harris, R. Rensink, and J. Beusmans. 1997. "Two strategies for learning a route in a driving simulator." *Journal of Environmental Psychology* 17:317–331.

Allen, G. L., K. C. Kirasic, A. W. Siegel, and J. F. Herman. 1979. "Developmental issues in cognitive mapping: The selection and utilization of environmental landmarks." *Child Development* 50:1062–1070.

Barry, C., C. Lever, R. Hayman, T. Hartley, S. Burton, J. O'Keefe, K. Jeffery, and N. Burgess. 2006. "The boundary vector cell model of place cell firing and spatial memory." *Reviews in the Neurosciences* 17:71–97.

Basten, K., and H. A. Mallot. 2010. "Simulated visual homing in desert ant natural environments: Efficience of skyline cues." *Biological Cybernetics* 102:413–425.

Baumann, O., and J. B. Mattingley. 2021. "Extrahippocampal contributions to spatial navigation in humans: A review of the neuroimaging evidence." *Hippocampus* 31:640–657.

Baumann, T., and H. A. Mallot. 2023. "Gateway identity and spatial remapping in a combined grid and place cell attractor." *Neural Networks* 157:226–339.

Bécu, M., D. Sheynikhovich, G. Tatur, C. P. Agathos, L. L. Bologna, J.-A. Sahel, and A. Arleo. 2020. "Age-realted preference for geometric spatial cues during real-world navigation." *Nature Human Behavior* 4:88–99.

Benedikt, M. L. 1979. "To take hold of space: Isovists and isovist fields." *Environment and Planning B* 6:47–65.

Bicanski, A., and N. Burgess. 2020. "Neuronal vector coding in spatial cognition." *Nature Reviews Neuroscience* 21:453–470.

Biegler, R., and R. G. M. Morris. 1993. "Landmark stability is a prerequisite for spatial but not discrimination learning." *Nature* 361:631–633.

Bilalić, M., T. Lindig, and L. Turella. 2019. "Parsing rooms: The role of the PPA and RSC in perceiving object relations and spatial layout." *Brain Structure and Function* 224:2505–2524.

Bloom, P., M. A. Peterson, L. Nadel, and M. Garrett, eds. 1996. *Language and space.* Cambridge, MA: MIT Press.

Carpenter, F., D. Manson, K. Jeffery, N. Burgess, and C. Barry. 2015. "Grid cells form a glocal representation of connected environments." *Current Biology* 25:1176–1182.

Cartwright, B. A., and T. S. Collett. 1982. "How honey bees use landmarks to guide their return to a food source." *Nature* 295:560–564.

Cartwright, B. A., and T. S. Collett. 1983. "Landmark learning in bees." *Journal of Comparative Physiology* 115:521–543.

Cheng, K. 1986. "A purely geometric module in the rat's spatial representation." *Cognition* 23:149–178.

Cheng, K., and C. R. Gallistel. 1984. "Testing the geometric power of an animal's spatial representation." In *Animal cognition,* edited by H. L. Roitblat, T. B. Bever, and H. S. Terrace. Hillsdale, NJ: Lawrence Erlbaum Associates.

Cheng, K., J. Huttenlocher, and N. S. Newcombe. 2013. "25 years of research on the use of geometry in spatial reorientation: A current theoretical perspective." *Psychonomic Bulletin & Review* 20:1033–1054.

Cheng, K., S. J. Shettleworth, J. Huttenlocher, and J. J. Rieser. 2007. "Bayesian integration of spatial information." *Psychological Bulletin* 133:625–637.

Collett, T. S., and J. Zeil. 2018. "Insect learning flights and walks." *Current Biology* 28:R984–R988.

Cornell, E. H., and C. D. Heth. 2006. "Home range and the development of children's way finding." *Advances in Child Development and Behavior* 34:173–206.

Credé, S., T. Thrash, C. Hölscher, and S. I. Fabrikant. 2020. "The advantage of globally visible landmarks for spatial learning." *Journal of Environmental Psychology* 67:101369.

Darwin, C. R. 1877. *The effects of cross and self fertilization in the vegetable kingdom.* New York: D. Appleton and Company.

Doeller, C. F., and N. Burgess. 2008. "Distinct error-correcting and incidental learning of location relative to landmarks and boundaries." *PNAS* 105:5909–5914.

Doeller, C. F., J. A. King, and N. Burgess. 2008. "Parallel striatal and hippocampal systems for landmarks and boundaries in spatial memory." *PNAS* 105:5915–5920.

Durier, V., P. Graham, and T. S. Collett. 2003. "Snapshot memories and landmark guidance in wood ants." *Current Biology* 13:1614–1618.

Ekstrom, A. D., M. J. Kahana, J. B. Caplan, T. A. Fields, E. A. Isham, E. L. Newman, and I. Fried. 2003. "Cellular networks underlying human spatial navigation." *Nature* 425:184–187.

Epstein, R. A. 2008. "Parahippocampal and retrosplenial contributions to human spatial navigation." *Trends in Cognitive Sciences* 12:388–396.

Epstein, R. A., and C. L. Baker. 2019. "Scene perception in the human brain." *Annual Review of Vision Science* 5:373–397.

Epstein, R., and N. Kanwisher. 1998. "A cortical representation of the local visual environment." *Nature* 392:598–601.

Evans, H. E., and K. M. O'Neil. 1991. "Beewolves." *Scientific American* 70 (2): 70–77.

Fodor, J. A. 1983. *The modularity of the mind.* Cambridge, MA: MIT Press.

Franz, M. O., B. Schölkopf, H. A. Mallot, and H. H. Bülthoff. 1998. "Where did I take that snapshot? Scene-based homing by image matching." *Biological Cybernetics* 79:191–202.

Fyhn, M., T. Hafting, A. Treves, M.-B. Moser, and E. Moser. 2007. "Hippocampal remapping and grid realignment in entorhinal cortex." *Nature* 446:190–194.

Gärdenfors, P. 2020. "Primary cognitive categories are determined by their invariances." *Frontiers in Psychology* 11:584017.

Gillner, S., A. M. Weiß, and H. A. Mallot. 2008. "Visual place recognition and homing in the absence of feature-based landmark information." *Cognition* 109:105–122.

Goodman, J. 2021. "Place vs. response learning: History, controversy, and neurobiology." *Frontiers in Behavioral Neuroscience* 14:598570.

Gootjes-Dreesbach, L., L. C. Pickup, A. W. Fitzgibbon, and A. Glennerster. 2017. "Comparison of view-based and reconstruction-based models of human navigational strategy." *Journal of Vision* 17 (9): 11.

Graham, M., M. A. Good, A. McGregor, and J. M. Pearce. 2006. "Spatial learning based on the shape of the environment is influenced by properties of the objects forming the shape." *Journal of Experimental Psychology: Animal Behavior Processes* 32:44–59.

Graham, P., and K. Cheng. 2009. "Ants use panoramic skyline as a visual cue during navigation." *Current Biology* 19:R935–R937.

Grieves, R. M., E. Duvelle, and P. A. Dudchenko. 2018. "A boundary vector cell model of place field repetition." *Spatial Cognition and Computation* 18:217–256.

Grieves, R. M., S. Jedidi-Ayoub, K. Mishchanchuk, A. Liu, S. Renaudineau, and K. Jeffery. 2020. "The place-cell representation of volumetric space in rats." *Nature Communications* 11:789.

Grieves, R. M., and K. J. Jeffery. 2017. "The representation of space in the brain." *Behavioural Processes* 135:113–131.

Gyselinck, V., and F. Pazzaglia, eds. 2012. *From mental imagery to spatial cognition and language: Essays in honour of Michel Denis.* London: Psychology Press.

Hamilton, D. A., I. Driscoll, and R. J. Sutherland. 2002. "Human place learning in a virtual Morris water task: Some important constraints on the flexibility of place navigation." *Behavioural Brain Research* 129:159–170.

Hermer, L., and E. S. Spelke. 1994. "A geometric process for spatial reorientation in young children." *Nature* 370:57–59.

Hort, J., J. Laczo, M. Vyhnalek, M. Bojar, J. Bures, and K. Vlcek. 2007. "Spatial navigation deficit in amnestic mild cognitive impairment." *PNAS* 104:4042–4047.

Høydal, Ø. A., E. R. Skytøen, S. O. Anderson, M.-B. Moser, and E. I. Moser. 2019. "Object-vector coding in the medial entorhinal cortex." *Nature* 568:400–404.

Hübner, W., and H. A. Mallot. 2007. "Metric embedding of view graphs: A vision and odometry-based approach to cognitive mapping." *Autonomous Robots* 23:183–196.

Jacobs, W. J., K. G. F. Thomas, H. E. Laurance, and L. Nadel. 1998. "Place learning in virtual space II: Topographical relations as one dimension of stimulus control." *Learning and Motivation* 29:288–308.

Janzen, G., T. Herrmann, S. Katz, and K. Schweizer. 2000. "Oblique angled intersections and barriers: Navigating through a virtual maze." *Lecture Notes in Computer Science* 1849:277–294.

Janzen, G., and M van Turennout. 2004. "Selective neural representation of objects relevant for navigation." *Nature Neuroscience* 7:673–677.

Jeffery, K. J., A. Jovalekic, M. Verriotis, and R. Hayman. 2013. "Navigating in a three-dimensional world." *Behavioral and Brain Sciences* 36:523–587.

Kant, I. 1781. *Critik der reinen Vernunft.* Riga: Johann Friedrich Hartknoch.

Kanwisher, N., and G. Yovel. 2006. "The fusiform face area: A cortical region specialized for the perception of faces." *Philosophical Transactions of the Royal Society B: Biological Sciences* 361:2109–2128.

Kelemen, E., and A. A. Fenton. 2010. "Dynamic grouping of hippocampal neural activity during cognitive control of two spatial frames." *PLoS Biology* 8 (6): e1000403.

Kravitz, D. J., K. S. Saleem, C. I. Baker, and M. Mishkin. 2011. "A new neural framework for visuospatial processing." *Nature Reviews Neuroscience* 12:217–230.

Lambalgen, M. van, and F. Hamm. 2005. *The proper treatment of events.* Malden: Blackwell.

Lambrinos, D., R. Möller, T. Labhart, R. Pfeifer, and R. Wehner. 2000. "A mobile robot employing insect strategies for navigation." *Robotics and Autonomous Systems* 30:39–64.

Learmonth, A. E., L. Nadel, and N. S. Newcombe. 2002. "Children's use of landmarks: Implications for modularity theory." *Psychological Science* 13:337–341.

Levinson, S. C. 2003. *Space in language and cognition.* Cambridge: Cambridge University Press.

Li, T., A. Arleo, and D. Sheynikhovich. 2020. "Modeling place cells and grid cells in multi-compartment environments: Entorhinal-hippocampal loop as a multisensor integration circuit." *Neural Networks* 121:37–51.

Lowry, S., N. Süderhauf, P. Newman, J. J. Leonard, D. Cox, P. Corke, and M. J. Milford. 2016. "Visual place recognition: A survey." *IEEE Transactions on Robotics* 32 (1): 1–19.

Mallot, H. A., and S. Lancier. 2018. "Place recognition from distant landmarks: Human performance and maximum likelihood model." *Biological Cybernetics* 112:291–303.

Mallot, H. A., S. Lancier, and M. Halfmann. 2017. "Psychophysics of place recognition." *Lecture Notes in Artificial Intelligence* 10523:118–136.

Marr, D. 1976. "Early processing of visual information." *Proceedings of the Royal Society (London) B* 275:483–519.

Michon, P.-E., and M. Denis. 2001. "When and why are visual landmarks used in giving directions?" In *Spatial information theory,* edited by D. Montello, 292–305. Berlin: Springer.

Mizunami, M., J. M. Weibbrecht, and N. J. Strausfeld. 1998. "Mushroom bodies of the cockroach: Their participation in place memory." *The Journal of Comparative Neurology* 402:520–537.

Möller, R. 2000. "Visual homing strategies of insects in a robot with analog processing." *Biological Cybernetics* 83:231–241.

Möller, R. 2002. "Insects exploit UV-green contrast for landmark navigation." *Journal of Theoretical Biology* 214:619–631.

Möller, R., and A. Vardy. 2006. "Local visual homing by matched-filter descent in image distances." *Biological Cybernetics* 95:413–430.

Morris, R. G. M. 1981. "Spatial localization does not require the presence of local cues." *Learning and Motivation* 12:239–260.

Moser, E. I., M.-B. Moser, and B. L. McNaughton. 2017. "Spatial representation in the hippocampal formation: A history." *Nature Neuroscience* 20:1448–1464.

Nardi, D., K. J. Singer, K. M. Price, S. E. Carpenter, J. A. Bryant, M. A. Hatheway, J. N. Johnson, A. K. Pairitz, K. L. Young, and N. S. Newcombe. 2021. "Navigating without vision: Spontaneous use of terrain slant in outdoor place learning." *Spatial Cognition & Computation* 21:235–255.

O'Keefe, J. 1991. "The hippocampal cognitive map and navigational strategies." In *Brain and space,* edited by J. Paillard, 273–295. Oxford: Oxford University Press.

O'Keefe, J., and J. Dostrovsky. 1971. "The hippocampus as a spatial map. Preliminary evidence from unit activity in the freely-moving rat." *Brain Research* 34:171–175.

O'Keefe, J., and L. Nadel. 1978. *The hippocampus as a cognitive map.* Oxford: Clarendon.

Pickup, L. C., A. W. Fitzgibbon, and A. Glennerster. 2013. "Modelling human visual navigation using multiview scene reconstruction." *Biological Cybernetics* 107:449–464.

Ramanoël, S., M. Durteste, M. Bécu, C. Habas, and A. Arleo. 2020. "Differential brain activity in regions linked to visuospatial processing during landmark-based navigation in young and healthy older adults." *Frontiers and Human Neuroscience* 14:552111.

Rolls, E. T. 2021. "Neurons including hippocampal spatial view cells, and navigation in primates including humans." *Hippocampus* 31:593–611.

Rosch, E. 1975. "Cognitive representations of semantic categories." *Journal of Experimental Psychology: General* 104:192–233.

Russell, B. 1915. *Our knowledge of the external world.* Chicago: The Open Court Publishing Company.

Shamsyeh Zahedi, M., and J. Zeil. 2018. "Fractal dimension and the navigational information provided by natural scenes." *PLoS ONE* 13 (5): e0196227.

Sheynikhovich, D., R. Chavarriaga, T. Strösslin, A. Arleo, and W. Gerstner. 2009. "Is there a geometric module for spatial orientation? Insights from a rodent navigation model." *Psychological Review* 116:540–566.

Skaggs, W., and B. McNaughton. 1998. "Spatial firing properties of hippocampal CA1 populations in an environment containing two visually identical regions." *The Journal of Neuroscience* 18:8455–8466.

Solstad, T., C. N. Boccara, E. Kropff, M.-B. Moser, and E. I. Moser. 2008. "Representation of geometric borders in the entorhinal cortex." *Science* 322:1865–1868.

Sperling, G., M. S. Landy, B. A. Dosher, and M. E. Perkins. 1989. "Kinetic depth effect and the identification of shape." *Journal of Experimental Psychology: Human Perception and Performance* 15 (4): 826–840.

Stankiewicz, B. J., and A. Kalia. 2007. "Acquisition and retention of structural versus object landmark nnowledge when navigating through a large-scale space." *Journal of Experimental Psychology: Human Perception and Performance* 33:378–390.

Steck, S. D., and H. A. Mallot. 2000. "The role of global and local landmarks in virtual environment navigation." *Presence: Teleoperators and Virtual Environments* 9:69–83.

Stürzl, W., A. Cheung, K. Cheng, and J. Zeil. 2008. "The information content of panoramic images I: The rotational errors and the similarity of views in rectangular experimental arenas." *Journal of Experimental Psychology: Animal Behavior Processes* 34:1–14.

Stürzl, W., and H. A. Mallot. 2006. "Efficient visual homing based on Fourier transformed panoramic images." *Robotics and Autonomous Systems* 54:300–313.

Tinbergen, N. 1958. *Curious naturalists.* London: Country Life Ltd.

Tinbergen, N., and W. Kruyt. 1938. "Über die Orientierung des Bienenwolfes (*Philanthus triangulum* FABR.) III. Die Bevorzugung bestimmter Wegmarken." *Zeitschrift für vergleichende Physiologie* 25:292–334.

Tolman, E. C., B. F. Ritchie, and D. Kalish. 1946. "Studies in spatial learning. II. Place learning versus response learning." *Journal of Experimental Psychology* 36:221–229.

Trullier, O., S. I. Wiener, A. Berthoz, and J.-A. Meyer. 1997. "Biologically based artificial navigation systems: Review and prospects." *Progress in Neurobiology* 51:483–544.

Vann, S. D., J. P. Aggleton, and E. A. Maguire. 2009. "What does the retrosplenial cortex do?" *Nature Reviews Neuroscience* 10:792–802.

Waller, D., J. M. Loomis, R. G. Golledge, and A. C. Beall. 2000. "Place learning in humans: The role of distance and direction information." *Spatial Cognition and Computation* 2:333–354.

Wenczel, R., L. Hepperle, and R. von Stülpnagel. 2017. "Gaze behavior during incidental and intentional navigation in an outdoor environment." *Spatial Cognition and Computation* 17:121–142.

Wiener, J. M., G. Franz, N. Rosmanith, A. Reichelt, H. A. Mallot, and H. H. Bülthoff. 2007. "Isovist analysis captures properties of space relevant for locomotion and experience." *Perception* 36:1066–1083.

Wiener, J. M., C. Hölscher, S. Büchner, and L. Konieczny. 2012. "Gaze behaviour during space perception and spatial decision making." *Psychological Research* 76:713–729.

Wilson, M. A., and B. L. McNaughton. 1993. "Dynamics of the hippocampal ensemble code for space." *Science* 261:1055–1058.

Wohlgemuth, M. J., C. Yu, and C. F. Moss. 2018. "3D hippocampal place field dynamics in free-flying echolocating bats." *Frontiers in Cellular Neuroscience* 12:270.

Yartsev, M. M., and N. Ulanowsky. 2013. "Representation of three-dimensional space in the hippocampus of flying bats." *Science* 340:367–372.

Yesiltepe, D., R. C. Dalton, and A. O. Torun. 2021. "Landmarks in wayfinding: A review of the existing literature." *Cognitive Processing* 22:369–410.

Yesiltepe, D., A. O. Torun, A. Courtrot, M. Hornberger, H. Spiers, and R. C. Dalton. 2021. "Computer models of saliency alone fail to predict subjective visual attention to landmarks during observed navigation." *Spatial Cognition and Computation* 21:39–66.

Zacks, J. M., and B. Tversky. 2001. "Event structure in perception and conception." *Psychological Review* 127:3–21.

Zeil, J., M. I. Hoffmann, and J. S. Chahl. 2003. "Catchment areas of panoramic snapshots in outdoor scenes." *Journal of the Optical Society of America A* 20:450–469.

Ziv, Y., L. D. Burns, E. D. Cocker, E. O. Hamel, K. K. Ghosh, L. J. Kitch, A. El Gamal, and M. J. Schnitzer. 2013. "Long-term dynamics of CA1 hippocampal place codes." *Nature Neuroscience* 16:264–266.

7 Spatial Memory

Memory comes in two major forms, called working and long-term memory, respectively. We define and discuss these concepts in general and in relation to spatial behavior in particular. Working memory is mostly concerned with the planning and monitoring of behavior, and we will describe behavioral paradigms addressing these issues. In humans, spatial imagery and the anisotropies of imagined space as well as the modal or amodal character of spatial working memory have received much attention. Neural substrates and computational models of spatial working memory are discussed as extensions of the hippocampal path integration system. In long-term memory, the most important idea is the distinction of route and map memories. Routes are sequences of stimulus–response or state–action schemata from which map knowledge can be built by connecting segments of multiple intersecting routes into networks. We discuss the notion of the cognitive map and suggest a simple definition as the declarative part of spatial long-term memory.

7.1 What Is Working Memory?

This chapter is not the first one in this book dealing with spatial memory. Indeed, most representations of space are not instantaneous but have a temporal duration and are therefore also forms of memory. This is clearly true for all representations of space built on temporal and spatial integration or extended exploration. Peripersonal space, for example, combines information from different viewing directions and vantage points into a stable representation. The mental transformations—perspective taking, mental rotation, and spatial updating—that were discussed in chapter 3 are processes carried out within this memory stage, which is therefore a form of working memory. Similarly, the home vector in path integration, as discussed in chapter 5, is a simple form of working memory held active during an ongoing excursion. Finally, in chapter 6, we defined landmarks as environmental cues stored in long-term memory as raw snapshots, more derived codes based on depth perception, or even elaborated object representations. All these examples involve memory but are also characterized by a strong perceptual component. In the present chapter, we turn to behavioral performances that are mainly driven by memory itself.

Table 7.1
Types of memory

	Iconic	Working memory	Long-term memory
Duration	< 1 sec	Seconds to minutes	Up to lifetime
Physiological mechanism	Receptor adaptation, propagation times, synaptic delays	Persistent and reverberating neural activity	Synaptic plasticity, adult neurogenesis
Function	Spatiotemporal processing of sensory information	Event monitoring, planning and plan execution, interaction with long-term memory (storage, recall, reconsolidation)	*Nondeclarative:* conditioning, habits, skills, routes *Declarative:* if–then rules, reference knowledge, cognitive maps
Capacity limitation	Several image frames	Miller's 7 ± 2 items, peripersonal space	No limit known

7.1.1 Distinctions

Working memory is among the most influential concepts introduced in the cognitive sciences during the past decades. It originated from the distinction of different memory types by their duration and for a while was thought to be sufficiently described as "short-term memory." The current view, however, is much broader and connects to all aspects of "thinking," including imagination, reasoning, problem solving, planning, executive function, and consciousness. Before we turn to working memory itself, we briefly review the most important memory concepts in context (see also table 7.1).

Iconic memory Perception and memory are often treated as well-defined, distinguishable faculties of the cognitive apparatus, but their delineation is to some extent arbitrary. Percepts always have a duration and persist after the stimulus vanishes. In this sense, the after-image visible after looking into a bright light source may be considered a memory, which in this case would originate already in the adaptation of retinal receptor cells. Sensory neurons often work as dynamical filters in the sense that their activity depends on, or integrates over, sensory input from a temporal or spatiotemporal receptive field with a time window on the order of tenths of seconds. An obvious example is visual motion detection, which in itself requires the consideration of past (i.e., remembered) input. Detectors for visual motion can be built on two input lines, a direct and a delayed one, which converge on a motion detector neuron. If this neuron detects the coincidence of the two inputs, a motion from the retinal origin of the delayed input toward that of the direct line is perceived. The delayed line thus holds a memory of the input to which the current image is compared. Memories based on the simple feed-forward dynamics of neural signal flow have been investigated psychophysically with groups of letters that are flashed briefly and have to be reported by the subject afterward (Sperling 1960). This type of memory is called "immediate" or "iconic" and is localized in the early sensory pathways.

Working memory Memories lasting for seconds or up to several minutes have traditionally been described as short-term memories. However, the definition by duration is artificial and has hence been replaced by functional and physiological criteria. When walking in a crowd, working memory may hold an image of a person as long as it takes the agent to pass without a collision. In a conversation, working memory lasts as long as the partner is present or at least as long as it takes a listener to comprehend a sentence or an argument. Working memory can be kept active by "rehearsal," for example if we keep uttering a telephone number until we enter it on a keypad and forget it shortly thereafter. By repeating the rehearsal longer, the item may also be actively memorized and transferred into long-term memory.

Working memory is thought to be based on persistent neural activity (Zylberberg and Strowbridge 2017). It is best studied in the prefrontal cortex but involves also a large number of other areas of the brain (Fuster 2001; D'Esposito and Postle 2015; Christophel et al. 2017). In a classical study on working memory in eye movement control, Goldman-Rakic (1995) investigated the representation of a visual target position in a delayed response task. In this task, a small target is flashed at some position in the visual field while a monkey is looking at a fixation cross. After a delay, a "go" signal is presented (e.g., a sound) and the monkey has to move its gaze to the remembered position of the now invisible target to obtain a reward. Goldman-Rakic (1995) and her colleagues identified neurons in the prefrontal cortex that kept firing during the entire duration of the delay period if the remembered target had occurred in a certain position for which the neuron was specific (the neuron's "memory field"). Persistent activity of the neuron thus represents the information of the required gaze position or movement. In general, persistent neural activity representing working memory contents can be based on various physiological mechanisms, including, among others, bistable firing behavior of individual neurons or dynamic attractors in neural circuits with feedback (Zylberberg and Strowbridge 2017). An example of the latter is the attractor dynamic thought to underlie the representation of heading and position in path integration as discussed in section 5.4 and figure 5.8. For a direct demonstration of a neural attractor representing the heading direction in *Drosophila*, see Kim et al. (2017).

Learning and long-term memory Learning can be defined as the adaptive change of behavior based on experience.[1] Such changes can be caused in many ways: for example, by the acquisition of new stimulus–response schemata or procedures (associative learning), the storage of facts and knowledge for informed choices, or even a more efficient interaction with working memory facilities to better evaluate the consequences of action alternatives. In any case, changing behavioral mechanisms or storing items into reference memory involves plastic changes of the brain such as the growth or degeneration of synapses or dendritic spines. This leads to changes in synaptic efficiency, as has been demonstrated in long-term potentiation (LTP). From the point of view of physiology, the hallmark of long-term memory is therefore gene expression and protein synthesis in the brain (for reviews, see Kandel, Dudai, and Mayford 2014; Minatohara, Akiyoshi, and Okuno 2016). Plasticity may even mean adult neurogenesis: that is, differentiation of novel neurons from stem cells and their incorporation into existing networks (Kempermann 2019). The plastic network changes caused by learning are also called the "memory trace."

Long-term memory serves different functions. We already mentioned "associative" learning, in which stimulus–response pairs are newly learned or modified, as in conditioning and reinforcement learning. The acquisition of knowledge such as facts, if-then rules, or memories of past events leads to "declarative" memories. Unlike associative learning, where the perception of the stimulus triggers the remembered response and is therefore also called "imperative" or "procedural," declarative memories provide information relevant for a behavioral decision but leave the decision itself to a separate planning stage. Planning is thought to happen in a subdivision of working memory known as the central executive. We will not go into detail here but come back to long-term memory types in the context of route and map learning below. For a classical discussion of long-term memory types, see Squire and Knowlton (1995).

7.1.2 Working Memory Function

In computer science, working memory is a temporary store holding the running program as well as input data or intermediate results required for the ongoing computation. If the computer stops, working memory is flushed and its contents are lost. Working memory is a component of the classical von Neumann architecture of a digital computer, although the term does not seem to have been used by von Neumann (1958) himself. The ideas have been picked up, however, in the functional account of working memory that has been given by Miller, Galanter, and Pribram (1960). In this book, the authors "speak of the memory we use for the execution of our plans as a kind of quick-access 'working memory'" (65). In particular, they discuss the analogy of psychological plans with computer programs, both of which are executed in a stepwise fashion and require intermediate storage of the state

1. The word "adaptive" is added here to distinguish learning from maladaptive changes of behavior such as addiction. Addiction is generally thought to be a disorder of learning.

of affairs, achieved subgoals, and the next things to do. Working memory is distinguished from a set-aside "storage" component: that is, the hard disk of a computer or the long-term memory in cognition, both with much larger capacities.

These considerations lead to a simple concept of a cognitive agent, which we discuss for the case of spatial cognition (see also figure 1.2 in the Introduction). It is composed of four parts: the *perception* unit provides information about egomotion and landmark cues; the *working memory (including executive functions)* controls ongoing maneuvers and generates novel plans; the *long-term memory*, which can only be accessed from working memory (both for storage and retrieval), provides supporting information; and the *action* system carries out motor behavior. Action and perception are coupled via the environment in the sense that the agent's motion will change its perception. Working memory is thus an essential part of the action–perception cycle (cf. section 4.4). Indeed, the prefrontal cortex, a structure whose involvement in working memory was already discussed above, is thought to control behavior as the top level of a subsumption architecture, as illustrated in figure 4.9c (see Fuster 2004). It should be noted, however, that working memory is defined by its function, not as a particular brain region. The von Neumann architecture applies to a *serial* computer in which a single working memory stage may be sufficient. The brain is clearly a parallel or distributed processor in which working memory function is realized in many different components.

In summary, working memory is much more than a storage of intermediate temporal duration. Rather, it has become a synonym for the machinery of thinking, imagery, prospection, planning, learning, text comprehension, and similar performances (Zimmer 2008; Carruthers 2013). Much work on working memory focuses on humans and specifically human abilities such as language. As compared to these, studies of working memory in the spatial domain lead to slightly different views that may be helpful for the understanding of the evolution of working memory at large.

7.1.3 Working Memory Capacity: The Multicomponent Model

Working memory is not only transient but also limited in storage capacity. If a subject is asked to memorize a list of words or a set of visual patterns, the number of remembered items is often quite small, and capacity has therefore been estimated as a discrete number as small as seven or even four. Capacity limitations can also be demonstrated for the distinguished levels of a continuous stimulus parameter such as color or line orientation, an effect described as probabilistic inference by Ma, Husain, and Bays (2014). Capacity depends on the type of encoding and can be increased by "chunking": that is, the coercion of sets of simple features into single, more complex ones. The total capacity is not a homogeneous or unstructured unity but contains "dissociable" components that can be identified in dual-task experiments. If a subject is imagining a visual scene and at the same time is asked to remember a sequence of words, the two tasks interfere less with each other than the combination of two visual or two sequence tasks would do. This type of experiment led Baddeley and Hitch (1974) to suggest that the working memory has (at least) two partially independent components, dubbed the "visuo-spatial sketchpad" and the "phonological loop," respectively. Both are coupled to the "central executive" in which processes are run that cannot be attributed to either of the other parts. The phonological loop is designed to explain experiments in language comprehension or word list repetition, which makes it hard to compare it to similar mechanisms in animal cognition. Sequences, however, do also occur outside the language (or otherwise phonological) domain (e.g., in the form of sequential actions or in route following). The visuospatial sketchpad is mostly a visual working memory, which may integrate over eye and head movements and would also be involved in visual imagery. The space mentioned in its name is mostly the two-dimensional image space, not the navigational space of spatial cognition.

Baddeley (2000) expanded the original two-component model by a third component, called the episodic buffer. The argument is based on the observation that in memory disorders, the performance in the recall of word lists is generally much lower than in the recall of meaningful sentences. In one patient, who could recall only about one word of a list, the ability to recall longer sentences was quite good, though not completely spared. This argues for a dissociation of the memory for arbitrary sequences (the phonological loop) and a memory for events or narratives. Baddeley (2000) therefore suggests the episodic buffer as a third component of working memory, especially interacting with episodic long-term memory.

In spatial working memory as discussed below, capacity limitations may also restrict the number of objects remembered in a given task or the accuracy of their remembered positions, a continuous quantity for which the probabilistic reasoning approach of Ma, Husain, and Bays (2014) may apply. Another type of limitation is spatial extent, for example, the "closure" of peripersonal space discussed in chapter 3. However, experimental evidence for these questions remains sparse.

7.2 Working Memory Tasks

Like other concepts in cognitive science, working memory is most usefully defined by experimental paradigms that operationalize the idea. In this section, we will therefore give an overview of the experimental tasks and performances used to study the concept. We start with rodents in which spatial behavior is probably the easiest, if not the only way to access working memory.

7.2.1 Rodents

Many experiments on place recognition discussed in the previous chapter (cf. figure 6.1) can be turned into working memory paradigms by varying the place cues between sessions or by changing the temporal sequence of the experimental procedure. Here we focus on two additional paradigms that relate to the discussion of working memory function given above. For a review, see Tsutsui et al. (2016).

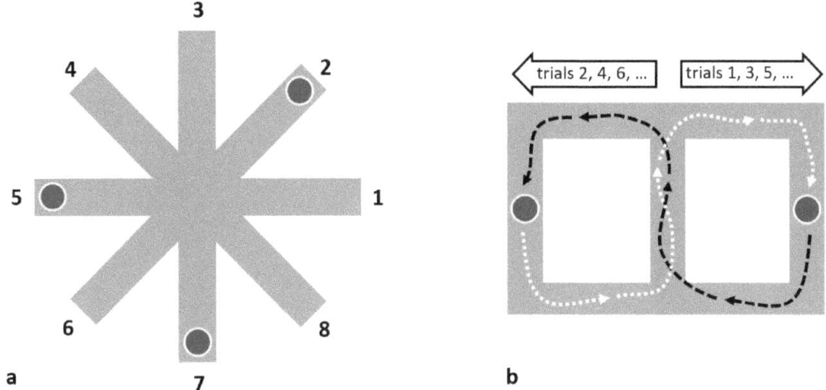

Figure 7.1
Two types of mazes used in the study of rodent working memory. (a) In a radial arm maze, a number of arms are selected for regular baiting (dark disks in arms 2, 5, and 7 in figure). Entering any of the other arms is counted as a reference memory error. Entering baited arms is correct when happening for the first time in each session. If a rat enters a depleted arm for a second time, however, this would be counted as a working memory error. (b) Figure-of-eight maze used in alternation tasks. The T-junction at the top center is the decision point at which the rat has to choose alternating arms at repeated trials. Reward for correct trials is provided in the outer arms.

A standard paradigm for working memory in rodents is food depletion. A rat is trained to receive food in a fixed subset of the arms of a radial arm maze such as the eight-arm maze shown in figure 7.1a. The same arms are baited at the beginning of every session and the rat will learn to check these arms while ignoring the others. If it does run into an arm that was never baited, this will be considered an error of long-term memory. If, however, it enters an arm for the second time after having taken the bait already earlier in the same session, this will be considered a working memory error. This protocol fits nicely to the above definitions: reentering a depleted arm should be prevented by memory of the ongoing process (the testing session) but is independent of long-term memory. Olton and Papas (1979) used this paradigm to demonstrate the role of the hippocampus in spatial working memory. Rats were trained in a radial seventeen-arm maze and then underwent a surgery in which the connections from the hippocampus via the fimbria and fornix were removed. After recovery, the animals were still able to navigate the maze and performed normally with respect to reference memory errors, while the working memory performance was lost and could not be recovered.

Another working memory task used with rodents is alternation in a figure-of-eight maze where the rat has to remember the last turning decision at a bifurcation of the path and choose the other one when it comes back to the decision point for the next time (figure 7.1b). Johnson and Redish (2007) established firing fields of hippocampal place

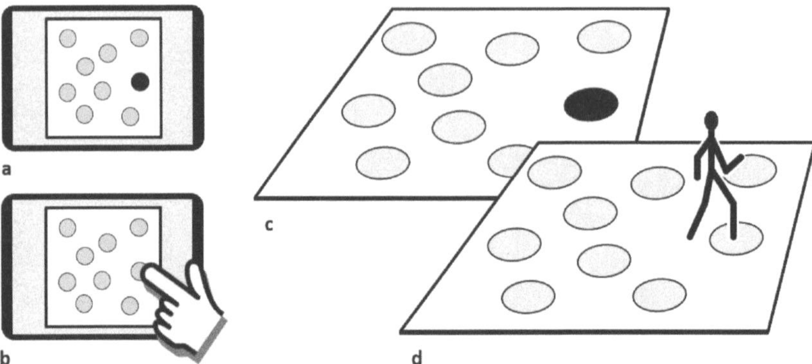

Figure 7.2
Screen and walking versions of the Corsi task. (a, b) Screen task: The "blocks" are marked on the screen and highlighted in a random sequence (only one step shown). (b) During recall, subjects point at the screen recalling targets and sequence. (c, d) Walking task. The "blocks" are laid out on the floor and the subjects watch the sequence from the side. During recall, the sequence is reproduced in walking.

cells during unrewarded exploration of a figure-of-eight maze. During the alternation task, in which a reward was given for correct decisions, some of these cells showed a previously unobserved pattern of activity: that is, they fired in places where they had previously been silent. In particular, neurons with firing fields in the initial sections after the bifurcation would now fire somewhat before the decision point. The position represented by this "out-of-place" firing was calculated by maximum likelihood estimation, which is an elaborated version of the population decoding procedure described in section 6.5.1. Results show that while the rat is sitting in front of the decision point, presumably negotiating the path alternatives, the population activity of place cells probes into the two possible directions. In this case, the hippocampal activity thus seems to reflect the planning of the next action.

Out-of-place firing of hippocampal place cells was also reported by Foster and Wilson (2006) in rats resting from a walk. In this situation, place cells that had been firing during the preceding motor activity are reactivated in a brief, inverted sequence lasting for some 100 ms. In simultaneous recordings of the hippocampal EEG (electro-encephalogram), short-wave events called "ripples" are seen at the time of replay. Replay is now thought to play a general role in the encoding and recall of information in the interaction of working and long-term memory; it is also found during sleep and may be involved in memory consolidation (Pfeiffer 2017).

7.2.2 The Corsi Block-Tapping Task

The place of spatial working memory in Baddeley's multicomponent model discussed in section 7.1.3 is unclear. Many tasks used to study the "visuospatial sketchpad" relate to visual, or image space, and cannot easily be generalized to navigational space. An example from grasp space is the "Corsi block-tapping task" (see Kessels et al. 2000), in which a subject is presented with an arrangement of nine blocks mounted on a board. The experimenter then touches a number of these blocks with the finger, and the patient is asked to repeat the indicated sequence. Before that, a retention phase of variable length is added, during which the subject has to hold a memory of the sequence. The dependent variable is called the Corsi span: that is, the length of correctly repeated sequences.

The type of working memory addressed in the Corsi task was investigated by adding a dual task during the retention phase (Della Sala et al. 1999). In a "spatial" dual task, subjects had to move their finger under haptic control (i.e., without vision) along a touchable edge; in a "visual" task, they just looked at a set of colorful paintings. Results show that the Corsi task is disrupted by the spatial task more than by the visual tasks. In another experiment, the Corsi task is replaced by a visual memory task not involving a memory of sequences. Subjects had to study a square grid with a number of squares filled in black color. After a retention phase, they were asked to reproduce the pattern of filled squares with a pen, which could be done in an arbitrary sequence. Again, dual tasks were carried out during the retention phase. In this experiment, the result was reversed: that is, the square filling task could be disrupted by the visual dual task but not by the spatial one. This result has been taken as evidence for a further separation of the visuospatial sketchpad into a spatial and a visual part (see also Baddeley 2003).

Corsi performance in peripersonal space is studied with the walking Corsi task (Piccardi et al. 2010 and figure 7.2c,d). In this case, the target locations are laid out as tiles on the floor of a walking arena, and sequences are presented by pointing with a laser pointer or flashlight. In the recall phase, subjects reproduce the sequence by stepping on each tile in the remembered sequence. By comparing the performance in the ordinary (grasp space) and walking Corsi tasks, Piccardi et al. (2010) demonstrated that patients with lesions in the temporal lobe performed differently in the small- and larger-scale spaces, indicating that memories for image and peripersonal space are not the same. The difference between these two memories was further investigated by Röser, Hardiess, and Mallot (2016) by testing all possible combinations of grasp space (computer screen) and peripersonal space (walking) for the encoding and reproduction phases of the experiment. The standard procedure would be to use the same medium for both sequence presentation and reproduction (see figure 7.2). In this case, performance is generally better in the screen–screen condition than in the floor–floor condition. Röser, Hardiess, and Mallot (2016) argue that the difference is due to a need for spatial updating during the reproduction by walking on the floor, since the egocentric position of the next tiles to step on is changing as the subjects moves around. If the sequence is presented on the floor but has to be reproduced on the screen, no such updating is necessary. However, in this case, the sequence has to be transformed and scaled, from the horizontal floor to the vertically mounted screen, which involves a perspective-taking task. Performance was below the screen–screen condition, comparable to the floor–floor condition. Finally, if the stimulus is presented on the screen but has to be reproduced in walking, both problems, perspective taking and spatial updating, arise. Indeed, performance in this condition was poorest.

7.2.3 Map Reading

Transformations between image spaces presented on a sheet of paper, say, and the environmental space of walking can also be tested directly with a map test. Subjects are presented with a simple map showing a few object positions in their peripersonal space. One position is marked and the subject is asked to indicate the object placed at this position. Dehaene et al. (2006) used this task in a cultural comparison study involving participants from North America and from an Amazonian indigenous group. While children at preschool age from both groups performed equally, North American adults showed a clear advantage over the age-matched indigenous group. This difference is likely due to the presence or absence of a formal school education in geometry. For our discussion, it emphasizes the point that memories for image and environmental spaces should not be considered the same thing.

7.2.4 The Anisotropy of Imagined Space

Judgment of relative direction When we plan an action in a remote space or simply think about it, we invoke a memory that contains various aspects of that space. These may be localized objects and landmarks belonging to the imagined space but also nonlocalized things such as smells or the emotions and events that we experienced during previous visits. In many cases, however, the memory will be image-like in the sense that it shows a configuration of objects as would be seen from a certain point of view. The imagery thus has an orientation that may be different upon repeated recall occasions and may change during imagined walking and perspective taking. Even if we try to imagine an aerial view or a map, an orientation has to be chosen, for example, the one in which the north direction is imagined upward or forward.

A large body of evidence shows that imagined space is not isotropic: that is, directional judgments are harder and more error prone if taken in some parts of the space or in others. This can be studied with a task called the "judgment of relative direction" (JRD), as illustrated in figure 7.3. In this task, a subject tested in some environment (left part of figure) is asked to imagine being in a place V (viewpoint) in some other environment, facing an object O, as shown in the right part of the figure. The direction VO is the imagined heading direction. Next, another object has to be recalled, shown in the figure as the target T. Finally, the subject is asked to report the pointing direction to T when facing O (angle φ in figure), either by physically pointing with the arm (i.e., relative to the true current heading) or by verbal report. The task is usually performed in the dark or with blindfolded subjects

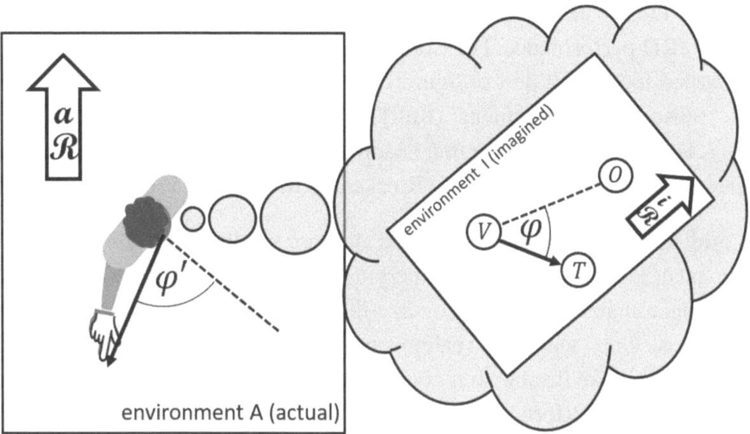

Figure 7.3
Judgment of relative direction (JRD). Blindfolded subjects are situated in an environment A and imagine being in a remembered environment I, at viewing position V. From there, they imagine looking at a remembered object O, such that the line VO defines the imagined heading or body orientation. They are then asked to imagine another remembered object, the target T, and the angle φ between O and T. Finally, they indicate this angle by actual pointing or verbal report. The pointing direction relative to the actual heading (dashed line in environment A) is the dependent variable φ'. $a\mathcal{R}$ and $i\mathcal{R}$ are the intrinsic reference directions of environments A and I, which also affect performance. In perspective taking, environment I is a remembered version of environment A.

such that actual and imagined environments will not interfere. The procedure is similar to perspective taking and has already been mentioned in section 3.4.2. Here we focus on cases where the imagined and actual environments differ. The imagery is therefore the result of a memory recall, not a perception persisting after closing of the eyes.

The best-studied anisotropy of imagined space is the intrinsic axis or reference direction ($i\mathcal{R}$ in figure 7.3; see Shelton and McNamara 2001; Mou and McNamara 2002). In the JRD task, it is determined by measuring the average pointing error as a function of the imagined heading direction (VO in figure 7.3). The heading direction with the best pointing performance, usually the long axis of the imagined room, is the intrinsic reference direction. Imagined headings orthogonal to this direction yield intermediate results while the performance for diagonal heading directions is poorest.

The intrinsic reference direction is a property of the imagined environment and may be either stored as a part of spatial memory or derived from the imagined spatial layout. In experimental setups, it is usually the longer midline of a rectangular room, but it may also be influenced by the position of doors, windows, or salient furniture such as a speaker's desk or the projection screen in a cinema hall. Indeed, in a circular lecture theater, there is a clear meaning of sitting in the front or in the back even if the room itself does not have an obvious axis.

In addition to the intrinsic reference axis of the imagined space, other axes have been found to affect JRD performance. For example, it may be advantageous to align the actual body axis (dashed line in left part of figure 7.3) and the imagined heading VO. This effect is known as "sensorimotor alignment" (Kelly, Avraamides, and Loomis 2007). Finally, we mention the role of the observer's actual heading relative to the intrinsic axis of the actual environment, which was investigated by Riecke and McNamara (2017).

Extended and nested spaces In extended outdoor environments, such reference axes also seem to exist. For example, when emerging from a subway station even in a familiar city, it may happen that we do not recognize the location because we are mistaken about the reference axis. Only when we realign our mental representation with the perceived views of the scene will we be able to recognize the site (e.g., May 2004). Such realignment may even take conscious effort. The maintenance of the reference direction, or "subjective north," over larger distances may be related to the sensorimotor alignment effect of Kelly, Avraamides, and Loomis (2007). It is immanent in path integration and is provided by the head direction system discussed in section 5.4: as long as the path integrator works, current heading will be represented relative to the reference direction that was defined when the integrator started. Mistakes are possible not only in the subway example, where continuous path integration is disrupted, but also in loop closing; that is, when we reach a familiar place from a previously unexplored direction. In summary, then, the sense of reference directions depends on two main components: the perceived or memorized intrinsic axis of a scene and the continuous update of its egocentric angle (angle between the intrinsic axis and current heading) in path integration.

Different environments may be represented with different intrinsic axes. Wang and Brockmole (2003) asked students at the campus of the University of Illinois at Urbana-Champaign to point from inside an unfamiliar, windowless room to various objects and locations inside and outside this room (see figure 7.4a). Inside the room, a number of everyday objects were placed and the participants studied these objects in advance. During pointing, the subjects were blindfolded. The outside objects were buildings on the same campus in which the test room was localized. Pointings to in-room objects were largely correct, without strong left or rightward biases for most subjects. In contrast, the pointings to on-campus objects or locations varied largely between subjects, with roughly correct interbuilding angles but large subject-specific biases. This indicates that the subjects took their directional judgments relative to different reference angles. That is, they represented the room and the campus with independent intrinsic orientations, which were roughly fixed for each subject but varied widely between them. Imagined space thus seems to be organized in a nested, hierarchical way.

The simultaneous existence and use of multiple reference frames was confirmed in a study by Strickrodt, Bülthoff, and Meilinger (2019). Subjects were trained in a corridor

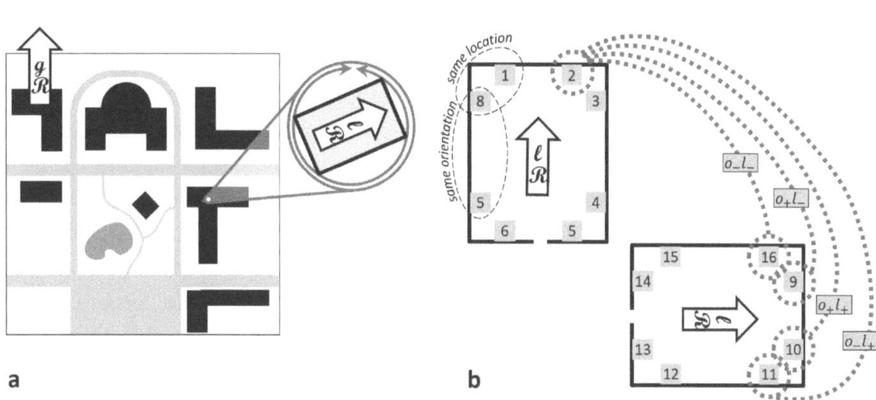

Figure 7.4

Representational axes of orientation. $g\mathcal{R}$: global axis of representation ("north"); $l\mathcal{R}$: local axis of representation. (a) Schematic of the nested environments experiment by Wang and Brockmole (2003). When pointing to targets inside or outside the lab room, the reference axes are not aligned. (b) Museum park experiment by Marchette et al. (2014). Only two of four museums are shown. Numbers 1 to 16: Objects on exhibition. Objects on the same wall share the same orientation; objects in the same corner are seen from the same viewpoint and thus share the same location. Across museums, sameness of orientation and location is defined in local coordinates. The dashed circles and lines illustrate some JRD tasks with object 2 as a start view (actual ego-position). For a JRD involving the imagined translocations $2 \mapsto 10$, orientation and location are locally the same (o_+l_+); for the translocation $2 \mapsto 16$ both orientation and location differ (o_-l_-). Translocations $2 \mapsto 9$ and $2 \mapsto 11$ are o_+l_- and o_-l_+, respectively. For further explanations, see text.

sectioned into two parts. Each part consisted of four straight segments connected by 90-degree turns left or right. The connection between the two parts, however, required a 45-degree turn. It may therefore be assumed that the main axes of the two parts were offset by the same angle. Subjects might also rely on a global orientation axis, for example, the initial heading when exploration started, which could be maintained by path integration. After training, subjects were asked to point to remembered out-of-sight targets (landmarks placed in each segment), assuming different body orientations. When pointing to targets within the current corridor section, performance was best when subjects were bodily aligned with the local axis: that is, looking along the corridor or straight toward the walls. However, when pointing across corridor parts, performance was best when subjects were bodily aligned with the global reference axis.

Judgments of relative direction depend not only on the imagined orientation but also on the imagined location (viewpoint V in figure 7.3). Marchette et al. (2014) designed a virtual environment with four large halls ("museums") arranged with different absolute orientation in a museum park (figure 7.4b). In this situation, each museum has its own local reference axis. Inside each museum, eight objects are exhibited in "alcoves" (i.e., between two blinds), such that subjects have to stand right in front of each object in order to see it. In the figure, the objects are numbered 1 to 8 for museum 1 and 9 to 16 for museum 2, starting with the far-left object, as seen from the museum door. Each room has a local reference axis ($l\mathcal{R}$ in the figure) defined by the view from the entrance door. With respect to this axis, four locations (corners) can be specified in the room reference frame: far left, far right, near left, and near right. Objects placed at the same museum wall induce the same imagined heading, which for objects 1, 2 in museum 1 and objects 9, 10 in museum 2 would be aligned with the respective room axis, while objects placed in the same corner induce a feeling of being in the same location, for example, the far-right corner for objects 2, 3 and 10, 11.

In a JRD task, subjects standing in front of one object would be asked to imagine facing another object that defines their imagined viewpoint (V) and orientation (O). They would then be asked to judge whether a target object (T) would be left or right of the imagined viewing direction. The dependent variable in this experiment is reaction time, following the idea that mental transformations between reference frames that might or might not be involved would require processing time and should thus lead to delayed responses. Reaction times were smallest when both objects shared the same orientation or the same location as defined in the local reference frames of each museum. Larger reaction times resulted for cross-directional and cross-locational judgments, irrespective of whether the actual and imagined observer positions are located in the same or in different museums. The rationale is illustrated for a few examples in figure 7.4b. Each JRD involves an imagined translocation from the actual egoposition to the imagined viewpoint V. A translocation from object 2 to object 10 (in short, $2 \mapsto 10$) does not require any mental rotation or updating since both orientation and locality are the same in the sense of the local reference frames; it is therefore marked $o_+ l_+$ in the figure. The translocations $2 \mapsto 9$ and $2 \mapsto 11$ require mental rotation or spatial updating, and the translocation $2 \mapsto 16$ requires both types of reference frame adjustment. Reaction time increases if more adjustments are required for the judgment. Interestingly, no effect of the global reference direction was found. The authors conclude that the representation of the local or imagined environment—namely, the spatial working memory—is coded relative to the local reference directions and location schemes (corners).

In addition to the behavioral data discussed so far, Marchette et al. (2014) also analyzed fMRI data. Activations in the retrosplenial cortex (RSC) were shown to reflect the layout of the museums. More specifically, voxel-based similarities between RSC activations (representational similarity analysis) resulting from different imagined views were larger, if the views shared the same implied orientation or location, as compared to cases where they did not. A multidimensional scaling algorithm applied to the similarity data even resulted in an approximate map of the objects within the museum.

A similar study in which the museums were replaced by rooms on different floors of a larger building was presented by Kim and Maguire (2018). No significant differences between tasks involving imagery of horizontally or vertically displaced rooms were found, indicating that the spatial working memory treats both cases in the same way.

Spatial working memory as a local chart Taken together, the spatial working memory appears as a local chart of the environment with a fixed reference frame in which the position and orientation (i.e., pose) of the observer are represented (see also Meilinger 2008). In this model, the anisotropies would result from a variable resolution of the chart, which must be assumed to be highest along the main axis of the reference frame and the center. If the observer moves without leaving the environment, its pose is continuously updated within the current frame. The local chart can also hold an imagined environment and supports imagined movements as well as jumps to novel positions. The extent of the chart is unclear, but it is clearly not unlimited. If the observer leaves the range of the current chart and reference frame (e.g., by walking into another room), a new chart with a new, environmentally fixed reference frame is formed or retrieved from long-term memory.

A much-debated issue in this context is the idea of allocentric versus egocentric memories. In the definition of Klatzky (1998), a representation is called allocentric if it does not change as the observer moves. But how does this apply to the local chart idea of spatial working memory? As long as we stay within a local environment, the representation of this environment would indeed be stable, using its fixed intrinsic reference frame. The only part that changes would be the representation of the observer's pose. As soon as we leave the local chart, be it in person or in imagination, a change of the representation will occur, showing the new environment in its own reference frame. In this sense, working memory would be locally allocentric; the selection of the local environment to be represented, however, is egocentric, since the chart will always contain the observer position.

7.2.5 Representational Neglect

A neurological disorder of spatial working memory was described in stroke patients by Bisiach and Luzzatti (1978). Patients who were familiar with the city of Milan (Milano, Italy) and especially with the central Cathedral Square (Piazza del Duomo) were asked in the lab to imagine being at the Piazza, looking east toward the cathedral doors from the far end of the square. They were then asked to name nearby places (offices, shops, museums, etc.) that came to their mind. They named a number of places that were all located to the right of their imagined heading direction (i.e., south of the square). Next they were asked to imagine walking across the square toward the cathedral doors, turn around, and imagine viewing the square from there. In the same task as before, they now named places north of the cathedral, which would appear again to their right.

This result leads to two important conclusions. First, all named objects were somehow stored in the patients' long-term memory, which thus seems unaffected by the disorder. Second, the stage mediating between long-term memory and speech production (i.e., the spatial working memory) works only for places located to the right of the subject, not to their left. A metaphor for this result compares working memory to a blackboard with egocentric representations of directions, on which the objects are represented upon retrieval from long-term memory and from which they are eventually read. In the neglect patients, the left side of this blackboard is damaged such that only rightward objects can be recalled. As soon as the subject performs a body turn, even in imagination, the blackboard also turns and other memories appear on the unaffected part.

Note that in the Milan experiment, imagined heading was aligned with the main axis of the Piazza. The result does therefore not exclude the possibility that the "blackboard" is oriented not by egocentric heading but with the local intrinsic reference direction, in agreement with the local chart idea discussed above.

The original paper by Bisiach and Luzzatti (1978) is a case study with just two patients. However, the condition seems to be quite common in patients with right-hemispheric lesions. Note that representational neglect differs from the more frequent phenomenon of perceptual or attentional neglect, in which patients, again with right-hemispheric lesions, ignore objects appearing in the left half of their visual field or features appearing on the left side of a larger object or pattern such as the marks for 7 to 11 hours on a clockface; for an overview, see Karnath and Dieterich (2006). In this case, no specific memory is involved. Guariglia et al. (2013) tested a large group of neglect patients and were able to delineate representational from perceptual neglect.

7.2.6 Position-Dependent Recall

In representational neglect, recall from long-term memory results in an egocentric image on the metaphorical blackboard discussed above. Even if the blackboard is intact (i.e., in the absence of neurological disorders), the recalled content will vary with recall position and imagined viewing direction. This idea was tested in a study by Röhrich, Hardiess, and Mallot (2014). Passersby in the downtown area of the German town of Tübingen were approached and asked to draw a sketch map of one of two well-known city squares. Interviews were carried out at a number of locations all in walking distance from the target square and distributed roughly equally around it. The target square itself, however, was not visible from any of the interview locations; the situation is illustrated in figure 7.5.

Participants produced good sketch maps but with various orientations. In each map, the "sketch orientation" was determined as the compass direction corresponding to the upward direction on the sketching paper. In the example of figure 7.5, we may assume that the main axis from the park to the mansion is north. In this case, we would say that the sketch map produced by the left person is oriented north, whereas the sketch map produced by the person to the right is oriented mainly west. The dependent variable of this experiment is the distribution of sketch orientations collected from many subjects and presented as a histogram with the frequencies for the four cardinal directions.

When data from all sketching locations are lumped together, a preferred orientation is found that depends on salient features in the environment. In the example of figure 7.5, this

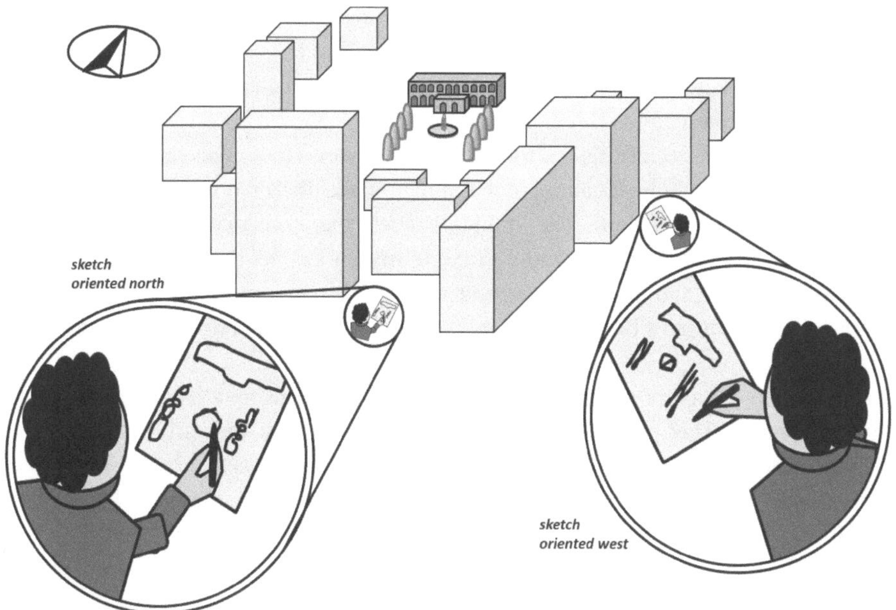

Figure 7.5
Position-dependent recall. Subjects are asked to produce sketch maps of an out-of-sight place (in the example, a mansion in a park). The light gray blocks symbolize buildings blocking the view to the target. The arrow in the upper left marks the north direction of the environment. Sketch map orientation varies with the production site and tends to align with the airline direction to the target square.

might be the axis of approaching the mansion from the park. In the Röhrich, Hardiess, and Mallot (2014) study, preferred orientations were uphill toward a landmark church building or toward the city hall dominating the market square. These directions might be called "canonical," in analogy of the canonical views known from object recognition (Bülthoff, Edelman, and Tarr 1995). When analyzing the histograms of individual interview locations, a second effect was found: that is, different orientations are preferred when producing the drawings at different locations. This shows that sketch-map orientation is a working memory performance, since long-term memories should not change as the observer walks around to another production site without acquiring new information about the target. The deviation of the local histograms from the global, "canonical" distribution was determined by calculating the differences between the local and the average histograms, which results in four numbers n, e, s, and w: that is, the corrected relative frequencies of sketch orientation for the four cardinal directions. Note that these numbers may take negative values. From the corrected frequencies, an average direction can be calculated by circular statistics, which, in this example, amounts to calculating the vectors with components $(e-w)$

and $(n - s)$, respectively. When plotted into a map of the interview locations, these vectors showed a clear component in the airline direction to the target square. This means that subjects produce sketch maps preferably in the orientation that they would see, could they fly from their current location directly to the goal.

Position-dependent recall was not found when interviews took place at a larger distance (about 2 km) from the target location; in this situation, the distribution of sketch orientation approaches the canonical view. The authors suggest therefore that sketch orientation is affected by both a position-independent long-term memory of the target place with the canonical orientation and a position-dependent working memory that is subject to spatial updating. The influence of this working memory component depends on the distance from the target.

The result has been reproduced by Meilinger et al. (2016) using a rebuilding task rather than sketch maps. This paper also shows an additional effect of body orientation. Virtual reality versions using the rebuilding task or an immersive sketching technique have been presented by Le Vinh, Meert, and Mallot (2020) and Grochulla and Mallot (2022), respectively.

7.2.7 Modality of Imagined Spaces

Spatial imagery can be generated not just from prior visual inspection or from memory recall but also from verbal descriptions. For example, one may describe a scene as a room with a window on the left, a desk next to the window, and an armchair in the right corner. One may then ask whether the representations generated by verbal description or by visual inspection differ from each other in systematic ways or not. If they do, they would be called "modal," since they would depend on the modality of acquisition; otherwise, they would be called amodal. Note that the term "modality," from which the word "modal" is derived, is used here in an extended sense. Normally, the sensory modalities are considered the five Aristotelian senses plus proprioception, as discussed also in section 2.2.2. Here, language is considered an additional modality, irrespective of whether it is presented as uttered speech or written text.

The issue of modality or amodality is of considerable theoretical and practical interest. If a sense of space precedes perception, as is assumed in empiricist philosophy (see, for example, the quote from Kant in chapter 1), perception would be merely "filled" into the preexisting spatial categories and should not affect the structure of spatial representations themselves. That is to say, the representations would be completely amodal. Understanding the degree of amodality in spatial representations is also an important issue in the understanding of spatial cognition in congenitally blind people who cannot have used vision to structure their representations of space (for review, see Schinazi, Thrash, and Chebat 2016).

Avraamides et al. (2004) tested the modality of spatial representations by comparing subjects' performance after learning an environment either by visual inspection or by verbal description. The visual image showed a room with a number of objects, while in the language condition, the objects were named and their position was specified in terms of direction and distance from the participant. In the test phase, subjects were blindfolded and given a pair of object names for which they had to judge their distance from each other as well as the allocentric orientation of the axis connecting the two objects. In addition to response accuracy, the variability and response latency were also analyzed. The experiment revealed no structural differences between the visual- and language-generated memories, hinting toward an amodal representation. Of course, the amount of information conveyed by visual inspection and by spatial language description is hard to balance. This may explain earlier findings of an advantage of visually generated memories.

Another paradigm used to address the issue is mental scanning. In this task, subjects are asked to imagine movements in an imagined environment. For example, one might ask a subject to imagine stepping out of the front door of their house, taking the first turn left, then the next turn right, then left again, and so on. After some time, the subjects are interrupted and report their current imagined position. The distance traveled to this position often correlates with the time spent scanning. This is taken as evidence for a maplike representation in which the representation of ego is moving with continuous speed.

Mellet et al. (2002) used a version of this paradigm to compare the performance after studying a map or a textual description of a novel environment. Rather than imagining to walk in this environment, subjects were asked to imagine a laser pointer moving between two specified locations. Mental scanning (i.e., the correlation between scanning time and traveled distance) was found in both conditions, indicating that a continuous, image-like representation has been formed in both acquisition conditions. However, an analysis of the brain regions recruited in each condition with positron emission tomography (PET) revealed substantial differences. In the language-based acquisition, mental scanning activated classical language areas such as the angular gyrus or Wernicke's area, whereas map-based visual acquisition resulted in increased activity in the medial temporal lobe. In addition, a frontoparietal network was activated in both conditions. This indicates that spatial memory involves different components, which may be modal or amodal to various degrees.

If spatial memory is at least partially dependent on language as an input modality, one might ask whether differences found between individual languages affect the structure of spatial representations in speakers of different languages. One important example is the cross-cultural comparison between speakers of languages with mostly egocentric reference systems based, for example, on the words "left" and "right" from languages using absolute, geocentric reference systems such as "north" and "south" or "uphill" and "downhill." Some such differences have indeed been reported; see for example, Majid et al. (2004).

A comprehensive discussion of this issue, however, is beyond the scope of this text. For a review, see Levinson (2003).

7.2.8 Working Memory Involvement in Wayfinding

Finding ways or repeating previously traveled routes is a complex task that besides working memory also involves long-term knowledge of places and connections. Working memory is involved in the acquisition of relevant information and its transfer into long-term memory, as well as in the planning of novel routes from long-term memory. We discuss three examples.

Meilinger, Knauff, and Bülthoff (2008) had subjects learn two routes in a virtual reality model of the downtown area of the city of Tübingen, Germany, in a dual-task paradigm. During learning, subjects were required to carry out one of three different secondary tasks designed to specifically address different subdivisions of working memory. In a visual task, subjects were verbally told a clock reading (ten minutes to ten, half past two) and had to decide whether the hands of the clock would be on the same half (upper, lower) or on different halves of the clockface. In the spatial task, subjects were presented with a sound and had to decide whether the source was left, right, or in front of them. Finally, the verbal tasks used common two-syllable words of the German language, some of which were turned into non-existing pseudo-words by changing the vowel of the first syllable: for example, "Mintag" instead of the correct "Montag" (Monday). The effort needed for each task was balanced by the frequency with which they had to be carried out.

Subjects who were engaged in a secondary task during the learning phase of the experiment made more errors during route reproduction than a control group. More interestingly, performance was affected differently by the three different tasks: the strongest effect was found for the verbal task, whereas the visual task led to a relatively low number of navigational errors. This result may indicate that subjects rely on verbal rehearsal during route learning, which would be impossible during the verbal secondary task. On the other hand, visual working memory seems to be of lower relevance for the encoding of route knowledge.

The situation seems to be different in the recall situation: that is, when an existing memory is to be used for planning. In a paradigm similar to the position-dependent recall described above, Basten, Meilinger, and Mallot (2012) asked subjects to produce sketch maps of familiar city squares after an imagined walk passing the square in one direction or another. In this case, the orientations of the produced sketch maps tended to show the square in the orientation in which the imagined route passed through the square. This may mean that visual memories of the location as seen with the respective heading direction are activated during planning.

The working memory contributions to wayfinding can also be accessed by retrospective verbal report. Spiers and Maguire (2008) had licensed London taxi drivers perform a number of routes in a virtual reality simulation of the City of London. After completing the travel, they were presented with a video recording of the travel and asked to report what they had been thinking at each instant. The reports were then classified by the authors into different types of thoughts. One interesting result is that for most of the time, drivers are not thinking about their travel at all; there are even long periods of "coasting" where the drivers engage in more or less automatic control loops. Thoughts that are related to navigation are employed with planning, which may or may not include picturing the upcoming locations and checking expectations about landmarks. Overall, the picture is quite complex but likely gives a realistic account of spatial thinking during wayfinding.

7.3 Models and Mechanisms for Spatial Working Memory

The properties of spatial working memory discussed in the previous section, and some additional ones that we did not discuss, have led Loomis, Klatzky, and Giudice (2013) to the formulation of their "spatial image" account of spatial working memory. It differs from visual images in the representation of three-dimensionality not only of object shapes but also of the goals of spatial action. It allows for simulated manipulation of objects as well as imagined observer motion. Resolution is coarse, and object representations are largely amodal.

For a summary of the neural structures involved in various aspects of spatial working and long-term memory in humans, see, for example, Vann, Aggleton, and Maguire (2009) and Julian et al. (2018). In brief, the hippocampus is considered to be concerned with the recognition of a scene or events within a scene, which leads back to its role in episodic memory. It interacts with the anterior thalamic nucleus, a structure known for its head direction cells. Also involved are a number of medial cortical areas related to visual input and imagery (occipital lobe), egocentric three-dimensional representations (parietal lobe), and allocentric three-dimensional representations in the retrosplenial cortex.

Quantitative modeling of spatial working memory is still largely limited to subsets of the phenomena discussed above. We will introduce two approaches to the representation of the environment in working memory: a map-based approach in which spatial working memory is seen as a local two-dimensional chart centered around the observer and a graph-based approach that treats spatial working memory as a structured collection of snapshots. In addition, we will briefly discuss a possible neural mechanism for path planning.

7.3.1 Representing the Current Environment

The allo-oriented chart The perception of peripersonal space, be it visual, auditory, or other, is initially egocentric and oriented relative to the current heading direction. Indeed, this is also what is needed on the motor side, for example, in order to grasp an object. If we look at a mug, then turn away to spot something elsewhere, and then reach for the mug outside the field of view, the reaching movement needs to take into account the previous body turn and use the information from an updated, egocentric representation of the mug.

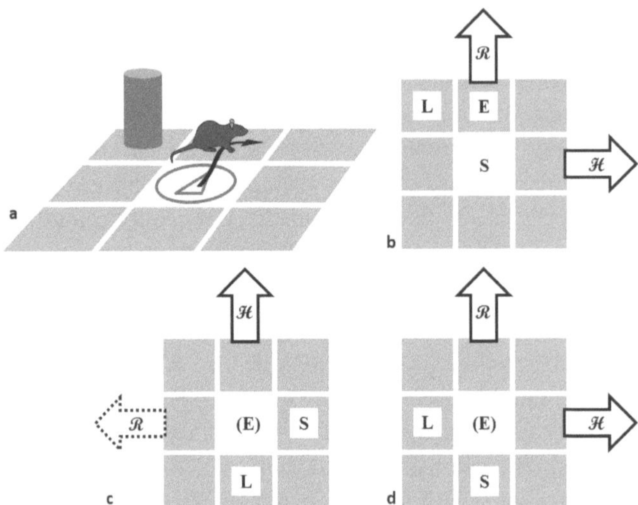

Figure 7.6
Reference frames for spatial working memory. (a) An agent moves from a start point forward and turns to the right. (b-d) Representation after the movement in different reference frames. S, E, L: are the positions of starting point, ego (the agent), and a landmark. $\mathcal{H}, \mathcal{R}$ heading and reference direction, which are initially identical. (b) True allocentric (or geocentric) scheme. S, L, and $\mathcal{R}$ remain unchanged while ego's position and heading are updated according to the performed movement. (c) True egocentric scheme. E is always in the center and $\mathcal{H}$ is represented upward. S and L are shifted and rotated accordingly. The reference direction $\mathcal{R}$ may or may not be represented. (d) The allo-oriented chart represents ego in the center, but the allocentric reference direction stays represented upward. As compared to the true allocentric scheme, (b), L and S are shifted, but the chart is not rotated.

The situation is different as soon as we want to build a map of a larger environment by scanning: that is, by connecting local patches of peripersonal space perceived and memorized in different places. In this case, all local patches that we stitch together need to have the same orientation. If this were the current heading, the entire representation built up so far would have to be counterrotated every time the observer turns. Alternatively, each newly added patch could be rotated into a common reference direction, which will be kept stable during observer motion. This common reference direction would be available from the heading component of path integration as described in section 5.2.3. Still, the observer position, or "ego," will stay in the center of the resulting chart, which means that the representation is egocentric in a literal sense. However, the orientation is not aligned with the current heading but with a fixed geo-centered direction. This is to say that the map is egocentric but "allo-oriented." This egocentric, allo-oriented representation is similar to that of an automotive navigation system operating in the fixed north-up mode.

The situation is illustrated in figure 7.6. Figure 7.6a shows a step of the observer in the forward direction followed by a right turn. In a truly allocentric representation in the sense of Klatzky (1998), the representation of the environment does not change upon observer motion. This is the case in figure 7.6b, where the starting point S and the landmark L are unchanged. Note, however, that observer position E and heading $\mathcal{H}$ have to be represented separately in this case, and that the observer representation has indeed moved out of the center. In the true egocentric case (figure 7.6c), the observer position is always in the center and does not require an explicit representation; it is therefore put in brackets in the figure. Map orientation is aligned with the current heading, and the intrinsic reference axis of the environment may or may not be represented separately. Finally, in the allo-oriented chart (figure 7.6d), orientation is along a fixed reference direction $\mathcal{R}$. Upon observer motion, the representation will be shifted but not rotated. For a discussion of reference frames in vision and working memory, see also Tatler and Land (2011).

Vector cell representation of the allo-oriented chart Byrne, Becker, and Burgess (2007) have argued that the egocentric representation of the visual environment (figure 7.6c) can be localized in a network of areas in the parietal cortex, summarized as the "parietal window." From this and additional input from the head direction system, the allo-oriented chart is thought to be constructed in the mediotemporal cortex, based on the activity of boundary and object vector cells found in the rodent brain (see also section 6.5.2). The term "vector" in the name of these cells refers to the fact that their activity encodes a distance and allocentric compass bearing from the position of the rat to either an extended boundary or a localized object. Each cell would thus correspond to a patch of the allo-oriented chart around the observer since its firing signals the presence of a boundary or an object at that patch. For example, in figure 7.6d, the cell marked "L" would be firing and thus indicate that there is an object east of the observer at distance 1. An array of such cells would be able to represent the space surrounding the agent (i.e., the allo-oriented chart). The content, or geometrical layout of objects and boundaries in that environment, is represented by the activity pattern on the chart and needs to be updated upon observer motion. Byrne, Becker, and Burgess (2007) use the attractor dynamics discussed in section 5.4 to achieve this updating: during rest, a stable attractor of activity forms on the vector cell array. As soon as the rat starts to move, input from the head direction system will shift the attractor in the appropriate direction. For elaborated versions of the model, see also Bicanski and Burgess (2018, 2020).

The view graph An alternative approach to spatial working memory in mazes is based on the idea to represent decision points by the views or local position information that can be perceived at each point (Schölkopf and Mallot 1995; Röhrich, Hardiess, and Mallot 2014). This idea is illustrated in figure 7.7a; it is motivated by the snapshot approach to spatial homing (section 6.2) and further supported by the existence of spatial view cells in the primate hippocampus (Rolls 2021). These cells have been shown to fire when a monkey

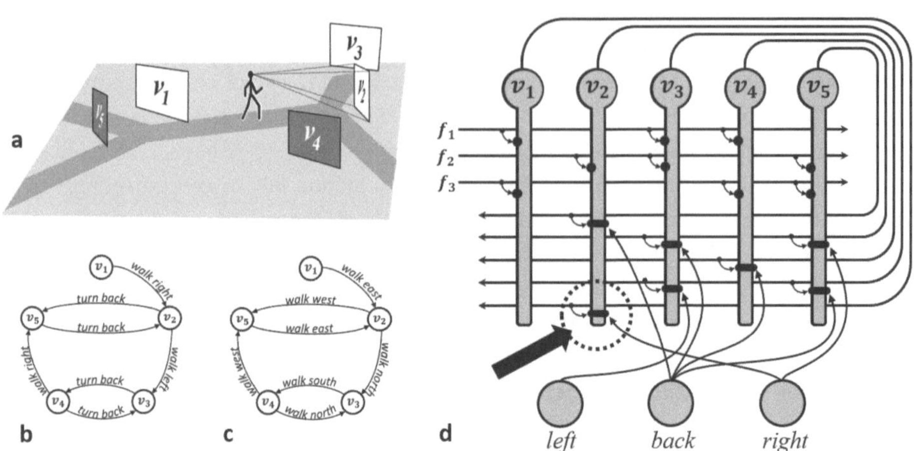

Figure 7.7

The view graph approach to spatial working memory. (a) An observer in a maze will perceive specific views at each decision point. Five views are shown as billboards; v_4 and v_5 are visible only from behind. The agent learns the maze by storing associations of the type "go right at view 1 to encounter view 2." (b) This information can be represented as a graph of the views and labeled connections. Required motions are coded in an egocentric scheme. (c) Same graph with allo-oriented encoding of motor actions. (d) Network model with view-units (v_1–v_5), feature inputs f_1–f_3 representing the local view information, and three action units for motor actions left, back, and right. The synapses from f_i to v_j learn local position information of the decision point (the view). The synapses in the lower part have two inputs, one from another view cell and one from an action unit. For example, the circled synapse in the lower left (arrow) fires unit v_2 if v_1 and action unit "right" are active. It thus encodes the association $(v_1, \mathit{right}) \rightarrow v_2$.

is looking at a particular object from a certain vantage point; see also section 6.5.4. View cells are similar to rodent boundary or object vector cells in that they have a firing field at a certain distance and world-centered direction from a viewed object or scene. If we describe the specificity of the spatial view cell by a snapshot, the firing field is simply the catchment area of this snapshot: that is, the region in space within which the scene looks roughly the same. Unlike boundary and object vector cells, view detectors defined in this way will also encode the look or identity of an object, not just the presence of an obstacle.

In addition to the view units themselves, the view graph encodes information about the bodily movements associated with view transitions. For example, a left turn may be required to change the view from a door to a window, say. Also, a smaller and a larger view of the same window will be connected by a forward or backward movement of the observer, respectively. These movements are represented in the view graph as labels added to the graph links. The links and their labels are not necessarily symmetric since the transitions back and forth between two views usually require different movements. Figure 7.7b,c shows the view graph with egocentric and absolute action labels, respectively. The original formulation by Schölkopf and Mallot (1995) was based on the egocentric approach, but extensions to the absolute scheme are straightforward (Mallot, Ecke, and Baumann 2020). This latter version contains the same information as the allo-oriented chart, plus information about object identities provided by the views.

The neural network dynamics of the view graph is shown in figure 7.7d. View units are activated by input from egocentric feature detectors conveying local position information and learn to recognize particular views. When the agent moves, the according action unit will be firing, and the three-way association between the current view, the action, and the next view will be learned. For this, special three-way synapses are assumed, as are shown in the lower part of figure 7.7d. After learning, coactivation of a given pair of view and action unit will result in the facilitation of the expected next view resulting from the motor action.

The view graph model nicely explains the results on position-dependent recall discussed in section 7.2.6. If an observer is asked to imagine a distant square, activity will spread through the view graph starting from the current position. The closest node belonging to the imagined square (i.e., the first one in the graph path showing this square) will feature a view in the approach direction, which may then be externalized as a sketch. The model also allows for different saliencies of different viewing directions or canonical views by assuming that view units representing a salient view axis are more frequent than other ones. In the recall from far-distant locations, these units would therefore be picked with higher probability.

View graph or vector cells? We conclude this discussion by noting that the maplike array of vector cells used in the Byrne, Becker, and Burgess (2007) model (see also figure 7.6d) is just a visualization. No such array has been observed in the brain, nor has this been claimed by the authors. Indeed, it is not necessary for the theory. All that counts is the connection pattern between neurons with different vector fields. In the maplike array, these are defined by adjacency and the reference direction in the map. In the brain, the units are thought to be irregularly arranged. They thus form a network, or graph, whose connections reflect the neighborhood in the map and are addressed by the appropriate movement directions defined by the head direction system. The computational structure, therefore, is again that of a labeled graph.

Summary: An atlas of allo-oriented charts The emerging picture of spatial working memory of large-scale environments is that it consists of a collection, or "atlas," of local charts, each with their intrinsic reference direction, which are "loaded" into working memory and used to monitor local motion. Each chart may cover areas that can be passed over in a few seconds or minutes. The representation of self ("ego") will always stay in the

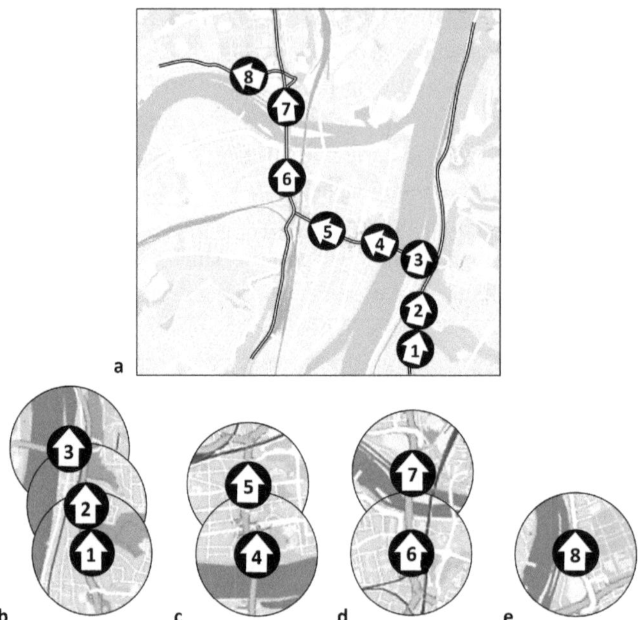

Figure 7.8

Summary of working memory contents during a complex route. (a) Map of a city at the junction of two rivers, superimposed with a route (steps 1–8). (b–e) Allo-oriented charts oriented by their reference directions. (b) Reference direction is pointing north while following the main river downstream. (c) Reference direction west while crossing the river and entering the city. (d) Reference direction north while approaching and crossing the tributary. (e) Reference direction west while following the tributary upstream. Each chart represents the local environment along a section of the route in which the reference direction does not substantially change. (Map of Koblenz, Germany: OpenStreetMap, Licence ODbL.)

center while the chart itself is "flowing" opposite to the current egomotion direction. The chart has a fixed reference direction, so that when we step sideways to the left, say, the chart would be shifted right but not rotated. Occasionally, the reference direction will change: for example, when we switch from thinking about an indoor space to the imagery of the surrounding city quarter (see figure 7.4a and Wang and Brockmole 2003) or when we perform substantial bodily turns (cf. the sensorimotor alignment effect of Kelly, Avraamides, and Loomis 2007). In these cases, a new allo-oriented chart together with its reference directions is retrieved from long-term memory and "loaded into" working memory for further use.

The idea is illustrated in figure 7.8 for a journey across a town with several intrinsic axes defined by two rivers meeting at about a right angle. Assume the driver of a car is approaching along the main river. The reference axis is then along this river (i.e., north in the example). Various stages of the allo-oriented chart with orientation north appear in figure 7.8b. As soon as the driver reaches the bridge and turns to the left to cross the river, a new allo-oriented chart with orientation west is activated, which leads into town (figure 7.8c). Reorientation will happen again when turning toward the tributary river (figure 7.8d) and after crossing this river and following it upstream (figure 7.8e).

7.3.2 Trajectory Planning with Place Cells

In section 6.5.1, we discussed the representation of place by the population activity of hippocampal place cells together with the out-of-place firing of place cells occurring at decision points of a route. This latter effect (i.e., the firing of place cells associated with spatial planning or imagery) leads to the idea that an array of place cells might also be used in planning. In the model by Ponulak and Hopfield (2013), for example, the planning of a trajectory to a goal would start with the activation of a group of place cells normally representing the goal location. If we assume that place cells representing adjacent places are synaptically connected, this activation will result in a wave spreading from the goal and thereby covering all possible paths in the network. All connections are considered to be bidirectional or reciprocal: that is, pairs of place cells will always be connected in both directions.

Ponulak and Hopfield (2013) hypothesize that synaptic strength will be modified by the activity wave according to a learning rule known as anti-spike-timing-dependent plasticity ("anti-STDP"). In this learning rule, synaptic strength is increased if the postsynaptic cell fires in a time window preceding the firing of the presynaptic cell. In the place cell array, this means that as the wave proceeds from a neuron A to a neuron B, say, the synaptic strength of the connection from B to A will be increased (i.e., the direction backward to the origin of the wave). After the wave has passed by, the modified synaptic strengths will therefore represent a field of directions, called the synaptic vector field, which can be followed back to the starting point of the wave.

The current position of the agent is represented by a group of active place cells again generated and stabilized by the attractor mechanism described in section 5.4. If the spreading wave from the goal passes this attractor, the synapses connecting its place cells to other ones closer to the goal will be strengthened, and the peak will move toward the goal. The algorithm is able to find efficient paths in simple environments. Note that for the functioning of this mechanism in the real brain, it must be assumed that out-of-place activity of place cells would cause the agent to move toward the regular firing fields of these cells. This is not implausible, but direct evidence for premotor activity in place cells seems to be missing.

Ponulak and Hopfield (2013) demonstrate the performance of the algorithm on lattices of spiking neurons in which some connections are cut to represent walls or obstacles. It can, however, also be applied to general graph search.

7.4 Routes

Besides the memory of places, the memory for routes is the simplest form of spatial long-term memory. Route navigation is a sequential behavior composed of steps such as landmark guidances or the reduction of a goal vector in path integration. The agent learns which mechanisms need to be applied and which snapshots, landmark cues, or vector parameters are to be used within each step of the route. During route production, navigational mechanisms are triggered by upcoming sensory cues and maybe by a representation of the current position obtained from working memory. Routes lead to a single goal, the endpoint of the route. Once initiated, the pursuit of a route can proceed automatically, without a need for intermediate planning steps as a stereotyped behavior; for classical accounts of the route concept, see chapter 2 of O'Keefe and Nadel (1978) and Kuipers (1978). Routes are also considered to mark an intermediate stage in the development of spatial knowledge in children, occurring between the learning of individual places and larger, two-dimensional survey knowledge (Siegel and White 1975). When exploring novel environments, initial excursions into unknown terrain are learned as individual routes before interconnections between different routes are discovered. In terms of neural mechanisms, route and map navigation are distinguished by their neural substrate. Routes are similar to habits and skills and recruit structures in the basal ganglia, whereas navigation by maps requires contributions from the hippocampus and other cortical areas. We will come back to the route-versus-map distinction below.

7.4.1 The Structure of Route Memories

The simplest way to generate some sort of route behavior is to remember a set of landmarks and approach each landmark as a beacon whenever it enters the field of view. With this strategy, we could walk from landmark 1 to landmark 2 and so on without a need for a memory of the landmark sequence. However, if we would accidentally turn around, we would be attracted to the previous landmark and start walking back. This problem may be relaxed by replacing the beacons by directed views taken in the forward direction of the route, as shown in figure 7.9a,b. If we now turn backward, the view will generally be unknown and will therefore not exert any attraction. Route memories composed of an unordered set or "bag" of directed snapshots have been suggested as a model of route navigation in insects by Baddeley et al. (2012).

In the "bag-of-snapshots" approach, the instantaneous motion of the observer is determined by its current sensory input alone. The snapshot to home to is the one stored in memory that most closely resembles the current visual input. This snapshot is used, together with the current visual input, to determine the direction of onward movement by snapshot homing (see section 6.2.3). The agent does not need additional information about its current position. In particular, it does not need to know its starting position in order to

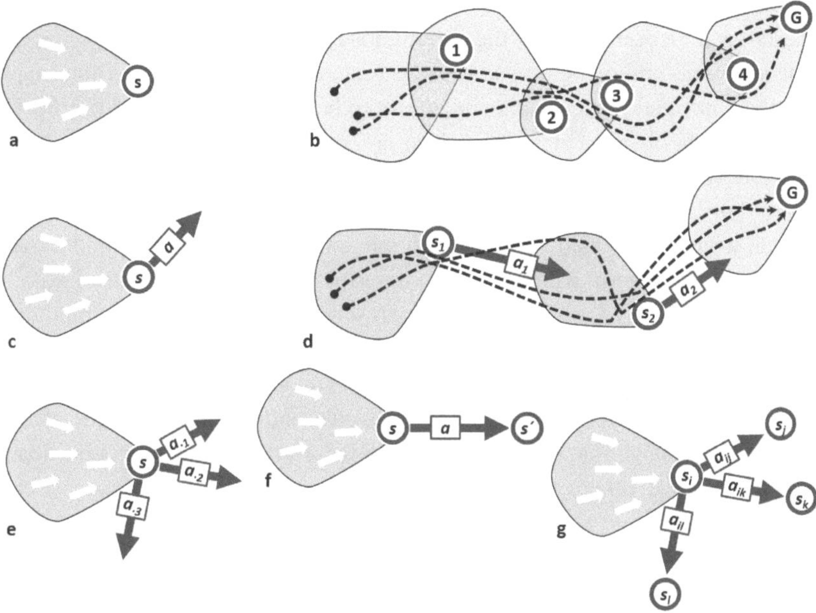

Figure 7.9
Types of knowledge represented in route memory. (a) A snapshot guidance with a directional snapshot $\boldsymbol{s}$ and a catchment area allowing one to approach the vantage point from one side. (b) A route generated by an unordered set of stored snapshot guidances activated when the agent enters the catchment area. The dashed lines show three possible paths. G marks the goal, and 1 to 4 are waypoints. (c) Direction or recognition-triggered response. In addition to the snapshot, memory contains an instruction for an action $\boldsymbol{a}$ to be carried out upon arrival at the vantage point. (d) A route generated from recognition-triggered responses where the action step uses path integration. When the vantage point is reached, the agent loads the associated vector into its path integrator and moves so as to reduce it to zero. This leads into the next catchment area. (e) Multiple responses $\boldsymbol{a}_{\cdot i}$ may be associated with one vantage point, for example, a path integration vector plus additional landmark guidances. (f) In addition to the action $\boldsymbol{a}$ executed at $\boldsymbol{s}$, memory may contain an expectation of the next vantage point to be approached. Such $(\boldsymbol{s}, \boldsymbol{a}, \boldsymbol{s}')$-associations represent the sequential ordering of the route steps as a chain. (g) Multiple three-way associations of the form $(\boldsymbol{s}_i, \boldsymbol{a}_{ij}, \boldsymbol{s}_j)$ constitute a graph memory.

initiate the route. All that is required is the identification of the best-matching snapshot from place memory.[2]

2. Note that this view differs from a classical definition of navigation, for example, by Levitt and Lawton (1990), where knowledge of the starting point is considered crucial. The selection of the best-matching snapshot implicitly contains the information that the agent is currently within the catchment area of this snapshot, but this information may be quite coarse, and it is not required but generated as a side effect from the first step of route navigation.

The situation is different at the next level of complexity, where the local movement decision depends on a representation of the current position, for example, a recognized place. The agent represents and recognizes waypoints (e.g., the vantage points of the snapshots) and remembers navigational decisions or actions to be taken at each point (figure 7.9c,d). Actions that can be associated with a place are guidances, such as "choose the right street and follow it until the next junction," "walk downhill to the river," or "approach the beacon," but also goal vectors specifying the position of the next subgoal. Distance and direction would be "loaded" into the path integrator, and the agent would move so as to reduce the vector to zero. Kuipers (2000) describes such rules as "control laws"; they are automatic behaviors, as discussed in chapters 4 and 5, but selected and modulated by memory mechanisms.

We can think of the represented current position as an inner *state* that the agent thereby assumes ("I am now at waypoint j") and denote this state and the snapshot associated with it by the letter s. The navigational action depends on this state, which is of course dependent on visual input but also on other factors such as the memory of places passed before. The $(s, \boldsymbol{a})$ association of a state and an action to take when the state is reached is a memory type known as direction or recognition-triggered response (Trullier et al. 1997). It is also feasible to store multiple associations at each place, for example, a vector for navigating the next section by path integration, together with bearing angles for various landmarks that the new heading should assume (figure 7.9e). In this case, a mechanism for cue integration is needed, as will be discussed in the empirical examples below.

At the next level of complexity, memory could also hold an expectation of the upcoming waypoint that should be reached if a particular action is carried out; the association will then take the form of a triple $(s, \boldsymbol{a}, s')$ or, in the case of multiple such triples, $(s_i, \boldsymbol{a}_{ij}, s_j)$, where j numbers the associations attached to node i; see figure 7.9f,g. In the words of Tolman (1932), $\boldsymbol{a}$ is a "means" to reach the "end" s', and $(s, \boldsymbol{a}, s')$ is a "means-ends pair" (see also figure 1.1 and the discussion there). The agent will carry out the remembered action: for example, walk into some direction until the expected next snapshot is recognized and can be used for homing. Triple associations of a recognized situation, actions to take in this situation, and the expected outcome are also known as "schemas"; see, for example, Arbib (2005).

The triple association contains also information about the sequence of route steps; the unordered set of guidances or recognition-triggered responses is thus transformed into an ordered chain, or a graph as shown in figure 7.10. For pure route memories, the graphs would be directed and free of loops, with one goal node that can be reached from any other node. Still, the whole memory will subserve only the navigation of just one route, ending at the sole goal location. Put differently, the motion decision at each waypoint does not depend on the goal, since the goal is fixed. We will see in the next section how this constraint can be relaxed in map memories.

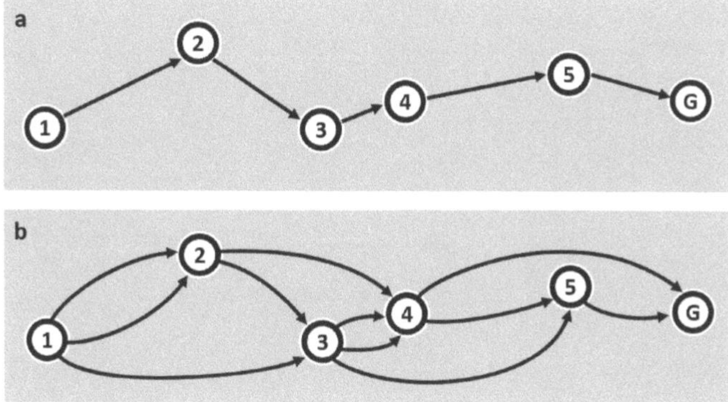

Figure 7.10
Graph models of route knowledge. Each arrow with its start and end node is a triple association (s, a, s'), as shown in figure 7.9g,f. G: goal; 1 to 5: waypoints. (a) Simple state–action chain. (b) Directed graph without cycles. Alternative paths are possible but will all end at the unique goal node. Navigational decisions at intermediate waypoints do therefore not depend on the goal.

Route memories are not strictly one-dimensional but have a nonzero extension to the sides. Indeed, if a navigator is slightly displaced sideways, successful navigation is generally still possible. Tolman (1948) therefore describes routes as "strip maps." One possible mechanism generating some width is the two-dimensional extent of the catchment areas: the route memory covers not just a single trajectory but a strip defined by the overlap of the catchment areas (total gray area in figure 7.9b). Additional factors widening the represented strip may be multiple, alternative action links between waypoints, as shown in figure 7.10b.

7.4.2 Empirical Findings

State–action associations Ants can be trained to routinely run back and forth between the nest and a feeding site. In section 5.2, we discussed this performance as controlled by path integration in an environment virtually void of landmark cues. In a study on the Australian ant species *Melophorus bagoti*, Kohler and Wehner (2005) studied the "commuting" behavior in an environment with sparse vegetation in which individual tussocks of *Cenchrus* grass provide good landmark information. The ants travel between the tussocks following well-defined paths using the same gaps between the tussocks most of the time.[3] Like the African desert ants discussed in chapter 5, *Melophorus* cannot use trail pheromones due to their high evaporation rate.

3. Methodologically, this statement relies on a suitable measure of the similarity of individual paths that allows one to judge their conformity with an assumed route. One such measure suggested by Hurlebaus et al. (2008) uses a Delaunay triangulation of the environment with the centers of the tussocks as vertices. Individual trajectories are transformed into sequences of touched triangles, which are then compared using a combinatorial approach.

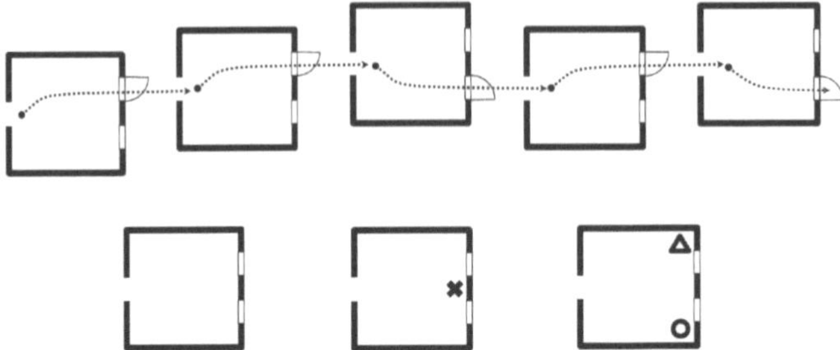

Figure 7.11
Sequence of binary decisions realized by rooms with two exit doors (modified after Waller and Lippa 2007). The top row shows a sequence of decisions that has to be learned by the subjects. The three single rooms in the bottom row show three experimental conditions—namely, an empty room, one landmark to be used by way of association, and two landmarks that can be learned as beacons.

In principle, the described route behavior might be expected from the combination of pursuit of a goal and obstacle avoidance, as discussed, for example, in section 4.3.4. However, Kohler and Wehner (2005) show that associations between landmark knowledge (snapshots) and actions also play a role. For example, if an ant is picked up at the nest, its home vector is zero. When released halfway along the route, these "zero-vector ants" run back straight to the nest following the regular route. The same is true for "full-vector ants": that is, ants picked up at the feeder and released along the route. Note that this result is quite different from that of the displacement experiments with *Cataglyphis* ants described in section 5.2, presumably due to the lack of landmarks in the *Cataglyphis* habitat. In contrast, *Melophorus* ants seem to recognize the release location by remembered landmark cues. They then know which way to take from there. This is precisely the behavior predicted by route memories of the type depicted in figure 7.9b.

In an additional experiment where ants were forced to use different routes for the outbound and inbound directions, ants displaced from their current way to a different position within the same route were able to "channel in," whereas displacement from the inbound to the outbound route or vice versa resulted in disorientation (Wehner et al. 2006). This may indicate that the current heading or the pursued goal influences the recognition of waypoints.

In humans, an experiment on simple route following has been performed by Waller and Lippa (2007); see figure 7.11. In a virtual environment, subjects walked through a flight of twenty square rooms each with two exit doors, only one of which would lead to the next room. In three conditions, rooms were either empty, marked by one discriminating landmark, or provided two landmarks placed next to the doors. In the empty-room condition,

---

Alternatively, Gonsek et al. (2021) suggest continuous measures based on dynamic time warping and the Fréchet distance between curves.

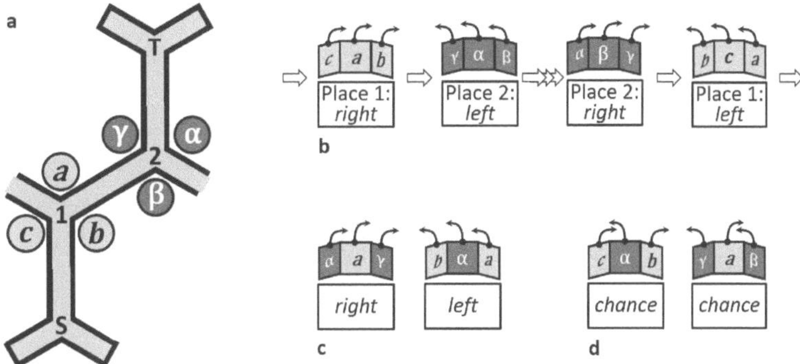

Figure 7.12
View voting experiment by Mallot and Gillner (2000). (a) Map of the experimental environment. Subjects learn a go-and-return route from the start point S via places 1 and 2 to the turning place T and then back via places 2 and 1 to the start. The decision points are marked by three landmarks each, denoted a, b, and c for place 1 (symbolized by black letters on bright background) and α, β, γ (bright letters on dark background) for place 2. (b) In the training phase, view–action associations are learned. Views of the same landmark encountered in the go or return part of the route appear in different parts of the visual field and are therefore distinguished as different views. The arrows indicate the motion decision associated with each view during learning. (c, d) In the test condition, views from different places are combined to form novel panoramas, either with equal action associations (c, consistent condition) or with different action associations (d, inconsistent condition). Subjects perform well in the consistent condition only but do not realize that panoramas combined views from different places.

subjects needed to learn the sequence of left or right decisions, which does not constitute a route memory in the strict sense. In the one-landmark condition, subjects could additionally learn to associate each landmark with a route decision (left or right), which constitutes a route memory of the direction type as depicted in figure 7.9c,d. In the two-landmarks condition, subjects would need to learn the landmark next to the correct door as a beacon, which corresponds to the situation depicted in figure 7.9a,b. Performance was poorest in the empty-room condition, intermediate in the direction condition with one landmark per room, and best in the beacon condition. This indicates that all strategies can be used but that there may be a small advantage for remembering beacons or snapshots rather than landmark–action associations.

Action voting for multiple cues In figure 7.9e,g we considered the case of multiple cue–action associations stored at a given place. This situation is actually quite common and likely leads to more robustness in route memories. It raises, however, two questions: what exactly is a cue, and how are the actions associated with each cue combined?

One study by Mallot and Gillner (2000) investigates route learning in an iterated Y-maze with landmark cues placed in each corner of the three-way junctions (figure 7.12). Subjects learn a go-and-return route with two decision points passed in reverse order when going back or forth. The main research question in this study is the object that is recognized in the recognition-triggered response: is it the decision point as a place, or is it just the view of the landmark cue? The views of a given landmark differ in the forward and backward direction of travel, both by the perspective that the observer takes on the three-dimensional buildings and by the position in the visual field in which the image occurs. Landmark views thus contain joint information about the position and heading of the observer and constitute a view graph as already discussed above (figure 7.7).

The view–action associations that subjects would learn in this situation are depicted in figure 7.12b. The overall view is split up into three parts, each depicting different landmark views. During training, each view and part of the visual field are stored separately and associated with the respective motion decision, which in the training phase is the same for all three parts. In the test phase, the views are recombined to novel panoramas containing views from different decision points. In this situation, the different, conflicting motion decisions will be recalled via the association from the three simultaneous views. Note that these panoramas could not occur during training, and the recognition of a place will therefore not be possible. Still, subjects make clear motion decisions as long as the motions associated with each of the views are consistent (figure 7.12c); as soon as the associated motions differ from each other, however, subjects get confused and make random decisions (figure 7.12d). This indicates that actions were associated with views, not with places, and that the eventual motion decision was derived from the individual associations by some sort of voting scheme or directional consensus.

A less symmetrical case of cue integration was reported for the interaction of global and local landmarks in a study by Steck and Mallot (2000); see figure 7.13. The training environment is again an iterated Y-maze with local landmark cues placed in each corner. These cues are visible only in the immediate vicinity of each junction and are therefore called "local." A second class of "global" landmarks was presented at elevated positions surrounding the maze and remained visible from within the entire maze. Such global landmarks can be used as a kind of compass (cf. section 2.2.2) but also as a reference for a remembered action to be carried out at a recognized decision point.

After training the subjects to perform a route in the maze, the ring of global landmarks surrounding the maze was rotated such that following the global or local landmarks would lead to different route decisions. The subjects roughly fell into two groups, one consistently following the local landmarks and ignoring the global ones and another one using the global landmarks throughout. Of course, this could reflect general preferences in the sense that subjects learned only one group of landmarks and ignored the other one from the beginning. However, when tested with just one type of landmark (figure 7.13c,d), all subjects performed well. This indicates that both local and global cues had been learned but that cue integration applied in the conflict case led to a decision for one landmark type but not the other.

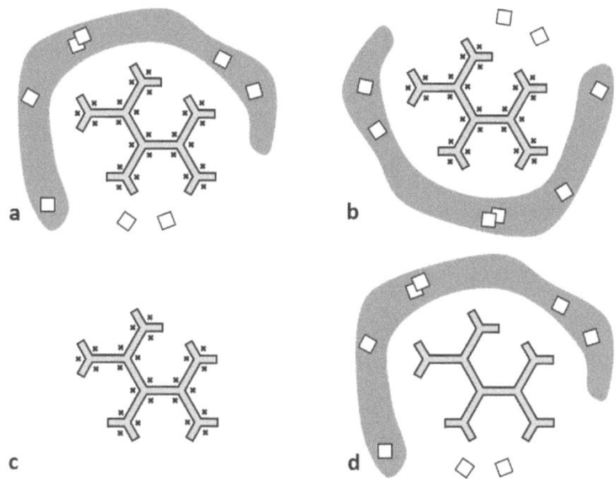

Figure 7.13
Local and global landmark experiment by Steck and Mallot (2000). (a) Subjects navigate an iterated Y-maze ("Hexatown") with local landmarks (marked "x") in all corners. Additionally, globally visible landmarks are provided on a surrounding mountain range or as high-rise buildings (squares). Subjects learn a route in configuration (a) and are then tested in configurations (b–d). In (b) the global landmarks are displaced by rotation; in (c) and (d) one group of landmarks is removed. For further explanation, see text.

Hoinville and Wehner (2018) suggest a quantitative model for the combination of multiple directional cues in ant navigation. These cues include "local vectors" associated with recognized landmarks or places and a "global" path-integration vector. Each vector suggests a movement direction but, by its length, also encodes the reliability of this information. Combination is then simply by summation, as in the cue accumulation scheme discussed in section 2.6. The model is able to explain a large number of quantitative behavioral data on ant navigation.

Sequentiality Route behavior is sequential by definition, but as we have seen in the discussion of the "bag-of-snapshots" approach to route memory, this may result simply from the spatial ordering of the waypoints without requiring a representation of the sequence in memory. In insects, some evidence to the contrary has been provided in a route-following experiment with honey bees, *Apis mellifera*. Chittka and Geiger (1995) trained bees to fly a route across a meadow, passing a row of identical tents that had been placed there as landmarks. The feeder was located between the third and the fourth tent. Of course, honey bees had no problem learning this route. In the test condition, the feeder position stayed the same, but the number of tents between the start and the feeder was varied. If tent spacing was closer than during training, the honey bees flew shorter distances ending up short of the feeder, while with larger spacing between the tents, bees would overshoot. Navigation is therefore not (exclusively) based on path integration, but also on the landmarks. Since these all look the same, the bee must know how many of the identical tends it has already passed and when the final tend is reached. Chittka and Geiger (1995) discuss this effect as "protocounting," suggesting that the bees were able to count the tents.[4]

A standard paradigm for studying route memories in humans is the recall of landmarks encoded during route learning; examples have already been discussed in section 6.4.2. Human subjects are indeed able to recall the sequence of landmarks appearing along a route. In addition, the position of occurrence influences the probability of recall, as do other parameters such as the relevance, information content, and salience of a landmark. As in classical sequence learning experiments based on word lists, items occurring initially or toward the end of a sequence are generally recalled better than items in the middle. These "primacy" and "recency" effects are also found in route memories. For a systematic comparison of route and list memories, see Hilton, Wiener, and Johnson (2021).

Direct evidence for the representation of sequentiality in human route memories comes from a priming experiment by Schweizer et al. (1998). Subjects learned a route through a corridor with a sequence of landmarks by watching a movie of the walk-through. They were then presented with images of objects that did or did not occur in the training environment and were asked to identify the known ones. Before the presentation of the test images, another image from the training sequence was presented as a prime. Reaction times were shorter when the test image had followed the prime image in the training sequence as compared to cases where the test image had preceded the prime image. The authors suggest that priming is caused by a directional spread of activation from the prime image in the forward direction of the learned route. A similar conclusion can be drawn from a study by Meilinger, Strickrodt, and Bülthoff (2018), in which subjects learned a route in an immersive virtual reality. When asked to point from one waypoint to another, performance was better if targets were ahead in route direction, at least for the initial trials.

In an imaging study by Wolbers, Weiller, and Büchel (2004), route learning in human subjects was tested with a version of the Hexatown environment described above (figure 7.13c). Only the landmarks in the direct approach direction were visible such that the resulting route was a strict sequence of landmarks and motion decisions. After learning, subjects were presented with two landmark views, a large one indicating a decision point and a small one that was to be imagined as a goal. They were then asked to indicate the motion decision required when facing the large view. Performance increased over the learning session, but only if the goal view was the direct successor of the decision point in the route sequence. This indicates that associations of the (s, a, s')-type have been learned.

4. In mathematics, the relation between sequentiality and counting is given by the "successor" function used in the definition of the natural numbers by means of the Peano axioms. The idea that spatial route knowledge may be the origin of higher-level cognitive performances has also been suggested for language production as a sequential performance by Hauser, Chomsky, and Fitch (2002).

Performance increase also correlates with fMRI activity in the posterior parietal cortex, which is therefore discussed as a possible site of sequential route knowledge.

7.5 From Routes to Maps

7.5.1 The "Cognitive Map" Metaphor

The term "cognitive map" was first introduced by Tolman (1948) to distinguish two types or theories of spatial behavior, stimulus–response associations as described in the previous section and purposive or goal-directed behavior. In this latter type of behavior, movement decisions are not determined by the current stimulus (the landmarks) or positional knowledge (recognized places, home vector) alone but also depend on the currently pursued goal. The term "cognitive map" replaces the older term "means-ends field" used by Tolman (1932), which is basically a network or graph of $(\boldsymbol{s}, \boldsymbol{a}, \boldsymbol{s}')$ associations as shown in figure 1.1. Conceptually, the two ideas, cognitive map and means-ends field, are pretty much the same, and we will see that the graph structure of the latter is more in line with modern thinking about cognitive space than the map metaphor. Indeed, the map metaphor is misleading in that it invokes the idea of a two-dimensional sheet or aerial image much as a printed geographical map. This seems to be far from Tolman's thinking, however. Rather, he sees the cognitive map as a body of knowledge about action alternatives applicable in each situation together with their respective outcomes. Learning and the addition of more action alternatives with a predictable outcome lead to a "widening" of the map, not necessarily in a geometrical sense, but first of all in the sense of a wider behavioral repertoire. A route would then be a particularly "narrow" map, specifying a sequence of automatic steps to a sole destination.

As a consequence of the sketched misinterpretation, the term "cognitive map" has become an issue of heated and sometimes polemic debates; for an example, see Bennett (1996). Of course, the term itself is just a metaphor, and the question of whether or not it is appropriate may be considered a matter of taste. However, two scientifically interesting questions can be asked: First, what is the role of numbers and coordinate systems in the description of space? And second, is this role the same for spatial working and long-term memories? Consider spatial working memory, which was described as a local chart of the current environment in the previous sections. This chart does possess some properties of a geographic map such as a coordinate origin (the observer), a main axis (e.g., heading or the intrinsic reference axis of a room), and maybe even place representations by coordinates (e.g., by object vector cells). It is, however, not a cognitive map in the sense of Tolman (1948) because it is limited in range and generally not sufficient to plan routes to distant goals. The information that is needed for long-distance planning can be expressed as a set of if-then statements of the type "if you take action $\boldsymbol{a}$ at place $\boldsymbol{s}$, you will reach place $\boldsymbol{s}'$." These statements do not depend on the current position of the observer and are therefore part of long-term memory. They are allocentric in the sense of Klatzky (1998), but they do not rely on coordinate systems or the representation of places by pairs of numbers.

Gallistel (1990) and others have tried to avoid these problems by defining the cognitive map as knowledge of the metric properties of space. This works nicely for path integration and other properties of spatial working memory but fails to include the knowledge of action alternatives and their consequences addressed by Tolman. We will therefore stick to Tolman's idea and allow for cognitive maps lacking metric information. The question of how much metric is included will be discussed in chapter 8.

In summary, we define the cognitive map as that part of spatial long-term memory that is required to plan and find ways to multiple distant goals. It generates a goal-dependent flexibility of behavior (i.e., the ability to navigate to varying goals) rather than just automatically repeating a route. As pointed out in Tolman's 1948 paper, the route versus map distinction is a distinction between automatic stimulus–response behavior, on the one hand, and planned and goal-dependent, or cognitive, behavior on the other. For a comprehensive discussion of the route versus map distinction, see also O'Keefe and Nadel (1978).

Another term used largely synonymously with the term "cognitive map" is "survey knowledge" (Siegel and White 1975). It stems from the idea that when overlooking an area from an elevated sight, spatial knowledge might be formed that would also support goal-dependent navigation. Aerial views are of course similar to geographic maps. The question, then, is whether cognitive maps that have been acquired piecemeal—i.e., by learning views of the scenery in a first-person perspective and combining them into $(\boldsymbol{s}, \boldsymbol{a}, \boldsymbol{s}')$-associations—can give rise to imagined aerial views as discussed, for example, in the section of mental scanning. At least when judging from sketch-map productions by human navigators, this does not seem to be a general case (see, e.g., Appleyard 1970).

Several authors have argued that cognitive maps do not exist in insects (Wehner and Menzel 1990; Bennett 1996; Cheung et al. 2014), although the question remains controversial (Cheeseman et al. 2014). We will not elaborate on this issue here but stick to evidence from mammals and humans, in which the situation is clearer. In any case, the appearance of a cognitive map in the sense discussed above should not be seen as a "Rubicon" that is or is not crossed and whose crossing immediately results in a full-fledged map behavior. Rather, transitions from "narrow" routes to wider maps (i.e., the representation of more or less elaborate action alternatives) can be gradual with respect to both logical distinctions and the emergence of navigational abilities in evolution and development.

In the remainder of this section, we will discuss evidence for behavioral mechanisms beyond stimulus–response behavior, following arguments from Tolman (1948) and O'Keefe and Nadel (1978). These are taken from three areas: (i) learning strategies (latent learning, trial-and-error behavior), (ii) mechanisms of wayfinding and planning, and (iii) neural substrates.

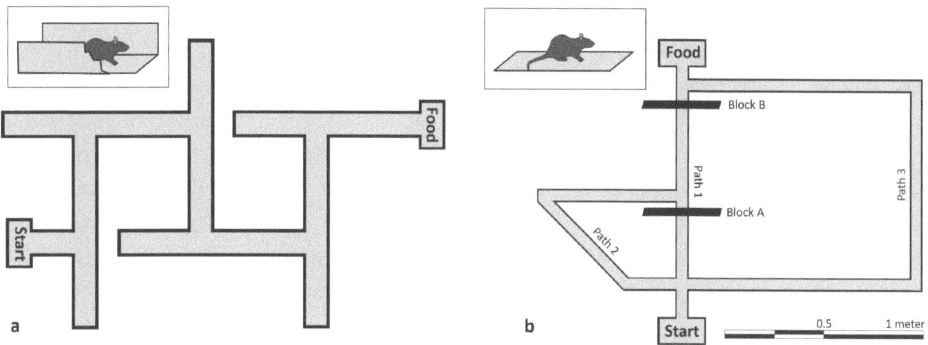

Figure 7.14
Two mazes used in classical experiments on map behavior. (a) Latent learning in an iterated T-maze with side walls (Blodget 1929). Rats receiving reward learn the direct route in about seven daily sessions. Rats trained initially without reward learn faster once reward is delivered. (b) Detour experiment using an elevated maze by Tolman and Honzik (1930). Without the blocks, rats choose the shortest path (path 1) from start to food site. After blocking at A, detour via path 2 is learned. When finally blocking at B, rats immediately switch to path 3, although the situation at the decision point is largely unchanged. Insets symbolize walled and elevated maze.

7.5.2 Learning

Latent learning Stimulus–response associations are usually learned by operand conditioning through reinforcement of the rewarded response. It is therefore interesting to look at the role of rewards in spatial learning. When learning a route with many steps, reward will be obtained only at the end, when the goal is eventually reached. In order to learn the earlier steps, some sort of reward propagation or "credit assignment" must be assumed that provides the "reward" for learning the earlier steps of the route. This problem is studied extensively in the theory of reinforcement learning (Sutton and Barto 2018). In addition, everyday experience tells us that we are able to pick up information also during free exploration of an environment without pursuing a particular goal. It can be argued, that in this situation rewards are not entirely absent but that they are intrinsically provided by the reward system of the brain. One might therefore say that learning is driven by curiosity and that the reward lies in the satisfaction of this curiosity. Learning without reward and reinforcement is called "latent," and if it exists, it must store stimulus–response or state–action associations not as procedures to be carried out whenever triggered by the stimulus but as options that the agent may choose to use as required.

Figure 7.14a shows an iterated T-maze used in a study on latent learning by Blodget (1929). The rat enters at the trunk of a T and then has to decide which arm to run into. The end of the correct arm is the trunk of the next T and so on. The final T leads to the reward. Rats were trained in this maze and learned to find the correct path after about seven training sessions. A second group of rats was also trained every day but without receiving a reward at the food site. Instead, they were taken out of the maze and put into a different box, where they received a reward after some waiting time. In this case, they ran around and explored the maze, but since there was no reward, they did not learn to run straight to the food box. However, if rewarding was started at day 3, rats did make progress faster than the original test group and caught up with their performance already at day 5. In a third group of rats, rewarding started only at day 7, and again the subsequent progress was faster than in the other groups. This indicates that the rats had learned something already during the trials where they were not rewarded and that they can make use of these memories once rewarding starts.

The idea of latent learning is that behavioral options are learned together with their respective outcomes but that learned options are not automatically executed every time their trigger stimulus appears. Rather, they are stored in some set-aside memory and activated as required in pursuit of the current goal. Voicu and Schmajuk (2002) present a neural network model of this performance and the memory store. It consists of a lattice of neurons representing a two-dimensional area with some resolution. Neurons representing adjacent places are reciprocally connected. During exploration and latent learning, knowledge about the environment is imprinted on this "canvas" by deleting connections (synapses) between places that have been experienced to be unconnected (i.e., separated by walls or gaps in the maze). After learning, activity is assumed to spread from both the current position of the agent and the goal, with some decay at each synapse. The agent moves toward the connected place with highest activity, (i.e., uphill in the activity landscape). Simulated trajectories nicely reproduce the latent learning effects reported by Blodget (1929).

Note that the planning procedure suggested by Voicu and Schmajuk (2002) is similar to the place cell planning model of Ponulak and Hopfield (2013) discussed in section 7.3.2. The latent learning component, however, is modeled by the deletion of synapses between neurons representing unconnected places (i.e., by synaptic changes), which makes the canvas a model of long-term memory.

Detour behavior In another experiment, Tolman and Honzik (1930) used a maze with three possible paths from start to goal; see figure 7.14b. The maze was built as an "elevated maze": that is, a set of bridges without side walls so that all paths were visible to the rat. Path 1 was straight and initially used by the rats. After introducing a barrier at position A, rats switched to path 2, which took a detour around the barrier, but joined path 1 for a common terminal section toward the reward. During these training phases without a barrier or with a barrier at position A, path 3 was ignored. In the test phase, the blockage was moved to position B, in the common terminal section of paths 1 and 2. Tolman and Honzik (1930) report that in this condition, all rats immediately chose path 3, which was the only one not affected by the block. They conclude that this path had been learned "latently" so it could be used later. Voicu and Schmajuk (2002) also test this behavior in their model and are able to reproduce the performance.

The experiment as described here also allows other conclusions. In order to be learned latently, path 3 must have been perceived by the rat while standing at the start or walking the other paths. It is therefore well possible that the choice in the test phase was based on the current perception and a planning process in working memory, not on previous learning. This type of detour behavior is not uncommon in many animal species, not only mammals but also amphibians or arthropods. Of course, in the heyday of behaviorism, when the paper was written, a planning process in working memory must have appeared just as revolutionary as a long-term memory learned without reward. For our argument about cognitive maps, however, the experiment would need further control.

Incentive learning and reward devaluation Strict stimulus–response behavior is automatic by definition: that is, it depends on the current stimulus only, not on a pursued goal or the motivational state of the agent. Of course, it can be doubted that this type of behavior at all exists. Even in the *Paramecium* discussed in section 4.1, inner states such as the nutrient level or the cell cycle will likely influence behavior. The resulting modulation, however, would not be the result of some sort of decision-making process but a direct reflection of the inner state. This type of behavioral modulation can be contrasted with the idea of "purposive" behavior, as addressed in Tolman's notion of the means-ends field (see figure 1.1). "Purposiveness" or goal-directedness of behavior requires the availability of action alternatives and a decision-making mechanism based on the evaluation of each expected outcome with respect to the pursued goal. The representation of action alternatives, their expected outcomes, and the value of these outcomes with respect to the current goal are crucial elements of the "map" level of spatial representations.

Experimental evidence for the role of the pursued goal comes from studies in which the incentive exerted by the goal is changed during the experiment. This can be done, for example, by providing a food reward, which is later devalued by offering the same type of food with some mild poison causing nausea. By this "taste aversion" training, the food is devalued and the rat will not further pursue it in the future. Experiments on incentive learning are mostly carried out in choice boxes with two types of levers or release mechanisms, but can also be done in T-mazes (e.g., Balleine and Dickinson 1998). For example, a rat might learn a T-maze with food and water rewards placed in the left and right arms of the "T," respectively. After water deprivation (i.e., when the rat is thirsty), it will turn left at the junction, while after food deprivation, it will choose the right arm. This already shows that the rat knows the location of water and food and does not simply perform an automatic motion decision at the junction. In addition, if the food is devalued by taste aversion training, the rat will no longer go for the food, indicating that the expected value of the action is taken into account in the decision-making process.

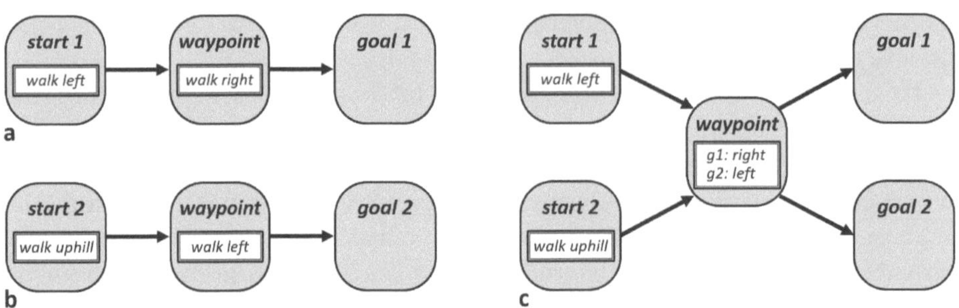

Figure 7.15
Knowledge transfer across routes. (a) A route from a start to a goal via a waypoint. At the start and the waypoint, the action required to continue on the route is stored as a recognition-triggered response. (b) A second such route. (c) If the agent realizes that the waypoint occurring in both routes is the same place (i.e., if the routes intersect), the knowledge required at the waypoint must be more specific. It is no longer an instruction or command but an if-then (if goal, then action) declaration. Such memories are called declarative.

7.5.3 Wayfinding

Knowledge transfer across routes The main difference between memories of the route and map type relates to the problem of knowledge transfer across routes. Assume an agent has learned two independent routes, each leading from a start point to a goal (see figure 7.15). Each step is characterized by a $(\boldsymbol{s}, \boldsymbol{a})$ or $(\boldsymbol{s}, \boldsymbol{a}, \boldsymbol{s}')$ association, as discussed in section 7.4. In the figure, the actions are specified by examples such as "walk left" or "walk uphill." Assume further that the two routes intersect, in which case they will share a common waypoint. If the agent now stores recognition-triggered responses, the decision at the common waypoint becomes ambiguous, since two responses would be activated and compete with each other. Even if one of them wins, the decision would not depend on the currently pursued goal; it might, for example, mislead an agent coming from route 1 to proceed to goal 2. The ambiguity can only be resolved if the logical structure of the memory associated with the waypoint is changed. It must not be a simple response, or command, that the agent carries out whenever the place is recognized but a declaration providing information about action alternatives: if heading for goal 1, walk right, and if heading for goal 2, walk left. In addition, the agent needs a representation of the pursued goal and a planning stage that reaches a motion decision based on the stored if-then clause and the current goal.

The representation of navigational knowledge as an if-then declaration has the additional advantage that it allows for the inference of novel routes, in our example the routes from start 1 to goal 2 and from start 2 to goal 1. Knowledge is thus transferred from known routes to other, novel routes. In addition, if longer routes are involved, any known place can become the goal as long as a sufficiently powerful planning device is available. O'Keefe and Nadel (1978) describe this property as the "flexibility" of map memories and contrast it to route memories, which work for one fixed goal only. The combination of known route segments into novel routes can also be considered a kind of shortcut behavior. It should be clear, however, that this is not the same as shortcut behavior based on path integration, as is also present in ants returning straight to their nest after curved excursions (see figure 5.1a). Shortcut behavior based on the recombination of known route segments is evidence for a cognitive map, whereas simple path integration is not.

Human performance The ability to transfer spatial knowledge from familiar to novel routes is an easily testable performance that can be taken as an operational definition of the cognitive map. Gillner and Mallot (1998) tested this idea in an iterated Y-maze ("Hexatown"); the layout is shown in figure 7.13c. All street junctions had 120-degree angles, and a model house was placed in each angle as a landmark. Subjects could view only one junction at a time since the view to distant junctions was blocked by trees. The environment was designed to require binary motion decisions at a regular spacing and to enforce navigation by landmarks and associated left/right decisions only. Subjects were trained in three phases: first a group of "excursions" starting from a fixed home location, then a set of "returns" to the home location, and finally a set of novel routes not touching the home location. Performance was measured by counting the errors defined as route decisions not reducing the graph distance (number of remaining decisions required) to the goal.

Without transfer, one would expect that the "novel" routes would be solved completely from scratch or by composing a return to home and a subsequent excursion, as had been learned in the first two phases. However, subjects immediately found the shortest route in the novel condition, indicating that knowledge transfer (graph short-cutting) had occurred. Overall, the error rate decreased with the number of routes traveled. A detailed analysis of the motion decisions also revealed a persistence effect in the sense that errors occurring at a particular place (view) were likely to occur again at the next visit of the same place, irrespective of the then pursued goal. This might indicate a transition from initial route learning with fixed view–decision associations to later map learning with goal-dependent flexibility in action selection.

In another study, Newman et al. (2007) used a virtual town with an irregular street network. Participants of the experiment were driving around in the environment and picked up (virtual) passengers who asked for a lift to some place in the town. Participants received reward for quick arrivals. Performance was measured as delivery path length relative to the shortest possible path ("percentage above optimal," PAO). The learning curve of excess path length plotted over trial number shows an exponential decrease: that is, knowledge acquired during one trip could be used also for later trips. If, however, the environment changed, learning had to start from scratch. This shows that the effect is not just a familiarization with the task but an instant of latent learning.

Neural substrate The distinction of route and map behavior is also reflected by the underlying neural systems. Indeed, a dissociation[5] of route and map memories of space was demonstrated in a study by Hartley et al. (2003). As in the previous studies, they used a virtual town in which subjects had to find routes to various goals. In a "route" condition, a sequence of goals was repeated several times such that subjects could store a fixed route. In a "wayfinding" condition, goals were presented in variable sequence, requiring subjects to plan routes from memory for each task. This is the situation in which map-type memory is most helpful. The experiment was carried out in an fMRI scanner, and neural activity was recorded. The authors report brain regions whose activity depends on the experimental condition. Subjects who perform better overall tend to activate their perirhinal cortex in the wayfinding task and the head of the caudate nucleus in route following. Other subjects did not show this effect.

The result clearly shows that the route versus map distinction is not just a logical construct but describes dissociable components of spatial cognition in the sense of cognitive neuroscience. It also is in agreement with findings from other domains indicating that basal ganglia (most notably the caudate nucleus) can be associated with nondeclarative forms of learning such as skill learning, while perirhinal cortex and other areas close to the hippocampus are associated with explicit or declarative forms of learning.

Further evidence for the role of the hippocampus in wayfinding was provided by a study on experts in spatial cognition: that is, taxi drivers in London (Maguire et al. 2000). Subjects had passed an exam as a licensed London taxi driver, which required a training time of 10 month to 3.5 years. In addition, they had professional experience for periods between 1.5 and 42 years. Structural MRI scans reveal that taxi drivers, as compared to age-matched controls, have bilaterally enlarged posterior hippocampi while the anterior hippocampi are reduced in size. The correlations with the time subjects had worked as taxi drivers was significant in the right hemisphere for both the posterior (positive correlation) and anterior hippocampus (negative correlation). This latter result indicates that posterior hippocampus increases in size as an effect of navigational practice. Thus, navigational practice seems to be causal for the increase of posterior hippocampal volume.

The discussed results should not be interpreted in the sense that the hippocampus is the cortical site of the cognitive map. Rather, it seems to be a structure involved in the encoding of information by plastic processes that take place in neocortical areas such as the posterior parietal or the retrosplenial cortex (Brodt et al. 2016; Boccia et al. 2016). The dissociation of route and map behavior into different parts of the brain, however, is confirmed by these more recent studies.

5. In neuropsychology, a *dissociation* between two behavioral competences is said to be present if they recruit different brain areas. The concept was originally developed for lesion studies. If the lesion of some brain area affects one behavioral performance but not another, the two competences would be dissociated. In double dissociations, a second lesion has been found in which the performance pattern is reversed.

7.5.4 The Cognitive Map as a Declarative Memory of Space

The distinction between routes and maps as formulated by O'Keefe and Nadel (1978) is an early example of the more general distinction between nondeclarative and declarative types of long-term memory, as introduced by Squire and Knowlton (1995). Declarative memory is a central concept of cognitive neuroscience: it includes explicit forms of memory such as general reference knowledge (e.g., "Paris is the capital of France") or knowledge of past events (e.g., "Yesterday, I had fish for dinner"). In humans, it can be expressed by verbal reports, as in the above examples, but it is also thought to exist in animals. Indeed, the protocols discussed above as evidence for map behavior—specifically, latent learning, detouring from memory, relearning of incentives, and transfer of knowledge between sequential tasks—may be used as procedural definition of declarative learning in animals.

Declarative memory has also been characterized as "knowing what," as opposed to nondeclarative memory, which is about "knowing how". This latter form of memory includes skills like shoelace binding or habits such as our daily routines. In the older literature, nondeclarative learning was also called procedural or even imperative learning, because at least some types of what is now summarized as nondeclarative memory result in immediate action demands. The following quote from Lorenz (1966) illustrates that route memories can indeed be seen as "imperative" memories in this sense:

> I once suddenly realized that when driving a car in Vienna I regularly used two different routes when approaching and when leaving a certain place of the city, and this was at a time when no one-way streets compelled me to do so. Rebelling against the creature of habit in myself, I tried using my customary return route for the outward journey, and vice versa. The astonishing result of the experiment was an undeniable feeling of anxiety so unpleasant that when I came to return I reverted to the habitual route.

As a conclusion of the route-versus-map debate, we state the definition of a cognitive map in the following way: like memories in other behavioral domains, spatial long-term memory comes in different types that can be classified as declarative or nondeclarative. The cognitive map is simply the declarative part of spatial long-term memory. Routes, remembered places, and guidances used for homing are also elements of long-term memory but belong to the nondeclarative side.

Key Points of Chapter 7

- Like other forms of memory, spatial memory can be divided into two major parts, working and long-term memory.
- Working memory is realized by sustained or reverberating neural activities such as the head and grid cell attractors discussed in chapter 5. It subserves the planning and monitoring of ongoing actions and events. Working memory is limited in duration, content, and spatial range.

- Long-term memory is associated with plastic changes such as gene expression, synaptic growth, or neurogenesis. It can be divided into a declarative part storing explicit knowledge and a nondeclarative part for skills, habits, stimulus–response associations, and so on.
- Working memory is experimentally studied with tasks requiring knowledge of the current context or situation. Typical paradigms include food depletion, the Corsi block-tabbing task, judgment of relative direction (JRD), spatial neglect and position-dependent recall, mental scanning, and route planning.
- Spatial working memory is modeled as a local chart with the observer in the center and a preferred axis along which representation is superior to other parts. In a neural network model based on boundary and object vector cells, the chart is egocentric but "allo-oriented": that is, the preferred direction aligns with the environmental geometry. Object and boundary positions are represented by vector cell firing. Alternatively, the view graph approach allows one to also include information about object identity.
- Spatial long-term memory contains representations of places and local views to which navigational actions (guidances) and their expected outcomes may be associated. Chains of place–action associations are called routes.
- If multiple actions are associated with individual places or views as if-then declarations, the representation allows the agent to infer novel routes from previously learned route segments. Such representations are called maps. They can be acquired by latent learning, without reinforcing reward.
- Map knowledge can be thought of as the declarative part of spatial long-term memory. In accordance with this idea, map behavior (planning novel routes) recruits areas in the medial temporal lobe (including the hippocampus) while mere reproduction of routes recruits networks in the basal ganglia.

References

Appleyard, D. 1970. "Styles and methods of structuring a city." *Environment & Behavior* 2:100–116.

Arbib, M. A. 2005. "Modules, brains and schemas." *Lecture Notes in Computer Science* 3393:153–166.

Avraamides, M. N., J. M. Loomis, R. Klatzky, and R. C. Golledge. 2004. "Functional equivalence of spatial representations derived from vision and language: Evidence from allocentric judgments." *Journal of Experimental Psychology: Learning, Memory, and Cognition* 30:801–814.

Baddeley, A. 2000. "The episodic buffer: A new component of working memory?" *Trends in Cognitive Sciences* 4:417–423.

Baddeley, A. 2003. "Working memory: Looking back and looking forward." *Nature Reviews Neuroscience* 4:829–839.

Baddeley, A. D., and G. J. Hitch. 1974. "Working memory." In *The psychology of learning and motivation: Advances in research and theory,* edited by G. A. Bower, 47–89. New York: Academic Press.

Baddeley, B., P. Graham, P. Husbands, and A. Philippides. 2012. "A model of ant route navigation driven by scene familiarity." *PLoS Computational Biology* 8 (1): e1002336.

Balleine, B. W., and A. Dickinson. 1998. "Goal-directed instrumental action: Contingency and incentive learning and their cortical substrates." *Neuropharmacology* 37:407–419.

Basten, K., T. Meilinger, and H. A. Mallot. 2012. "Mental travel primes place orientation in spatial recall." *Lecture Notes in Artificial Intelligence* 7463:378–385.

Bennett, A. T. D. 1996. "Do animals have cognitive maps?" *The Journal of Experimental Biology* 199:219–224.

Bicanski, A., and N. Burgess. 2018. "A neural-level model of spatial memory and imagery." *eLife* 7:e33752.

Bicanski, A., and N. Burgess. 2020. "Neuronal vector coding in spatial cognition." *Nature Reviews Neuroscience* 21:453–470.

Bisiach, E., and C. Luzzatti. 1978. "Unilateral neglect of representational space." *Cortex* 14:129–133.

Blodget, H. C. 1929. "The effect of the introduction of reward upon maze performance in rats." *University of California Publications in Psychology* 4:113–134.

Boccia, M., C. Guariglia, U. Sabatini, and F. Nemmi. 2016. "Navigating toward a novel environment from a route or survey perspective: Neural correlates and context-dependent connectivity." *Brain Structure & Function* 221:2005–2021.

Brodt, S., D. Pöhlchen, V. L. Flanagan, S. Glasauer, S. Gais, and M. Schönauer. 2016. "Rapid and independent memory formation in the parietal cortex." *PNAS* 113:13251–13256.

Bülthoff, H. H., S. Y. Edelman, and M. J. Tarr. 1995. "How are three-dimensional objects represented in the brain?" *Cerebral Cortex* 5:247–260.

Byrne, P., S. Becker, and N. Burgess. 2007. "Remembering the past and imagining the future: A neural model of spatial memory and imagery." *Psychological Review* 114:340–375.

Carruthers, P. 2013. "Evolution of working memory." *PNAS* 110:10371–10378.

Cheeseman, J. F., C. D. Millar, U. Greggers, K. Lehmann, M. D. M. Pawley, C. R. Gallistel, G. R. Warman, and R. Menzel. 2014. "Way-finding in displaced clock-shifted bees proves bees use cognitive map." *PNAS* 111:8949–8954.

Cheung, A., M. Collett, T. S. Collett, A. Dewar, F. Dyer, P. Graham, M. Mangan, et al. 2014. "Still no convincing evidence for cognitive for cognitive map use by honeybees." *PNAS* 111:E4396–E4397.

Chittka, L., and K. Geiger. 1995. "Can honey bees count landmarks?" *Animal Behaviour* 49:159–164.

Christophel, T. B., P. C. Klink, B. Spitzer, P. R. Roelfsma, and J.-D. Haynes. 2017. "The distributed nature of working memory." *Trends in Cognitive Sciences* 21:111–124.

D'Esposito, M., and B. P. Postle. 2015. "The cognitive neuroscience of working memory." *Annual Reviews in Psychology* 66:115–142.

Dehaene, S., V. Izard, P. Pica, and E. Spelke. 2006. "Core knowledge of geometry in an Amazonian indigene group." *Science* 311:381–384.

Della Sala, S., C. Gray, A. Baddeley, N. Allamano, and L. Wilson. 1999. "Pattern span: A tool for unwelding visuo-spatial memory." *Neuropsychologia* 37:1189–1199.

Foster, D. J., and M. A. Wilson. 2006. "Reverse replay of behavioural sequences in hippocampal place cells during the awake state." *Nature* 440:680–683.

Fuster, J. M. 2001. "The prefrontal cortex—An update: Time is the essence." *Neuron* 30:319–333.

Fuster, J. M. 2004. "Upper processing stages of the perception-action cycle." *Trends in Cognitive Sciences* 8:143–145.

Gallistel, C. R. 1990. *The organization of learning.* Cambridge, MA: MIT Press.

Gillner, S., and H. A. Mallot. 1998. "Navigation and acquisition of spatial knowledge in a virtual maze." *Journal of Cognitive Neuroscience* 10:445–463.

Goldman-Rakic, P. S. 1995. "Cellular basis of working memory." *Neuron* 14:477–485.

Gonsek, A., M. Jeschke, S. Rönnau, and O. J. N. Bertrand. 2021. "From paths to routes: A method of path classification." *Frontiers in Behavioral Neuroscience.* doi:10.3389/fnbeh.2020.610560.

Grochulla, B., and H. A. Mallot. 2022. "Presence and perceived body orientation affect the recall of out-of-sight places in an immersive sketching experiment." *bioRxiv.* doi:10.1101/2022.10.25.513723.

Guariglia, C., L. Palermo, L. Piccardi, G. Iaria, and C. Incoccia. 2013. "Neglecting the left side of a city square but not the left side of its clock: Prevalence and characteristics of representational neglect." *PLoS ONE* 8 (7): e67390.

Hartley, T., E. A. Maguire, H. J. Spiers, and N. Burgess. 2003. "The well-worn route and the path less traveled: distinct neural bases of route following and wayfinding in humans." *Neuron* 37:877–888.

Hauser, M. D., N. Chomsky, and W. T. Fitch. 2002. "The faculty of language: What is it, who has it, and how did it evolve?" *Science* 298:1569–1579.

Hilton, C., J. Wiener, and A. Johnson. 2021. "Serial memory for landmarks encountered during route navigation." *Quarterly Journal of Experimental Psychology* 74:2137–2153.

Hoinville, T., and R. Wehner. 2018. "Optimal multiguidance integration in insect navigation." *PNAS* 115:2824–2829.

Hurlebaus, R., K. Basten, H. A. Mallot, and J. M. Wiener. 2008. "Route learning strategies in a virtual cluttered environment." *Lecture Notes in Computer Science* 5248:105–120.

Johnson, A., and A. D. Redish. 2007. "Neural ensembles in CA3 transiently encode paths forward of the animal at a decision point." *Journal of Neuroscience* 27:12176–12189.

Julian, J. B., A. T. Keinath, S. A. Marchette, and R. A. Epstein. 2018. "The neurocognitive basis of spatial reorientation." *Current Biology* 28:R1059–R1073.

Kandel, E. R., Y. Dudai, and M. R. Mayford. 2014. "The molecular and systems biology of memory." *Cell* 157:163–186.

Karnath, H.-O., and M. Dieterich. 2006. "Spatial neglect—a vestibular disorder?" *Brain* 129:293–305.

Kelly, J. W., M. N. Avraamides, and J. M. Loomis. 2007. "Sensorimotor alignment effects in the learning environment and in novel environments." *Journal of Experimental Psychology: Learning, Memory, and Cognition* 33:1092–1107.

Kempermann, G. 2019. "Environmental enrichment, new neurons, and the neurobiology of individuality." *Nature Reviews Neuroscience* 20:236–245.

Kessels, R. P. C., M. J. E. van Zandvoort, A. Postma, L. J. Kappelle, and E. H. F. de Haan. 2000. "The Corsi block-tapping task: Standardization and normative data." *Applied Neuropsychology* 7:252–258.

Kim, M., and E. A. Maguire. 2018. "Hippocampus, retrosplenial and parahippocampal cortices encode multicompartment 3D space in a hierarchical manner." *Cerebral Cortex* 28:1898–1909.

Kim, S. S., H. Rounault, S. Druckmann, and V. Jayaraman. 2017. "Ring attractor dynamics in the *Drosophila* central brain." *Science* 356:849–853.

Klatzky, R. L. 1998. "Allocentric and egocentric spatial representations: Definitions, distinctions, and interconnections." *Lecture Notes in Artificial Intelligence* 1404:1–17.

Kohler, M., and R. Wehner. 2005. "Idiosyncratic route-based memories in desert ants, *Melophorus bagoti:* How do they interact with path-integration vectors?" *Neurobiology of Learning and Memory* 83:1–12.

Kuipers, B. 1978. "Modeling spatial knowledge." *Cognitive Science* 2:129–153.

Kuipers, B. 2000. "The spatial semantic hierarchy." *Artificial Intelligence* 119:191–233.

Le Vinh, L., A. Meert, and H. A. Mallot. 2020. "The influence of position on spatial representation in working memory." *Lecture Notes in Artificial Intelligence* 12162:50–58.

Levinson, S. C. 2003. *Space in language and cognition.* Cambridge: Cambridge University Press.

Levitt, T. S., and D. T. Lawton. 1990. "Qualitative navigation for mobile robots." *Artificial Intelligence* 44:305–360.

Loomis, J. M., R. L. Klatzky, and N. A. Giudice. 2013. "Representing 3D space in working memory: Spatial images from vision, hearing, touch, and language." In *Multisensory imagery: Theory and applications,* edited by S. Lacey and R. Lawson, 131–156. New York: Springer.

Lorenz, K. 1966. *On aggression.* London: Routledge.

Ma, W. J., M. Husain, and P. M. Bays. 2014. "Changing concepts of working memory." *Nature Neuroscience* 17:347–356.

Maguire, E., D. Gadlan, I. Johnsrude, C. Good, J. Ashburner, R. Frackowiak, and C. Frith. 2000. "Navigation-related structural changes in the hippocampi of taxi drivers." *PNAS* 97 (8): 4298–4403.

Majid, A., M. Bowerman, D. B. M. Haun, and S. C. Levinson. 2004. "Can language restructure cognition? The case for space." *Trends in Cognitive Sciences* 8:108–114.

Mallot, H. A., G. A. Ecke, and T. Baumann. 2020. "Dual population coding for path planning in graphs with overlapping place representations." *Lecture Notes in Artificial Intelligence* 12162:3–17.

Mallot, H. A., and S. Gillner. 2000. "Route navigation without place recognition: What is recognized in recognition–triggered responses?" *Perception* 29:43–55.

Marchette, S. A., L. K. Vass, J. Ryan, and R. A. Epstein. 2014. "Anchoring the neural compass: coding of local spatial reference frames in human parietal lobe." *Nature Neuroscience* 17:1598–1606.

May, M. 2004. "Imaginal perspective switches in remembered environments: Transformation vs interference accounts." *Cognitive Psychology* 48:163–206.

Meilinger, T. 2008. "The network of reference frames theory: A synthesis of graphs and cognitive maps." *Lecture Notes in Artificial Intelligence* 5248:344–360.

Meilinger, T., J. Frankenstein, N. Simon, H. H. Bülthoff, and J.-P. Bresciani. 2016. "Not all memories are the same: Situational context influences spatial recall within one's city of residency." *Psychonomic Bulletin & Review* 23:246–252.

Meilinger, T., M. Knauff, and H. H. Bülthoff. 2008. "Working memory in wayfinding—A dual task experiment in a virtual city." *Cognitive Science* 32:755–770.

Meilinger, T., M. Strickrodt, and H. H. Bülthoff. 2018. "Spatial survey estimation is incremental and relies on directed memory structures." *Lecture Notes in Artificial Intelligence* 11034:27–42.

Mellet, E., S. Bricogne, F. Crivello, B. Mazoyer, M. Denis, and N. Tzourio-Mazoyer. 2002. "Neural basis of mental scanning of a topographic representation built from a text." *Cerebral Cortex* 12:1322–1330.

Miller, G. A., E. Galanter, and K. H. Pribram. 1960. *Plans and the structure of behavior*. New York: Holt, Rinchart and Winston, Inc.

Minatohara, K., M. Akiyoshi, and H. Okuno. 2016. "Role of immediate-early genes in synaptic plasticity and neuronal ensembles underlying the memory trace." *Frontiers in Molecular Neuroscience* 8:78.

Mou, W. M., and T. P. McNamara. 2002. "Intrinsic frames of reference in spatial memory." *Journal of Experimental Psychology: Learning, Memory and Cognition* 28:162–170.

Neumann, J. von. 1958. *The computer and the brain*. New Haven, CT: Yale University Press.

Newman, E. L., J. B. Caplan, M. P. Kirschen, I. O. Korolev, R. Sekuler, and M. J. Kahana. 2007. "Learning your way around town: How virtual taxicab drivers learn to use both layout and landmark information." *Cognition* 104:231–253.

O'Keefe, J., and L. Nadel. 1978. *The hippocampus as a cognitive map*. Oxford: Clarendon.

Olton, D. S., and B. C. Papas. 1979. "Spatial memory and hippocampal function." *Neuropsychologica* 17:6669–682.

Pfeiffer, B. E. 2017. "The content of hippcampal 'replay'." *Hippocampus* 30:6–18.

Piccardi, L., A. Berthoz, M. Baulac, M. Demos, S. Dupont, S. Samson, and C. Guariglia. 2010. "Different spatial memory systems are involved in small and large-scale environments: Evidence from patients with temporal lobe epilepsy." *Experimental Brain Research* 206:171–177.

Ponulak, F., and J. J. Hopfield. 2013. "Rapid, parallel path planning by propagating wavefronts of spiking neural activity." *Frontiers in Computational Neuroscience* 7:98.

Riecke, B. E., and T. P. McNamara. 2017. "Where you are affects what you can easilys imagine: Environmental geometry elicits sensorimotor interference in remote perspective taking." *Cognition* 169:1–14.

Röhrich, W., G. Hardiess, and H. A. Mallot. 2014. "View-based organization and interplay of spatial working and longterm memories." *PLoS ONE* 9 (11): e112793.

Rolls, E. T. 2021. "Neurons including hippocampal spatial view cells, and navigation in primates including humans." *Hippocampus* 31:593–611.

Röser, A., G. Hardiess, and H. A. Mallot. 2016. "Modality dependence and intermodal transfer in the Corsi spatial sequence task: Screen vs. floor." *Experimental Brain Research* 234:1849–1862.

Schinazi, V. R., T. Thrash, and D.-R. Chebat. 2016. "Spatial navigation by congenitally blind individuals." *WIREs Cognitive Science* 7:37–58.

Schölkopf, B., and H. A. Mallot. 1995. "View-based cognitive mapping and path planning." *Adaptive Behavior* 3:311–348.

Schweizer, K., T. Herrmann, G. Janzen, and S. Katz. 1998. "The route direction effect and its constraints." *Lecture Notes in Computer Science* 1404:19–38.

Shelton, A. L., and T. P. McNamara. 2001. "Systems of spatial reference in human memory." *Cognitive Psychology* 43:274–310.

Siegel, A. W., and S. H. White. 1975. "The development of spatial representations of large-scale environments." *Advances in Child Development and Behavior* 10:9–55.

Sperling, G. 1960. "The information available in brief visual presentations." *Psychological Monographs: General and Applied* 74:1–29.

Spiers, H. J., and E. A. Maguire. 2008. "The dynamic nature of cognition during wayfinding." *Journal of Environmental Psychology* 28:232–249.

Squire, L. R., and B. J. Knowlton. 1995. "Memory, hippocampus, and brain systems." In *The cognitive neurosciences,* edited by M. S. Gazzaniga, 825–837. Cambridge, MA: MIT Press.

Steck, S. D., and H. A. Mallot. 2000. "The role of global and local landmarks in virtual environment navigation." *Presence: Teleoperators and Virtual Environments* 9:69–83.

Strickrodt, M., H. Bülthoff, and T. Meilinger. 2019. "Memory for navigable space is flexible and not restricted to exclusive local or global memory units." *Journal of Experimental Psychology: Learning, Memory, and Cognition* 45:993–1013.

Sutton, R. S., and A. G. Barto. 2018. *Reinforcement learning: An introduction.* 2nd ed. Cambridge, MA: MIT Press.

Tatler, B. W., and M. F. Land. 2011. "Vision and the representation of the surroundings in spatial memory." *Philosophical Transactions of the Royal Society (London) B* 366:596–610.

Tolman, E. C. 1932. *Purposive behavior in animals and men.* New York: The Century Company.

Tolman, E. C. 1948. "Cognitive maps in rats and man." *Psychological Review* 55:189–208.

Tolman, E. C., and C. H. Honzik. 1930. "Insight in rats." *University of California Publications in Psychology* 4:215–232.

Trullier, O., S. I. Wiener, A. Berthoz, and J.-A. Meyer. 1997. "Biologically based artificial navigation systems: Review and prospects." *Progress in Neurobiology* 51:483–544.

Tsutsui, K.-I., K. Oyama, S. Nakamura, and T. Iijima. 2016. "Comparative overview of visuospatial working memory in monkeys and rats." *Frontiers in Systems Neuroscience* 10:99.

Vann, S. D., J. P. Aggleton, and E. A. Maguire. 2009. "What does the retrosplenial cortex do?" *Nature Reviews Neuroscience* 10:792–802.

Voicu, H., and N. Schmajuk. 2002. "Latent learning, shortcuts and detours: A computational model." *Behavioural Processes* 59:67–86.

Waller, D., and Y. Lippa. 2007. "Landmarks as beacons and associative cues: Their role in route learning." *Memory & Cognition* 35:910–924.

Wang, R. F., and J. R. Brockmole. 2003. "Human navigation in nested environments." *Journal of Experimental Psychology: Learning, Memory, and Cognition* 29:398–404.

Wehner, R., M. Boyer, F. Loertscher, S. Sommer, and U. Menzi. 2006. "Ant navigation: One-way routes rather than maps." *Current Biology* 16:75–79.

Wehner, R., and R. Menzel. 1990. "Do insects have cognitive maps?" *Annual Review of Neuroscience* 13:403–414.

Wolbers, T., C. Weiller, and C. Büchel. 2004. "Neural foundations of emerging route knowledge in complex spatial environments." *Cognitive Brain Research* 21:401–411.

Zimmer, H. D. 2008. "Visual and spatial working memory: From boxes to networks." *Neuroscience and Biobehavioral Reviews* 32:1373–1395.

Zylberberg, J., and B. W. Strowbridge. 2017. "Mechanisms of persistent activity in cortical circuits: Possible neural substrates for working memory." *Annual Review of Neuroscience* 40:603–627.

8 Maps and Graphs

In this chapter, we summarize the graph theory of spatial representation as is needed to explain the simple wayfinding behaviors discussed in the previous chapter. More complex spatial behavior based on spatial planning and reasoning will require richer representations of space. We discuss two possible amendments of basic graph models of cognitive space, metric information and the hierarchy of places and regions. Behavioral experiments show that metric information in spatial long-term memory is fragmentary and consists mostly of "local" measurements such as the lengths of place-to-place transitions or the angles at which two connections meet. Global metric embeddings of all known places into a common coordinate frame as are used in robotics (SLAM) do not seem to play a major role in human spatial cognition. In addition to local metric, memory contains a notion of regions: that is, it is organized in a hierarchical, multiscale network whose nodes represent spatial entities with various extent or granularity. Multiscale graphs containing regional nodes in addition to a basic layer of places simplify long-distance planning and may even emulate some properties of global metric maps.

8.1 Spatial Problem Solving

In the previous chapter, we discussed the spatial information needed for the reproduction of known routes and the inference of novel routes from intersecting known ones. As a result, the ideas of wayfinding and "topological navigation" appeared as special cases within the larger domain of performances known as problem solving or "qualitative reasoning" (Kuipers 1978, 2000). Consider as an example the well-known Tower of Hanoi game where disks with various diameters have to be stacked on rods, such that smaller disks are always supported by larger ones or by the ground plane. The goal is to collect all disks on one rod while observing the size rule. In this game, every allowed arrangement of disks (i.e., all arrangements with smaller disks on top of larger ones) is a "state," and every allowed move will lead to another such state. For a Tower of Hanoi game with three disks and rods, $3^3 = 27$ possible states exist that are connected by the moves in a characteristic pattern known as the Hanoi graph. The problem can be formulated as searching a path in this graph or "problem space," leading from the current state to the goal state. Problem spaces can be formulated for many other fields ranging from cooking recipes to assembly plans for furniture or machinery and on to complex games such as chess or Go.

In the logic of the previous chapter, the problem space or state–action graph is a reference or long-term memory of action alternatives and their expected outcomes. In order to actually produce a behavior, the agent will need a stage for planning a path in the graph and for executing the planned actions (i.e., a working memory). The whole machinery then becomes similar to a finite automaton in machine theory where the state transitions are executed by the stored $(s, \boldsymbol{a}, s')$ associations together with the planning stage. For an elaboration of these ideas in terms of the state–action network as a "world graph" interacting with a vector of drives, see Arbib and Lieblich (1977).

Navigational graphs may occasionally have the clear and well-defined structure as found in the problem space of the Tower of Hanoi game, but they will generally deviate from this pattern in a number of ways. Important problems include directedness, granularity and fuzziness of state definitions, metric, and dimension.

Directedness The Hanoi graph is undirected, meaning that every move can be reverted. In navigational graphs, this is not generally the case, since scenes and landmarks may look quite different when seen from one direction or another. Knowing a route in outbound direction does therefore not necessarily imply the knowledge of the return route. This problem was already discussed in connection with the view graph (figure 7.7), which is a directed graph. Of course, directed graph links also exist in nonspatial problem spaces, for example in a cooking recipe where a step like mixing the ingredients of a dough, say, can not be undone.

Granularity and the fuzziness of states Another difference between navigational and other problem spaces is that the states are often not well defined. If we identify states with recognized places, the ambiguities of place definition discussed in chapter 6 apply. This is in part a problem of granularity and resolution: that is, the representation of individual street junctions, neighborhoods, or entire cities by nodes of the graph. From a neurobiological perspective, the population coding of place by hippocampal place cells adds an element of fuzziness to the notion of a state, which can therefore not be simply identified with the firing of a single place cell. Again, ambiguities of state definitions are not a specialty of navigational problem spaces; in the cooking domain, an example would be the question of whether a steak is still in the state "rare" or already "well done."

Metric One property that does distinguish navigational problem spaces from graphs in other domains is *metric*. Graphs have their own definition of distance, which is basically the minimal number of steps needed to move from one node to another. The "graph distance" between nodes or states is usually an integer number (the number of steps) and is well defined in all problem domains. In spatial problems, it also plays a role, for example, as the depth of planning: that is, the number of steps needed to reach a goal. In addition to graph distance, however, space has another measure of distance—namely, the geometric (usually Euclidean) distance between places. Metric distance is a continuous quantity and obeys the laws of geometry, as will be discussed below. It is easy to see that in labyrinthine alleys, two places may be metrically close but connected only via a winding path such that their graph distance may be large. When treating wayfinding as just another example of problem solving, the possible role of metric distance is neglected. The extent to which metric information is relevant in human wayfinding performances is an issue of ongoing research.

Dimension A final property of navigational spaces that is closely related to the existence of metric distances is *dimension*. Navigational spaces have a dimension, usually two or three, an attribute that graphs are lacking altogether.[1] Indeed, all graphs can be plotted on sheets of paper without losing any information, whereas metric relations in three-dimensional space would be lost in such drawings. Dimension can only be defined in normed vector spaces; it is the size of the largest set of linear independent vectors contained in a space.

To sum up, spatial memories do have properties of mathematical graphs but also other types of structure related to geometry. In this chapter, we will discuss the basic idea of graph-based memories of space as well as possible extensions required to account for aspects of fuzziness, granularity, metric, and hierarchy.

8.2 Graphs: Basic Concepts

8.2.1 Graph Representations

In mathematics, a graph is defined as a pair of two sets, a set of n vertices or nodes, and a set of edges (a.k.a. links, arcs) between the vertices that can be conceived of as a subset of the set of pairs of vertices (i.e., the Cartesian product of the set of vertices with itself). Since we deal with state–action graphs, we denote the vertices as "states" ($\boldsymbol{s}_i$) and the links as "actions" ($\boldsymbol{a}_{ij}$); see figure 8.1a. The case shown in the figure is a directed graph since the link from node $\boldsymbol{s}_i$ to $\boldsymbol{s}_j$, denoted as $\boldsymbol{a}_{ij}$, may exist while the link in the reverse direction, $\boldsymbol{a}_{ji}$, is missing. Note that the means-ends field shown in figure 1.1 is a directed multigraph, since multiple links between the same pair of vertices may exist.

A graph is completely described by its adjacency matrix $\boldsymbol{A}$ (figure 8.1b), which is a square ($n \times n$) matrix with components $A_{ij} = 1$ if link $\boldsymbol{a}_{ij}$ exists and $A_{ij} = 0$ otherwise. The adjacency matrix is symmetric if the graph is undirected. It can be used to derive properties

1. The dimension of a graph is sometimes defined by considering the graph as a polyhedron with the nodes as vertices and the graph links as edges. Its dimension is then the ordinary dimension of the smallest Euclidean space containing this polyhedron. For place graphs, this measure is not related to the number of dimensions of the space containing the places but will mostly reflect their number.

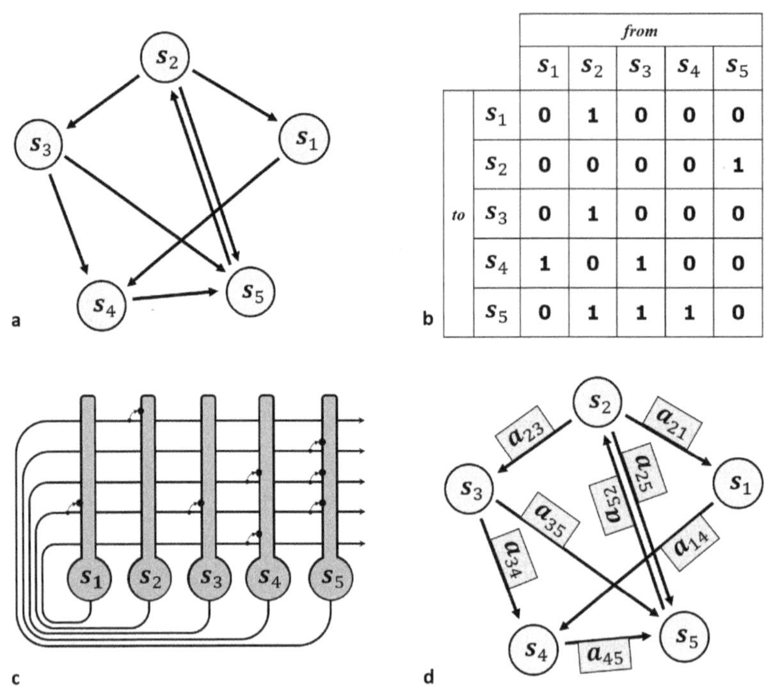

		from				
		s_1	s_2	s_3	s_4	s_5
	s_1	0	1	0	0	0
	s_2	0	0	0	0	1
to	s_3	0	1	0	0	0
	s_4	1	0	1	0	0
	s_5	0	1	1	1	0

Figure 8.1
Graphs and graph representations. (a) A simple directed graph with five nodes and eight arcs or links. (b) Adjacency matrix of this graph shown as a table. Each cell corresponds to one possible link. An entry of 1 means that the link is present, while 0 means that the link is absent. (c) An associative neural network representing the graph. Each node corresponds to a neuron, and the links are represented by synapses. If a neuron is active, activity will spread to the other neurons connected in the graph. (d) A labeled graph. Each link is associated with some additional information such as a motor action required to move from one node (state) to another.

of the graph with the methods of matrix algebra ("spectral graph theory"; e.g., Luxburg 2007). For example, the iterated adjacency matrix $\boldsymbol{A} \times \boldsymbol{A} = \boldsymbol{A}^2$ has nonzero entries for all vertex pairs connected via two steps, a property that generalizes to k-step paths that are represented by the matrix $\boldsymbol{A}^k$. If the agent is at a node i, this can be represented by a canonical unit vector $\boldsymbol{x} = (0, ..., 1, 0, ...0)^\top$ where the 1 occurs at position i. The vector $\boldsymbol{Ax}$ then has nonzero entries for all nodes at which the agent may end up by moving one step while $\boldsymbol{A}^k\boldsymbol{x}$ gives the number of ways in which each node can be reached in exactly k steps. If the start node i is part of a densely connected subset of the graph with only few external connections, the vector $\boldsymbol{A}^k\boldsymbol{x}$ for large k will have nonzero entries for all nodes in the highly interconnected area. Put more generally, the eigenvectors of the adjacency matrix can thus be used to infer highly interconnected subsets or "cliques" of the graph.

A common way to represent graphs as neural networks is shown in figure 8.1c. Each vertex is identified with one neuron and connects to the other neurons as specified by the graph links. Neural networks are per se directional networks; for the representation of an undirected graph, a pair of reciprocal synapses between any two connected neurons would therefore be required. Neural networks implementing directed or undirected graphs have been suggested by various authors (e.g., Schmajuk and Thieme 1992; Schölkopf and Mallot 1995; Muller, Stead, and Pach 1996); for review, see Trullier et al. (1997) and Madl et al. (2015). They operate as forward models: that is, they can make predictions about what is going to happen if the agent proceeds along some link from a given node. In this case, activity will spread to the connected units, which can be interpreted as a prediction of what places would be reached. The backward question, which asks for the places that need to be visited in order to reach a prescribed goal, is not easily answered with this kind of network. We will come back to this planning problem below.

Strictly speaking, the graph representations of figure 8.1a–c are not state–action graphs since the links are not distinguishable actions. All that is known is that a connection exists, and the agent may find it ad hoc, for example, by searching. The logical structure of a full state–action graph requires "labels": that is, specifications of the actions, whose execution results in the transition of each link (figure. 8.1d). The labeling can be thought of as a mapping from the set of links (a subset of the pairs of vertices) into a set of possible actions, such as "move left," "move uphill," or "search for a bridge." One possibility to include labels in neural network models of a graph, again in the logic of a forward model, is shown in figure 7.7d.

8.2.2 Ontology of Graph Vertices

There is no single or unique way to turn knowledge of an environment into a graph. Consider a simple street map as shown in figure 8.2a. At first glance, it might be obvious to use the street junctions as vertices. This approach leads to a graph of places, as illustrated in figure 8.2b. It requires a reliable mechanism of place recognition even when reaching the junction from different sides. Place graphs based on snapshot recognition have been used in biomorphic robotics; see, for example, Kuipers (2000), Franz et al. (1998), Hübner and Mallot (2007), and Differt and Stürzl (2021). However, obtaining directional invariance in snapshot recognition is computationally expensive. It may even be detrimental since snapshots are recorded in an egocentric viewing direction and therefore contain implicit information about the agent's pose, which is lost if invariance of recognition is obtained. Rather than deriving direction-invariant place information from directed views, one might therefore use the directed views themselves as graph nodes. This results in a view graph representation, as shown in figure 8.2c (Schölkopf and Mallot 1995). The view graph and

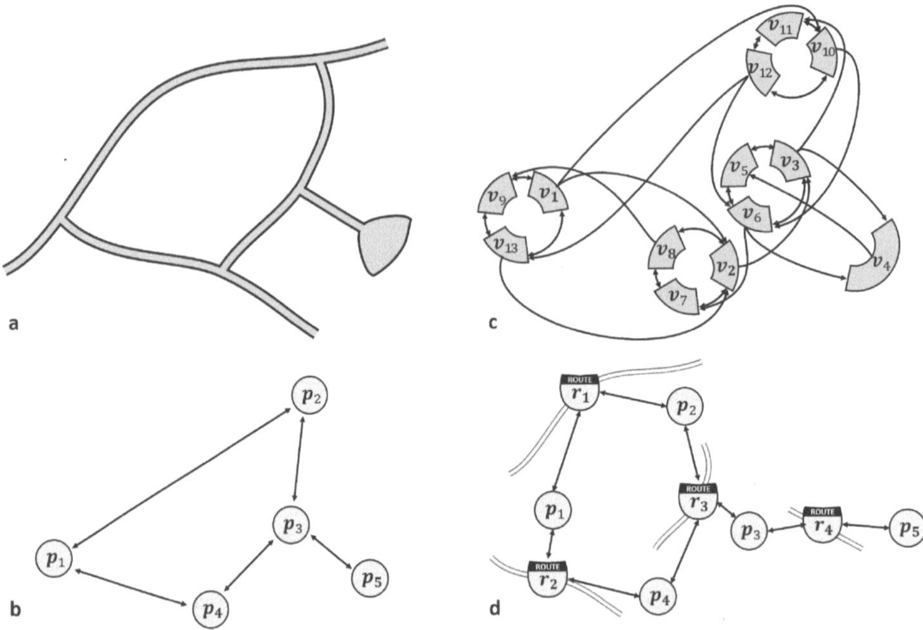

Figure 8.2
Representing environments as graphs. (a) A simple road map with three street junctions and one dead-end place. (b) Representation as a place graph with bidirectional links. Actions must be specified in a world-centered scheme. (c) In the view graph, each node implicitly encodes location and approach direction. Links are directional and actions can be specified in ego- or geo-centric schemes. (d) Bipartite graph of routes and places. Actions are mostly the entering and exiting of the streets.

its possible implementation in a neural network was already discussed as a model of spatial working memory (see figure 7.7) but may also underlie spatial long-term memories. A robot application using directed views as a basis of spatial memory is the "RatSLAM" algorithm by Wyeth and Milford (2009).

The idea of the view graph leads to the question of what might be the minimal visual information on which spatial graphs can be based. Views may be panoramic and would then comprise relatively large packets of information. They may also be restricted to a smaller angular range and, in the extreme case, be reduced to mere image features. In this case, however, the number of vertices in the graph will get very large, and the problem of aliasing (i.e., the confusion of similar features occurring in different places) will loom. Mallot, Ecke, and Baumann (2020) designed an algorithm based on image features as "mini-snapshots" that solves the aliasing problem by treating each graph node only as partial information of the navigational state. Place is thus represented in a population code over the set of graph nodes. Each activated node then casts a vote on the required motion decision, and these votes determine the final action by a second population coding step, as described in section 7.4.2.

Besides the difficulties with the concepts of snapshot and view, another critique of using "places" as the vertices of space graphs comes from the discussions in chapters 3 and 6. The notion of a place is not well defined and other structures, such as known routes and guidances, region boundaries, junctions and street hubs, or landmarks, might also be included in the graph. This list leads back to the discussion of the legibility of an environment discussed in section 4.3.3 and the work of Kevin Lynch (1960). One example for a graph structure using different types of elements as its nodes is the "skeleton" map suggested by Kuipers, Teucci, and Stankiewicz (2003). It is a bipartite graph: that is, a graph with two types of nodes, in this case places and ways, with connections only between places and ways; see figure 8.2d. The idea is that way nodes may represent easily navigated highways, and route planning mostly amounts to reaching the highway and leaving it again at the appropriate junction. The skeleton reduces the depth of planning required for a given navigation and adds an element of hierarchy to the representation of space. For example, when planning a trip from Chicago to Los Angeles, it might suffice to find the start of Route 66 and follow it to the end without considering the many places along the route as individual navigational steps.

So far, nodes of the graph have always corresponded to places, views, or ways that have been visited by the agent during map learning. Landmarks would show up by their images in a snapshot or as objects in some other form of place code but not as separate vertices of the graph. This is different in robotic SLAM algorithms, which are based on a metric embedding of the landmarks themselves. The result is a plan of landmark positions (occupancy map) from which a graph of navigable space must be derived as a second step. We come back to metric embeddings below.

8.2.3 Planning

Computationally, path planning in a space graph is an instance of a graph search, usually solved by algorithms such as the well-known one by Dijkstra (1959). Intuitively, one might think about the problem by imagining the graph as a network of water pipes. When water is poured into a start node, it will flow along all connected pipes and reach the connected nodes. We may assume that the speed of flow is the same for all tubes (plain graph) or that tube lengths differ as would be specified by distance labels attached to each link. For each node, we record the time taken by the water to arrive (its distance from the start) and the pipe through which it first did so. The node at the origin of this pipe is called the "predecessor" of the current node for the particular search problem. When the water reaches the goal, we can construct the shortest route by concatenating the predecessors backward from the goal.

In the discussed example, we do not need to know in advance how far the goal is away from the start point. If we had a reasonable estimate for this distance, we could also use an alternative approach known as the A^*-algorithm (Hart, Nilsson, and Raphael 1968). For example, we could use the graph distance itself as the distance estimate. Graph distance is the number of links in the shortest possible path between two nodes or, if the links have distance labels, the minimum of the sum of the link distances along all possible paths. In the above water-pouring analogy, graph distance can be measured without knowledge of the optimal path: it simply corresponds to the arrival time of the water wave at the goal. The A^*-algorithm can then be visualized by considering the neighbors of the start node one by one. We pour water into the first of the neighboring nodes and measure its distance to the goal. Next, we pour water into the second node, measuring its distance to the goal and so on. Finally, we pick the one neighboring node whose graph distance to the goal, augmented by the distance from the start to this node (1 in unlabeled graphs), is smallest. We move to this node, make it the new start node, and repeat the whole procedure from there.

In the technical literature, many versions and improvements of these basic ideas have been described. Particularly in the field of spatial information theory, procedures for route planning are studied intensively for traffic and transportation networks; see, for example, Bast et al. (2016). These issues, however, are beyond the scope of this book.

In biological systems, graphs might be identified with neural networks, as depicted in figure 8.1c. In this case, the intuition of water pouring is replaced by waves of activity propagating through the network, either from the start, from the goal, or from both sides. Schölkopf and Mallot (1995), for example, used a version of the A^*-algorithm based on the propagation of neural activity. Other approaches have already been discussed in sections 7.5.2 (Voicu and Schmajuk 2002) and 7.3.2 (Ponulak and Hopfield 2013). In the latter approach, a mechanism called anti-spike-timing-dependent plasticity was postulated to take the role of the predecessors in a Dijkstra-like planning scheme.

8.3 Metric Maps

From Euclid to Minkowski and up to this day, the understanding and modeling of space has been a major driving force in the advancement of mathematical theory. It may therefore come as a a surprise that in a text on spatial cognition, metrics as a central concept of geometry is discussed only in the second to last chapter. The reason for this is, however, that cognitive space is not natural space and that spatial cognition is about the organization of spatial behavior, spatial knowledge, and spatial thinking but not about space itself.

Issues of geometry have been discussed in the context of space perception (including the non-Euclidean structure of visual space) and in path integration, both of which are based on the perception of continuous quantities such as visual distance (depth) or the speed and direction of egomotion. When it comes to wayfinding, however, places are treated as discrete, recognizable entities rather than as mathematical points in a continuous manifold.

Similarly, "ways" are not just continuous trajectories but contain discrete decision points that are indeed their crucial elements. Therefore, topological or graph-theoretic modeling as summarized in the previous section is the natural approach in this case. Metric knowledge of space, as far as it at all exists, is an addition to this topological knowledge, not its basis. In a sense, we thus follow the "Bourbaki" axiomatic foundation of mathematics in which set theory, algebra, and topology logically precede the study of geometry. We now turn to the question of how and to what extent metrical information is included in cognitive representations of space.

8.3.1 Metric and Normed Spaces

In mathematics, a "metric space" is a set M of points p for which a distance function can be defined. A distance function is a function that assigns to each pair of points p, q a nonnegative number $D(p, q)$, which satisfies the following conditions:

1. *Reflectivity:* The distance of a point to itself is zero, and two points are identical if their distance is zero; $D(p, q) = 0 \Longleftrightarrow p = q$ for all $p, q \in M$.
2. *Symmetry:* The distance between two points does not have a direction; $D(p, q) = D(q, p)$ for all $p, q \in M$.
3. *Triangular inequality:* The distance from one point to another cannot be larger than the summed distances measured via a third point; $D(p, q) \leq D(p, s) + D(s, q)$ for all $p, q, s \in M$.

Distance functions may exist for continuous as well as for discrete spaces. For example, in a simple graph structure as discussed in the previous section, a possible definition of the distance of two points (nodes) is the number of steps in the shortest path connecting the two nodes.

Distance functions always imply a system of nested neighborhoods. For a point q, the neighborhood of radius r consists of all points whose distance to q is smaller than r, $N_q(r) = \{p \in M \mid D(p, q) < r\}$. Such neighborhood systems are also called "topologies." A space with a nested neighborhood structure is called a topological space. As pointed out before, graphs have a distance function and are therefore topological spaces. They are not, however, metric spaces, which will be defined below. Therefore, graph navigation is also called "topological" navigation.

The distance function describes one important property of space: that is, the notion of relative nearness and distance ordering in nested neighborhoods. Another property not automatically reflected by this notion, however, is isometry or congruence. In a metric space, we might pick two points p and q and construct a ruler spanning the distance from p to q: that is, a ruler with length $D(p, q)$. If we now move the ruler to a new position, its ends mark two new points, u, v, of which we assume that they are also included in our metric space. Isometry would then obtain if $D(u, v)$ equals $D(p, q)$. We could also produce multiple copies of the ruler and lay them out in a row, thus marking points at distances

$D(p, q)$, $2D(p, q)$, $3D(p, q)$, and so on. This idea is the basis of analytical geometry and the notion of a coordinate system.

Isometry is an obvious part of our standard intuitions about space but is generally not realized in graph representations of space or in maps with variable resolution. It requires first of all that the space contains the endpoints of the ruler at every possible position that it can be moved to. This condition is generally satisfied in continuous spaces but not in graphs. More important, a notion of additivity of points and multiplicity of distances must be provided: when we move the ruler from its original position p to a new position u, the vector addition ("+") and the multiplication with a number ("×") should be defined in a way that allows us to express the endpoint of the ruler, v, as $v = u + q + (-1) \times p$. Spaces in which the addition of elements and the multiplication of an element with a real number are well defined, are called vector spaces. We will therefore write the elements of M as vectors from here on.

If the operation $\times$ is defined for the multiplication with real numbers (i.e., if we can compress or stretch our ruler by arbitrary factors) the resulting vector space can be identified with a number space, $M = \mathbb{R}^n$. The dimension n is the largest number of linearly independent vectors that can be found in M: that is, the size of the largest possible set of vectors $\boldsymbol{v}_i \neq 0$ for which the linear combination $\sum_{i=1}^{n} a_i \boldsymbol{v}_i$ cannot yield 0 as long as none of the coefficients vanishes ($a_i \neq 0$ for all i). This condition implies that for any set of $n+1$ vectors, coefficients can be found such that the sum vanishes. In this case, each vector can thus be expressed as a weighted sum of the n others. For example, in a three-dimensional space, each vector can be expressed as a linear combination of three others, which may thus be chosen as a basis of the vector space.

A metric space with the described isometric structure is called a normed vector space. The distance function can be replaced by the so-called norm, which does not take two points as its arguments, as was the case for the distance function, but merely the difference of these points, which is now well defined by our ruler movements. The norm is thus a function assigning a number $\|\boldsymbol{v}\|$ to each vector $\boldsymbol{v} \in M$; it must satisfy the following conditions:

1. Nonnegativity: $\|\boldsymbol{v}\| \geq 0$ for all $\boldsymbol{v}$ and $\|\boldsymbol{v}\| = 0$ if and only if $\boldsymbol{v} = 0$.
2. Multiplication with a scalar: $\|\lambda \boldsymbol{v}\| = |\lambda| \|\boldsymbol{v}\|$ for all $\lambda \in \mathbb{R}$ and all $\boldsymbol{v} \in M$.
3. Triangular inequality: $\|\boldsymbol{u} + \boldsymbol{v}\| \leq \|\boldsymbol{u}\| + \|\boldsymbol{v}\|$ for all $\boldsymbol{u}, \boldsymbol{v} \in M$.

The distance function is then defined as the norm of the difference, $D(\boldsymbol{u}, \boldsymbol{v}) := \|\boldsymbol{u} - \boldsymbol{v}\|$. In addition to the properties of a general distance function defined above, the norm accounts for the described additivity and multiplication properties.

The most intuitive normed vector space is the well-known Euclidean space with the Euclidean norm

$$\|\boldsymbol{v}\|_E = \left(\sum_{i=1}^{n} v_i^2 \right)^{\frac{1}{2}} \tag{8.34}$$

where n is the dimension of the vector space and the v_i are the components of the vector $\boldsymbol{v}$. Other norms are possible, however, and may even be relevant for spatial cognition. For example, the city block, or Manhattan norm,

$$||\boldsymbol{v}||_C = \sum_{i=1}^{n} |v_i| \tag{8.35}$$

captures path distances in "gridiron" (orthogonal) street plans.

One final concept defined in Euclidean space is the angle between two intersecting lines or between three points, marking an apex and two legs. Some special angles can be defined solely by the norm. For example, two vectors $\boldsymbol{u}, \boldsymbol{v} \neq 0$ are said to be orthogonal if they satisfy the Pythagorean theorem $||\boldsymbol{u}||^2 + ||\boldsymbol{v}||^2 = ||\boldsymbol{u}+\boldsymbol{v}||^2$ or the isosceles triangle condition $||\boldsymbol{u}+\boldsymbol{v}|| = ||\boldsymbol{u}-\boldsymbol{v}||$.[2] For a full-fledged definition of continuous angles, a notion of the cosine function or of an inner product is required (see, e.g., Balestro et al. 2017), which is of course provided in Euclidean space.

In analytical geometry, vectors in a vector space are generally expressed as lists of numbers (i.e., by their coordinates). It is important to note, however, that the definitions of a normed vector space and of the distance and angles given therein are independent of the choice of the coordinate system and that the question of where the "true" coordinate origin should be put is arbitrary. This is obvious for geographical conventions such as placing the zero meridian of the globe in Greenwich or orienting north up in a map. Cognitive maps may also have centers or hubs: for example, the navigator's home place, where spatial knowledge is particularly rich and detailed, but these need not be coordinate centers in the sense of analytical geometry. As was discussed already in section 1.3.3, it is unlikely that the brain represents places by coordinates (i.e., as lists of numbers); rather, known places are entities in their own right just as familiar objects or remembered events. Place recognition is based on the place codes discussed in chapter 6 and on the nearness or adjacency of other recognized places, not on the measurement of coordinates. Cognitive representations, even if they do contain metric knowledge such as distances and angles, might therefore be coordinate free.

8.3.2 Partial Metric Knowledge

The mathematical definitions of metric and normed spaces, as given above, are very clear and allow rigorous distinction between spaces with and without metric information. In cognitive modeling, however, we need to allow for incomplete information in the sense that distances between some points may be known while distances between other points are not. This can be modeled by considering labeled graphs (figure 8.1d), where the labels

2. The Pythagorean notion of orthogonality underlies the use of knotted cords for the construction of right angles practiced already in ancient Egypt. Cords form a loop with twelve equal sections marked by knots. If three sections are laid out along one leg of a triangle and four along the second leg such that the remaining five knots are straightened out, the angle between the three- and four-knot sections is 90 degrees, since $3^2 + 4^2 = 5^2$.

attached to each link contain information about distance, required time of travel, or bearing angles of other points or landmarks as seen when traveling along the link. Such information might be stored separately for each link, without checking for geometric requirements such as the triangular inequality. For example, distance labels in a triangular subgraph might read "1 ; 2 ; 4," although no triangle with these side lengths can exist, because the sum of the two shorter sides is less than the longer side. The storage of metric information without embedding into a global metric space has been called "local metric" by Foo et al. (2005).

Before we turn to the empirical evidence for the existence of (local or global) metrics in cognitive space, we discuss the idea of the metric embedding of a labeled graph. Metric embedding means that coordinate values are computed for each graph node, such that the distances and maybe angles resulting from these computed coordinates most closely equal the originally measured distances and angles. This is a question studied in several areas of mathematics and computer science, including distance geometry, multidimensional scaling (MDS), or simultaneous localization and mapping (SLAM). In coordinate-free representations, these methods may be used to infer missing distance measurements from known ones or to correct noisy measurements based on metric constraints.

Multidimensional scaling MDS (Borg and Groenen 1997; Mardia, Kent, and Bibby 1979) is based on a subset of a set of $n \times n$ similarities or distances d_{ij} between a set of n points. The distances should be symmetric or, if they are not, can be replaced by their average, $(d_{ij} + d_{ji})/2$. Similarities are considered the inverse of distances and allow the computation of a metrically ordered representation of objects, say, from a matrix of confusion probabilities of all object pairs in an object recognition task. If confusion is likely, similarity is high and distance should be short. Here, we assume that the d_{ij} are perceived distances between places, as might be obtained from path integration.

In the global metric embedding, each place i is assumed to have coordinates in p-dimensional Euclidean space, $\boldsymbol{x}_i = (x_{i1}, \ldots, x_{ip})^\top$, which are initially unknown. The coordinates of all points are jointly referred to as the configuration. In metric MDS,[3] the so-called stress function s_p

$$s_p^2(\boldsymbol{x}_1, \ldots, \boldsymbol{x}_n) = \sum_{i \neq j} (\|\boldsymbol{x}_i - \boldsymbol{x}_j\|_p - d_{ij})^2 \tag{8.36}$$

is minimized in the $p \times n$ coordinate values of all points in the configuration. $\|\cdot\|_p$ denotes the Euclidean norm in p-dimensional space. The resulting values of the $\boldsymbol{x}_i$ are the sought coordinates in the embedding. Note that stress s can always be made to vanish if all triples of distance measurements (d_{ij}, d_{jk}, d_{ik}) satisfy the triangular inequality and $p \geq n-1$. Of

3. In addition to metric MDS, there is also a procedure known as classical MDS (Mardia, Kent, and Bibby 1979) for which an analytical solution is known. It minimizes a somewhat different error function known as "strain." Classical MDS is related to principal component analysis and finds axes of largest extension (variance) in the configuration.

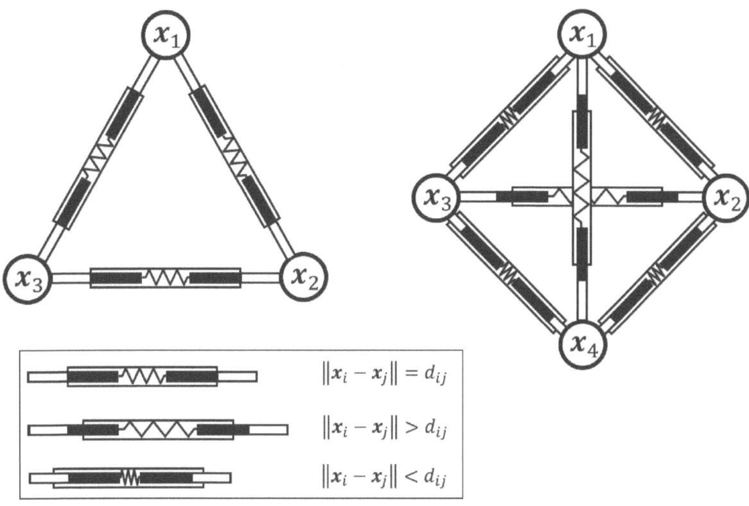

Figure 8.3
Multidimensional scaling. Measured distances d_{ij} are shown as extendable telescopic bars working both ways against a spring. For simplicity, we assume that all d_{ij} are identical. Left: With just three points, a perfect embedding is possible and leads to an isometric triangle. Right: If a fourth point is added, a distance-true embedding in the two-dimensional plane is no longer possible. MDS thus stretches two connections (the diagonal ones) and compresses the others. Note that other solutions with equal stress are obtained by permuting the nodes. This is an idiosyncrasy of this example and goes away if the d_{ij} are not strictly equal. Note also that a true embedding would be possible in three-dimensional space, resulting in a regular tetrahedron. For an elaboration of the spring analogy, including an extension to rotational springs, see Golfarelli, Maio, and Rizzi (2001).

course, for our discussion of representations of space, we will assume $p=2$ or $p=3$: that is, embeddings will always be in two- or three-dimensional space. In general problems of representation, such as the example of object representation mentioned above, the most interesting question is usually how many dimensions are needed to allow for a reasonable embedding.

Figure 8.3 illustrates the optimization process in MDS. Measured distances are shown as extendable telescopic bars connected to the points between which distances have been measured. If only three points are considered, a planar embedding is possible as long as the triangular inequality holds. In this case, the telescopic bars are neither extended nor compressed, and the stress would be zero. If a configuration of four points is forced into a planar arrangement, some of the telescopic bars need to be extended while others are compressed. The configuration will be characterized by a minimized, but not vanishing, stress. This minimum might be local: that is, the optimization process might be caught in a nonoptimal configuration.

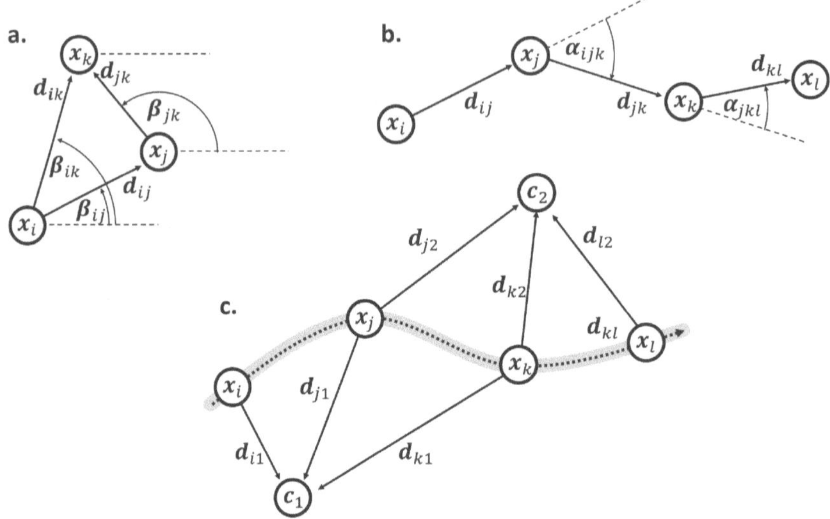

Figure 8.4
Other types of metric embedding. (a) Compass bearings β_{ij} and distances d_{ij} of target locations $\boldsymbol{x}_j$ are measured at various viewpoints $\boldsymbol{x}_i$. Bearings are defined relative to a global compass direction indicated by the dashed lines (true east). The reference direction is chosen as the positive x-direction, in agreement with equations 8.37 and 8.39 and the mathematical angle convention. (b) Local angles and traveled distances. The agent travels from $\boldsymbol{x}_i$ to $\boldsymbol{x}_l$ and measures turns (heading changes) α_{ijk} together with traveled distances d_{ij}. The procedure is similar to path integration without a compass but allows better metric embeddings if places are recognized and visited repeatedly. See equation 8.44. (c) Simultaneous localization and mapping (SLAM). At various positions $\boldsymbol{x}_i$ along its trajectory, the robot measures distances d_{ij} to a set of recognizable cues appearing at locations $\boldsymbol{c}_j$. The unknown cue locations and points of measurement are determined by error minimization.

MDS embeddings are not unique but are defined only up to a shift and an orthonormal transformation (i.e., rotation and mirroring). They can be calculated also if distance measurements for some pairs of points are missing. Even for a plain, unlabeled graph, MDS embeddings can be calculated by setting the distance between connected points to 1 and all other distances to some large number, or by using the graph distance as an estimate.

Multidimensional scaling in based on distances between visited and recognizable points. In other embedding procedures, additional types of local metric information can be taken into account, in particular compass bearings, angles formed in triples of points without reference to a compass (e.g., heading changes), and the distance and bearing of nonvisited landmarks. Angular information with and without compass plays a role in experiments on human map behavior, while mapping by multiple off-trajectory cues is a standard approach in robotics. We will briefly go through these procedures in the sequel.

Including compass bearings Assume an observer endowed with a compass walks around in an environment, measuring the bearings of a set of recognized landmarks as angles relative to true east[4] as provided by the compass (see figure 8.4a). If enough measurements are collected, it will be possible to localize the landmarks and viewpoints by way of triangulation. In psychological measurements, subjects may be mistaken about the true east direction, but bearing judgments may still be obtained. The mental map on which these judgments are based, however, may be distorted in the sense that assumed east may actually be different in different parts of the representation. Even in this case, metric embeddings from bearing data may be interesting. For example, they can tell us if mental representations, however distorted, are at least locally consistent and can be "unwrapped" without overlaps.

To obtain such embeddings from bearing estimates, we name all points (i.e., viewpoints and landmarks) jointly as a list of vectors $(\boldsymbol{x}_i, i = 1, ..., n)$. For some subset S of the set of pairs $\{(i,j)\}$, we assume to have bearing measurements β_{ij}, taken relative to the participant's subjective compass. In the final embedding, the coordinates of the points involved in a bearing measurement β_{ij} should satisfy the condition

$$\frac{\boldsymbol{x}_j - \boldsymbol{x}_i}{\|\boldsymbol{x}_j - \boldsymbol{x}_i\|} = \begin{pmatrix} \cos \beta_{ij} \\ \sin \beta_{ij} \end{pmatrix} =: \boldsymbol{b}_{ij} \tag{8.37}$$

where $\boldsymbol{b}_{ij}$ denotes the unit vector in direction β_{ij}. An objective function f for angular scaling can thus be formulated as

$$f(\boldsymbol{x}_1, \ldots, \boldsymbol{x}_n) = \sum_{(i,j)\in S} \left(\frac{\boldsymbol{x}_j - \boldsymbol{x}_i}{\|\boldsymbol{x}_j - \boldsymbol{x}_i\|} - \boldsymbol{b}_{ij} \right)^2 = \sum_{(i,j)\in S} 2 - 2\frac{((\boldsymbol{x}_j - \boldsymbol{x}_i) \cdot \boldsymbol{b}_{ij})}{\|\boldsymbol{x}_j - \boldsymbol{x}_i\|}, \tag{8.38}$$

where $(\cdot)$ in the second sum denotes the dot product. Minimization of f in the configuration $(\boldsymbol{x}_1, \ldots, \boldsymbol{x}_n)$ is unique up to overall scaling and two-dimensional shift. These ambiguities can be removed by choosing $\boldsymbol{x}_1 = 0$ and $\|\boldsymbol{x}_2\| = 1$. Minimization of f in equation 8.38 has been suggested by Wender, Wagener-Wender, and Rothkegel (1997) and used to model pointing data. An extension considering also measurements of point-to-point distances (d_{ij}) has been given by Waller and Haun (2003).

It seems to have been overlooked so far that the problem of joint distance and bearing scaling has a simple analytical solution based on linear regression, which does not require a minimization procedure. To see this, we observe that in the presence of distance estimates d_{ij}, the constraints from equation 8.37 become

$$\boldsymbol{x}_j - \boldsymbol{x}_i = d_{ij}\boldsymbol{b}_{ij} = (d_{ij} \cos \beta_{ij}, d_{ij} \sin \beta_{ij})^\top. \tag{8.39}$$

4. We express angles with respect to east (positive x-direction in a coordinate system) rather than north to satisfy the mathematical angle convention in which the vector $(\cos \phi, \sin \phi)^\top$ is said to point in direction ϕ. Angles are thus measured from the positive x-axis in the counterclockwise sense. By coincidence, taking east as a reference is in agreement with the etymology of the word "orientation," which is derived from medieval maps in which east ("the Orient") was represented up. In figure 8.3 and elsewhere, we still draw the east direction rightward.

Assume now that we have K measurements for pairs (i, j). Let the indices of the two points involved in the kth measurement be $(i(k), j(k))$. We renumber the measurements as d_k and β_k and collect them into a $K \times 2$ matrix

$$\boldsymbol{B} = \begin{pmatrix} d_1 \cos \beta_1 & d_1 \sin \beta_1 \\ \vdots & \vdots \\ d_K \cos \beta_K & d_K \sin \beta_K \end{pmatrix}. \tag{8.40}$$

Next we define a $n \times 2$ matrix $\boldsymbol{X}$ whose ith row holds the sought point coordinates $\boldsymbol{x}_i^\top = (x_{i,1}, x_{i,2})$. Finally, we need a $K \times n$ design matrix $\boldsymbol{G}$ with the coefficients

$$g_{kl} = \begin{cases} 1 & \text{if } l = j(k) \\ -1 & \text{if } l = i(k) \\ 0 & \text{otherwise} \end{cases}, \tag{8.41}$$

and $l = 1, ..., n$. That is, each row of the matrix has exactly one coefficient +1 marking the target of the measurement and one coefficient –1 marking the start point of the measurement. With these conventions, we can write the constraints from equation 8.39 as

$$\boldsymbol{B} = \boldsymbol{G}\boldsymbol{X}. \tag{8.42}$$

In order to find the least squares solution for $\boldsymbol{X}$, we note that it will be defined only up to a global shift. We therefore choose $\boldsymbol{x}_1 = (0, 0)^\top$ and delete the first row of matrix $\boldsymbol{X}$ and the first column of matrix $\boldsymbol{G}$. With the reduced matrices $\boldsymbol{G}$ and $\boldsymbol{X}$, we can now calculate the best-fitting configuration with the pseudo-inverse as

$$\boldsymbol{X}^* = \left(\boldsymbol{G}^\top \boldsymbol{G}\right)^{-1} \boldsymbol{G}^\top \boldsymbol{B}. \tag{8.43}$$

This approach can also be used in the absence of distance estimates by setting all $d_{ij} = 1$. The algorithm will then try to find a configuration with equal distances.

Including local angles The scaling of compass bearings as described in equation 8.38 depends on the assumption of a global reference direction relative to which bearings are measured. The estimate of the reference direction may be prone to errors, but it is included as a global element in the scaling procedure. If it is omitted, the only remaining way to include angular information is by considering three points of the configuration as a triangle with its included angles. An algorithm for joint angular and distance embedding based entirely on this type of local information has been presented by Hübner and Mallot (2007). In this approach, error measures and constraints are based not just on two points but on three, forming a local triangle. As the agent proceeds through a sequence of points numbered (i, j, k), it takes local measurements of the distances d_{ij} and d_{jk} for the first and second steps, as well as of the required heading change α_{ijk}; see figure 8.4b. We allow $i = k$ (i.e., walking back and forth between two points) in which case the angle α_{iji} is set to 180 degrees. These measurements can be based on pure “idiothetic” path integration as described in chapter 5 and do not rely on any kind of compass or global reference system. The embedding is then based on the isometry of all triangles for which the local measurements are available.

For a formal description, we consider all triplets $T = \{(i,j,k)\}$ that the agent has explored, together with the related measurements $(d_{ij}, d_{jk}, \alpha_{ijk})$. Each point may appear many times in this list. The cost function can then be written as

$$
\begin{aligned}
& f(\boldsymbol{x}_1, \ldots, \boldsymbol{x}_n) = \\
& \sum_{(i,j,k)\in T} \lambda_1 \left[((\boldsymbol{x}_j - \boldsymbol{x}_i) \cdot (\boldsymbol{x}_j - \boldsymbol{x}_k)) - d_{ij} d_{jk} \cos \alpha_{ijk}\right]^2 + \\
& \qquad \lambda_2 \left[((\boldsymbol{x}_j - \boldsymbol{x}_i) \otimes (\boldsymbol{x}_j - \boldsymbol{x}_k)) - d_{ij} d_{jk} \sin \alpha_{ijk}\right]^2 .
\end{aligned}
\tag{8.44}
$$

As before, $(\cdot)$ denotes the dot product, while $(\otimes)$ is used here to denote the third component of the cross-product, $(\boldsymbol{a} \otimes \boldsymbol{b}) := a_1 b_2 - a_2 b_1$; it equals twice the area of the triangle (i,j,k). The terms $d_{ij} d_{jk} \cos \alpha_{ijk}$ and $d_{ij} d_{jk} \sin \alpha_{ijk}$ are estimates of the same quantities based on the local measurements. The constants λ_1 and λ_2 are used to weight the two components of the cost function based on their variances. Note that the triplets of the type (i,j,i) will show only in the first component of the sum, since the included angles were set to 180 degrees such that $\sin \alpha_{iji} = 0$ in these cases, and the cross-product involving two parallel vectors will also yield zero.

Simultaneous localization and mapping Metric embedding is most intensively studied in robotics as part of the theory of simultaneous localization and mapping (SLAM; for an introduction, see Durrant-Whyte and Bailey 2006). While the general idea is similar to the procedure discussed so far, there are a number of differences that are mostly due to the requirements arising in robot applications.

In the simplest case depicted in figure 8.4c, SLAM is based on a set of distance measurements taken from the agent to environmental features or cues that can be recognized from multiple positions along the agent's path. These distance measurements are often obtained using laser range finders and are therefore much more accurate than comparable estimates made by human observers. The unknowns are given by the set of cue locations together with the positions along the trajectory from which the measurements are taken. Since all measurements are between the robot and a landmark, the set of distance estimates is not complete. Still, a cost function can be formulated in the style of the MDS stress given in equation 8.36. The solution specifies both the landmark positions and the agent's trajectory (hence the name of the procedure) and can thus be used also for path integration or, to use the technical term from robotics, odometry.

This simple scheme has been elaborated in many ways. As in the biological examples discussed above, landmark bearings and egomotion cues can be included as additional information. Statistical modeling of the errors in the various measurements and in the recognition of the landmarks can be used to further optimize the embedding. Of course, in robotic applications, the goal will be to obtain the best possible representation of the environment. This strength of the SLAM theory may turn into a weakness when it comes to cognitive modeling of human metric navigation. Biological models need to be able to accommodate the substantial errors that humans make and to understand the mechanisms whose application leads to the observed suboptimal performance.

8.3.3 Metric Embedding in Humans?

Metric information is clearly present in path integration as well as in some aspects of spatial working memory: for example, those that are probed with the judgment of relative direction task. Here we address the role of metric in other representations: namely, the spatial problem graphs discussed in the previous section. It is important to note that metric information may in principle be present in some spatial representations while it is lacking in others. Therefore, the ability of an agent to perform path integration does not imply that all its representations of space will automatically share this metric structure. Indeed, we will see in this section that the metric knowledge in spatial long-term memory is incomplete in systematic ways.

Asymmetric distance estimates One required property of a metric representation is the symmetry of its distance function: that is, the distance from A to B should always equal the distance from B to A. McNamara and Diwadkar (1997) tested symmetry of remembered distances by having subjects study and memorize a sheet of paper showing an irregular maplike arrangement of dots, each labeled with the name of an object, as shown in figure 8.5a. Four of the location names were printed in uppercase; these could be used as "landmarks" structuring the other places into four regions. Subjects learned the arrangement by filling in the landmark names on sheets that showed only the four landmark locations and subsequently filling in positions and names of the nonlandmarks (figure 8.5c). In the test phase, subjects were given two place names as the start and the target of a distance estimate, using the known size of the study sheet as a scale. Distances were reported in writing using the inch (2.54 cm) as a unit. If the pairs of names were a landmark together with a nearby place in the same region, distance estimates were unsymmetrical in the sense that the distance from a landmark to a nonlandmark (e.g., SOAP to ring) was consistently reported to be shorter by about 20 percent than the reverse distance (ring to SOAP). The effect was not found for distance estimates between different landmark regions or for long-distance pairs within regions. In an additional experiment, the same task was applied to remembered landmarks and minor buildings on the university campus. Again, the distance from "Great Hall" to "Minor Annex," say, was judged smaller than the distance from the Annex to the Hall, although the results were less clear than in the paper-and-pencil version of the experiment.

The authors discuss the effect in the light of the "implicit scaling theory" of Holyoak and Mah (1982); see also figure 8.5d: the naming of the first place is thought to set the scale of

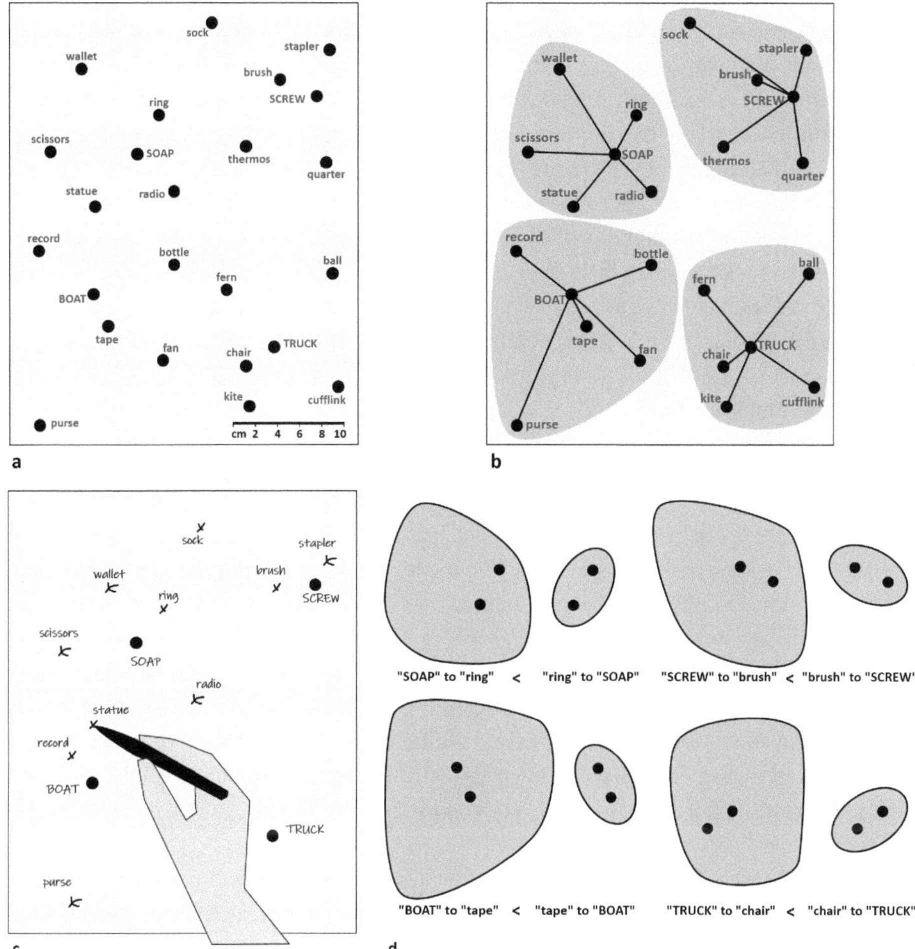

Figure 8.5
Schematic of the experiment on the symmetry of remembered distances by McNamara and Diwadkar (1997). (a) Study sheet with points and place names. (b) Implicit structure of "landmarks" printed in uppercase letters and "nonlandmarks" printed in lowercase. This sheet is not shown to the subjects. (c) Training procedure. Subjects are given a sheet with the landmark positions and are asked to fill in the landmark names. In a second step, they also fill in the positions and names of the nonlandmarks. Training is repeated until positions are correct up to .95 cm. (d) Implicit scaling hypothesis: Landmark to nonlandmark tasks are judged in larger contexts (gray areas), nonlandmark to landmark tasks in smaller ones. Part (a) redrawn from McNamara and Diwadkar (1997).

the representation into which the two places are loaded in order to estimate the distance. If the task starts with a landmark name, this representation is thought to contain the entire region whose diameter is large as compared to the target distance. This is indicated by the large gray areas in figure 8.5d, which are the regions shown also in figure 8.5b. If, however, the task starts with a nonlandmark, the representation will cover only a small part of the map containing just the start and the target (gray ellipses in figure 8.5d), in which the target distance is relatively large. This idea is reminiscent of the wellknown Ebbinghaus (or Titchener) illusion of visual perception in which a circular disc surrounded by a wide ring of larger disks looks smaller than the same disk surrounded by a narrow ring of smaller disks. This interpretation is also consistent with the lack of distance asymmetries between distant places or places in different landmark regions. If the "implicit scaling theory" is correct, it leaves open the possibility that the spatial long-term memory itself does indeed satisfy the symmetry axiom, since the asymmetry results only from the conditions under which this memory is downloaded into a working memory in which the distance judgment is made.

The sum of angles of a triangle The fact that the sum of angles of a triangle is 180 degrees relies on the parallel postulate of Euclidean geometry and may be violated in metric spaces with non-Euclidean distance functions. As an example, consider the surface of a sphere and the geodesic distance function given by the lengths of arcs (sections of great circles) connecting two points on the sphere. The sum of angles of spherical triangles exceeds 180 degrees. The sum of angles of a triangle is therefore less fundamental than the symmetry of distance, which is part of the definition of a distance function and is therefore preserved even in non-Euclidean geometries. Still, its status in human spatial cognition is of great interest since it might indicate a deviation from the Euclidean normed vector space model of space.

Moar and Bower (1983) addressed this question with pointing experiments involving familiar targets in the city of Cambridge, UK. Subjects were instructed to picture themselves at one of three familiar places connected by straightly running streets that roughly form a rectangular triangle. They were then asked to imagine the view toward one of the other corners and mark on a sheet of paper the angle between the so-defined forward direction and the third corner of the street triangle. This procedure was repeated for all corners and a number of other street triads, and the average reported angles were added together. The results show that the sum of angles in the triangles is between about 210 and 250 degrees: that is, significantly larger than the expected 180 degrees. This result may indicate a deviation of spatial long-term memory from Euclidean geometry, but it can also be interpreted as a perceptual bias toward orthogonality in the perception of angles together with local metric encoding in a labeled graph.

Completing learned triangles Clear evidence for the presence of at least some metric knowledge in human spatial long-term memory comes from an experiment by Foo et al. (2005). In a virtual environment, subjects learn a configuration of three places, a home place and two target points. The connections between home and each target were 8 m long, and both ways were trained extensively. Also known from exploration is the angle between the two targets when looking from home. After training, subjects have thus learned two legs of a triangle plus the included angle. If they did compute a metric embedding, they should be able to calculate the third leg connecting the two targets, as well as the angles of this connection with the learned legs, and should therefore be able to find the metric shortcut between the ends of the two legs. Note that the design is similar to triangular completion in path integration. It differs, however, in the learning procedure, which is a repeated and independent study of the two outward legs and the included angle, and the testing task, which is a shortcut between the two target places, not a return to home.

Performance in the shortcut task is clearly not random but also much poorer than performance in the outward legs, which were used as controls. Subjects missed the goal by several meters. Performance can be improved if landmarks are added. In this case, however, subjects start with the same initial error as in the landmark-free condition but are able to correct later when landmarks are recognized. This is clearly visible from abrupt changes of direction in the trajectories as soon as the catchment area of landmark homing is reached. In summary, the results show that some knowledge of the shortcut is present, but it is less than would be expected from true triangulation. The results are thus consistent with the idea of local metrics in a labeled graph: metric labels for the outbound legs and the included angle are available, but overall metric embedding is poor.

Accumulation of local information into global maps The distinction of local and global metric is related to the notion of vista and environmental spaces of Montello (1993), which was briefly mentioned in chapter 1. Vista space is the space that can be overlooked from a given point of view. The objects therein can be localized by stereo vision or other mechanisms of visual depth perception, as described in section 3.2. In contrast, the spatial information in Montello's environmental space is obtained piecemeal by actively walking across and exploring visually disconnected environments such as the rooms and corridors of an apartment or the streets and squares of a city quarter. The construction of environmental space might stop on the level of plain graph structures, or it might collect pieces of metric information from each local vista space into a global map by some sort of metric embedding. We discuss two experiments addressing this issue.

Meilinger, Strickrodt, and Bülthoff (2016) designed two virtual environments with the structure of a vista and an environmental space, respectively. In the vista space condition of the experiment, seven objects were placed on the floor of a large room that was otherwise empty. In the environmental space condition, the objects remained at the exact same position, but the room was partitioned by walls placed around and between the objects such that they now came to lie within a meandering corridor in which only a small subset of the objects was visible at any one time. Subjects explored the environment either by walking through the corridor or by inspecting the open room from a fixed point of view. In the test phase, the corridor walls and objects were removed. Learning success was tested by indicating a position on the floor and asking the subjects to report which object had been placed there. If the criterion was not reached, training was repeated in the same vista or environmental space used also for the initial training. Spatial knowledge as tested by the "what was here?" task was thus the same in both conditions. Subjects were then teleported to one of the object locations and asked to point to one of the other locations, as indicated by an object name. Performance was clearly better (but not faster) in the open room situation, in which the subjects had studied the configuration from a fixed point of view only. The perception of visual space in the vista space condition thus provides a better metric information than the putative embedding of multiple view and egomotion cues obtained in the environmental space condition. Still, performance in the environmental space condition was not random, indicating that some metric structure has been recovered even in this situation.

Another prediction that can be made for metric embedding is that it should improve with an increasing number of measurements available. If an environment is explored repeatedly over an extended period of time, one might expect that metric judgments get more accurate. However, this does not seem to be the case. Ishikawa and Montello (2006) had subjects learn two routes in the suburbs of Santa Barbara, California, by driving them around in a car. The routes were some 2 to 3 km in length and initially unknown to the subjects. During and after each trip, a number of measures of metric knowledge were recorded, including estimates of distance between waypoints, pointings, and sketch-map drawings. The results over ten training sessions show a remarkably low amount of improvement; indeed, the "learning curves" are virtually flat.

The relation of cognitive and actual distance in urban environments is a longstanding issue of research in cognitive geography; see, for example, Golledge (2002) and Montello (2018). Systematic distortions of distance have been reported and are generally compatible with the local metrics idea. In a recent study, Manley, Filomena, and Mavros (2021) compare cognitive distance with Euclidean and route distances in a large sample of major cities worldwide and find a clear correlation even for true distances as large as 2 km. Distances are generally overestimated, and overestimation increases for larger true distances. Again, these results support the idea of partial metric knowledge.

The role of the third dimension was briefly discussed in the context of surface slope and the firing specificities of grid cells in animals exploring three-dimensional volumes or scaffolds (section 6.5.3). Stored knowledge of elevation has been reported in a study by Gärling et al. (1990), who asked subjects to rate the absolute elevation of familiar places in the city of Umeå, Sweden. The places were between 500 m and 4 km apart and were presented in pairs. Subjects were asked to judge which one is more elevated and to estimate the elevational difference in meters. Overall performance was good but subjects with less navigational experience of Umeå (but still knowledgeable of the tested places) consistently overestimated the elevational differences.

Surface slope can also be used as an ordering constraint or compass cue, in particular if the uphill direction marks a fixed compass bearing throughout the environment. This would, for example, be the case at the slopes of a mountain valley. Linguistic evidence for the inclusion of uphill and downhill directions in spatial memory has already been discussed in section 7.2.7. In a navigational study, Restat et al. (2004) tested the wayfinding abilities of human subjects in virtual environments with the same street and landmark pattern appearing either on a flat plane or with slopes in different directions. Navigation was improved in the sloped conditions: that is, subjects were able to build better representations if the environment provided the constant uphill direction as a compass cue. This might be a result of improved metric embedding; however, pointing performance from memory was not improved.

Goal vectors for path planning One performance often taken as evidence for a dense metric map is the straight-line approach of a distant goal from an arbitrary starting point. Straight-line approaches are common in path integration when the goal is represented as a goal vector and the path is found by moving in the direction of this vector while constantly updating it with the actually performed egomotion; see section 5.2. In the case of ant navigation studied there, the goal vector is simply the current content of the path integrator, which the ant decides to reduce by homeward walking as soon as a food item is discovered and picked up. In more elaborate navigational systems, vectors may be associated with recognized places as a type of direction or recognition-triggered response (i.e., as a long-term memory). Storing metric vectors for each pair of known places, however, is not very efficient. It is rather more plausible to assume that metric vectors are inferred from graph memories with local metric knowledge, simply by concatenating the place-to-place vectors along the required path. The resulting vector could then be "loaded" into the path integrator, and the agent may walk so as to reduce it until the goal is reached. The resulting path would not follow the individual graph links but rather a straighter line. It may also go wrong, for example, if a detour required to reach a bridge is left out. In this case, the agent would need to resort to a more step-by-step navigation scheme. In any case, the sketched procedure for vector navigation explains straight-line approaches without assuming a continuous metric map.

In one study, Normand and Boesch (2009) tracked trajectories of wild chimpanzees moving in a forest area of several kilometers in diameter. The chimpanzees regularly visit trees for harvesting various kinds of fruit as soon as these have ripened. They do not follow a stereotyped route but visit the trees in an irregular sequence. The data show straight-line trajectories over several of hundreds of meters across the tropical rain-forest.

Human navigators are very good at finding paths to distant goals in urban environments. In a study evaluating more than half a million trajectories of pedestrian walks recorded via

GPS tracking in Boston and San Francisco, Bongiorno et al. (2021) show that actual paths are somewhat longer than the optimal paths and, more important, that different paths are chosen if origin and destination are swapped. This latter result is not predicted if planning is assumed to take place in a place graph labeled with local distances or in a continuous map. The authors therefore suggest another planning scheme based on a continuously updated estimate of the direction to the goal as described above. Paths are then chosen so as to keep the walking direction close to the direction of the goal. This vector-based strategy does indeed predict the different paths chosen for forward and backward walks since local street directions will force the agent away from the vector and may do so differently when walking a given connection forward or back. In addition, it might also explain the so-called initial segment strategy observed by Bailenson, Shum, and Uttal (2000) in path choice in printed maps: if two paths of equal length are shown on the map, subjects prefer the one starting with a straight initial section (see also section 8.4.2).

Impossible virtual environments Another line of research into the relation of global and local metrics uses impossible environments, in which locally perceived metric relations cannot consistently be embedded into an overall metric map. Of course, experiments of this type cannot easily be carried out with real-world exploration but require virtual reality technology. One real-world example reminiscent of this approach might be a house in which the door of a small room has been bricked and painted over long ago. The room behind the bricked door may be forgotten and overlooked, although it leaves a gap in the plan of the house, but this gap is not apparent to the inhabitants as long as they do not actually start to construct a measured plan. This observation shows an incompleteness of metric cognitive maps that can be studied more systematically with virtual reality technology.

Ruddle et al. (2000) constructed a virtual environment as a five-by-five array of square chambers, connected by "hyperlinks" whose activation promotes the navigator either from one room to an adjacent one or to another room further away, which can otherwise be reached only via several intermediate steps. The experiments were carried out in a desktop setting: that is, the subjects did not actually move and could judge simulated egomotion from their visual input only. When searching for objects distributed across the chambers or repeating routes through a given sequence of chambers, subjects were able to use the long-distance hyperlinks for efficient performance. The results show that metric consistency is not required for the buildup of a spatial representation and thus generally support the graph approach of spatial memory.

As pointed out above, multidimensional scaling and similar methods of metric embedding can be applied to inconsistent distance data and will then try to capture the relations as close as possible in a distorted representation. The idea that human observers embed local metric knowledge by some sort of optimizing process can therefore be tested by the study of shortcuts performed in consistent and metrically impossible virtual environments. Warren et al. (2017) designed a complex virtual maze consisting of a central place, a number of arms radiating from there into all directions, and a small number of peripheral corridors connecting arm endpoints. Two of these peripheral connections could include "wormholes"; that is, hyperlinks that would teleport the subjects over a long distance and thus shorten the perceived distance between two peripheral points. In an immersive virtual reality setup (video goggles and free walking in tracked arena), subjects explored mazes with wormholes or with ordinary corridors between peripheral points. They were then released at one peripheral point. The maze disappeared and subjects were asked to walk on a textured ground plane to the assumed position of a target. The paradigm is similar to the completion of learned triangles discussed above (Foo et al. 2005): learned route segments have to be combined into a metric shortcut. Results show that shortcuts in the presence of wormholes differ from shortcuts produced in the ordinary mazes in the sense that subjects tended to walk closer to the wormhole entries when these were present. This effect is indeed predicted if shortcuts were planned in a distorted representation in which places connected by a wormhole are drawn closer together than their metric distance implies.

To sum up, metric information in human spatial long-term memory seems to be partial and incomplete and can be modeled as a labeled graph in which local distances and angles can be represented but are not checked for global consistency (Mallot and Basten 2009; Warren 2019). The representation does not contain a global coordinate system, which would of course imply a consistent embedding. Still, the distances themselves may be Euclidean, although the overall resulting representation of space is not. Note that this is different from the non-Euclidean properties of visual and peripersonal space discussed in chapter 3. Visual space is continuous and possesses a distance function that, however, deviates from the Euclidean definition. Spatial long-term memory is discrete and does not possess a globally consistent distance function at all.

8.4 Regions and Spatial Hierarchies

In the graphs discussed in section 8.2, nodes may be distinguished by the spatial category they represent (i.e., as views, places, or ways). They are, however, all on the same level of hierarchy such that a place represented by one node will not be part of a region represented by another node. This may be considered a shortcoming since mental representations of space clearly do contain objects such as regions, gateways, or hubs that hierarchically structure large spaces and support wayfinding and spatial planning. In this section, we discuss evidence and models for a hierarchical organization of spatial representations.

8.4.1 Hierarchy in Graphs

Figure 8.6 summarizes different ways in which spatial hierarchies can be encoded in graphs. The most obvious case is the graph-theoretic tree: that is, a graph without cycles in which one node is the "root." This root node is connected to "branches" marking the next lower level in hierarchy. The branches split up into twigs and so on until the lowest level of

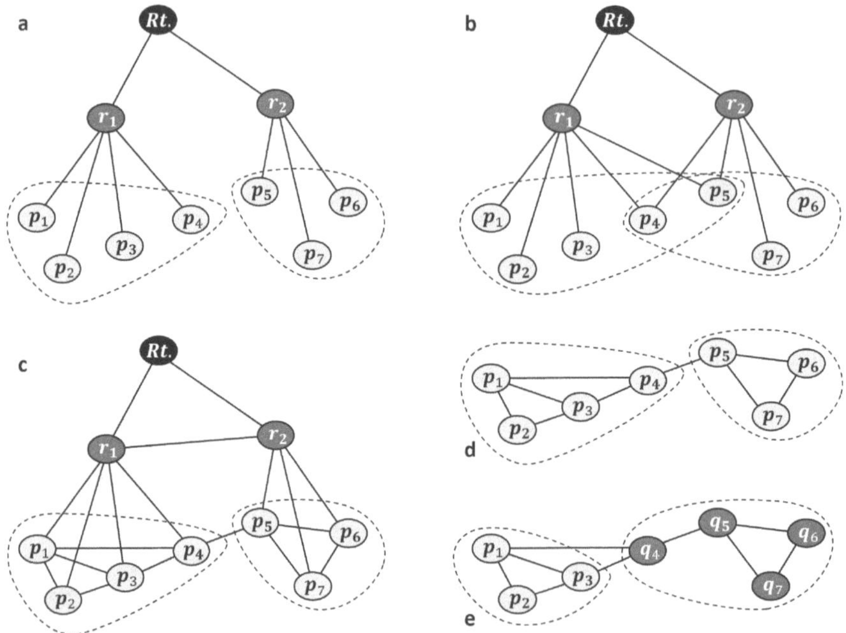

Figure 8.6
Representation of spatial hierarchies in graphs. (a) A tree with root node (Rt.), two region nodes ($\boldsymbol{r}_1, \boldsymbol{r}_2$), and seven leaf nodes (places $\boldsymbol{p}_1, \ldots, \boldsymbol{p}_7$). The dashed lines indicate the regions. (b) A polyhierarchy with overlapping regions. (c) If place-to-place and region-to-region links are considered, the original tree can be used as a "spanning tree." (d) If region nodes are not explicitly represented, regions can still be defined by connectedness. The nodes $\boldsymbol{p}_4$ and $\boldsymbol{p}_5$ are gateways that have to be touched by all inter-region paths. (e) Regionalization may also be based on place similarities (common function, spatial nearness, similar building style, etc.). For example, the places $\boldsymbol{p}_i$ might be locations on a university campus while places $\boldsymbol{q}_j$ might belong to a shopping area. Note that the connectivity in (d) and (e) is the same.

the hierarchy is reached in the "leaves"; see figure 8.6a for a tree with three levels of hierarchy. All connections are between nodes at subsequent levels of the hierarchy and indicate inclusion of the lower-level object in the higher-level category. Paths from one leave node to another are not direct but will always lead through high-level nodes. In cognitive maps, we could interpret the root node as an agent's home range, the branches and twigs as major and minor regions, and the leaves as places or views. Spatial representations organized as strict hierarchical trees have been called "strong hierarchies" by McNamara (1986).

While the hierarchy in a graph-theoretic tree with known root is uniquely defined, spatial hierarchies will often show overlapping regions. In figure 1.4d, we discussed a place specified as being in the region "riverfront" as well as in the region "downtown" with the idea that these regions are overlapping but not contained in one another. Ordering structures with partial overlap are known as polyhierarchies; an example appears in figure 8.6b, where the places $\boldsymbol{p}_4$ and $\boldsymbol{p}_5$ are both included in either region. Polyhierarchies are a weaker form of hierarchy in which the inclusion relation is not unique (one place may belong to multiple regions) but still monotonic: if a place belongs to some region, it also belongs to all larger structures that the region is included in. If the inclusion direction is represented by directed graph links, polyhierarchies correspond to directed graphs without cycles.

Figure 8.6c shows a graph with "vertical" links indicating hierarchical inclusion and "horizontal" links for spatial adjacency or connectedness. Horizontal links exist on all hierarchical levels such that places are connected to neighboring places and regions to adjacent regions and so on. For example, horizontal links on the place level may be used in walking while links on the region level indicate railway or airline connections. The vertical links will be mostly relevant for path planning. In graph theory, the vertical links define a spanning tree: that is, a subgraph without cycles containing all graph nodes. The case of figure 8.6c is discussed as "partially hierarchical" by McNamara (1986).

Graph hierarchies may also be implicit: that is, without explicit representation of regions by higher-level nodes. The clustering of graphs based on the density of connections is a well-studied problem in graph theory and machine learning; see, for example Aggarwal (2010) and Luxburg (2007). In the example shown in figure 8.6d, two regions with dense internal connections and sparse region-to-region connections would be found. Note, however, that spatial regions need not always be defined by connectivity. In figure 8.6e, the same graph is shown again but with a different interpretation: now, two regions are defined by properties of the nodes. These regions need not coincide with the connectivity clusters. Explicit region representations are of course able to capture perceived regionalizations independent of the connectivity.

8.4.2 Evidence for Hierarchical Organization

Inheritance of spatial relations In hierarchical structures, subordinate nodes are often initialized with properties of their superordinate node as a default. In object-oriented programming, this is known as "inheritance" by a "child node" from its "parent." In spatial cognition, inheritance of spatial relations between superordinate nodes by their subordinates leads to systematic distortions in directional estimates, as has been demonstrated by Stevens and Coupe (1978) and Tversky (1981).

The study by Stevens and Coupe (1978) uses directional judgments between geographical points in North and Central America. Subjects are asked to imagine a map with a circle around some start location and indicate on that circle the allocentric or compass direction to the goal. In one example, subjects rated the direction from the Atlantic mouth of the Panama Canal to its Pacific mouth as being east-to-west, when in fact it is northwest to southeast. This result reflects the spatial relation of the two oceans connected by the canal that are also the containing regions of the two mouths. The judgment of the direction of the

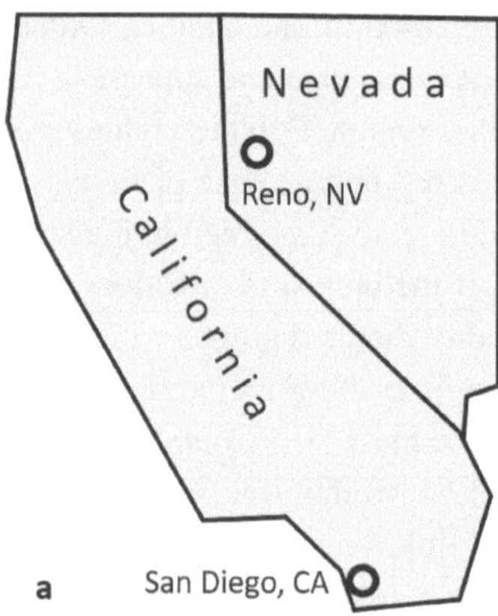

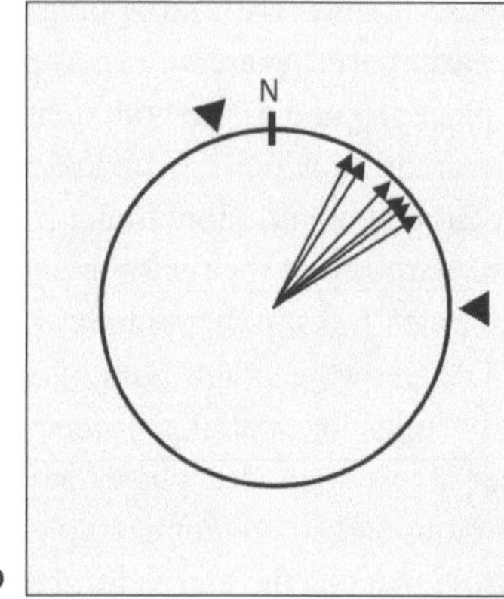

Figure 8.7
Inheritance of metric relations from superordinate spatial entities. (a) One of the experimental conditions used by Stevens and Coupe (1978); the map was not shown in the experiment. Subjects are asked to judge the compass direction from San Diego, California, to Reno, Nevada. The true direction is north–northeast, while Nevada as the superordinate entity is west or northwest of California. (b) Schematic data sheet as used in the experiment. In a circle with a marking for north, subjects draw a line from the center of the circle in the judged compass direction. On average, the reported direction is northeast. The triangles (not shown to the subjects) mark the true direction and the assumed direction between the superordinate regions (i.e., California to Nevada).

Panama Canal is therefore "inherited" from the spatial relation of the containing, superordinate spatial entities (i.e., the Pacific and the Atlantic oceans). In another example, the direction from Toronto, Canada, to Seattle, Washington, is judged to be southeast, when in fact Seattle has a higher northern latitude than Toronto. Again, the explanation would be that Canada is north of the United States and the perceived relation of the subordinate entities, the cities of Toronto and Seattle, is inherited from the superordinate ones.

The experiment was repeated with a number of additional relations, and the directional judgments were compared with the true directions between the named points and the centers of gravity of the containing "regions" (figure 8.7). In all cases, the judgments were biased away from the true subordinate relations and toward the superordinate relations. This indicates that the superordinate entities and their directional relations are represented in memory and that these relations bias the directional judgments between their contained subregions.

In a second experiment, Stevens and Coupe (1978) asked subjects about the relations of points marked on a sheet of paper. The points were introduced as "cities" on a map which additionally could contain a boundary between two "counties." City X was slightly left ("west") of city Y, which was clearly visible if no boundary was shown. In the learning phase, the subject studied the map in one of three conditions: either without a boundary, with a vertical ("north-south") boundary passing between cities X and Y, or with a curved "north-south" boundary putting city X into the "eastern" county and city Y into the "western" one. After studying the arrangement, subjects had to report from memory which of the two cities was further to the east. The results are veridical in the control condition, or with the straight border putting the eastern (right) city into the eastern (right) county. If, however, the curved boundary is used (i.e., if the eastern city is included in the western county), subjects make substantial errors. This indicates that belonging to one of the counties and the relative position of these counties are taken into account when making the relational judgment.

It should be noted, however, that the results reported above may also be due to visual grouping, without the involvement of a representation of large-scale space. This is even more likely for the comparisons between North and South American cities presented by Tversky (1981). The geographical outlines of the Americas are perceived as visual shapes whose relative position is only schematically represented. Generally, they are thought to be offset only in the north–south direction, while the more eastern position of South America is ignored. As a result, Lima, Peru, is often judged as being west of Miami, Florida (i.e., further left on the standard map) when in fact it is located about 3 degrees east of the Miami meridian.

Clustering geographical space In the tradition of Lynch (1960) and Appleyard (1969), Golledge and Spector (1978) developed their "anchor point" hypothesis of spatial knowledge of urban environments. According to this idea, cities are perceived through a number of salient or well-known locations, called anchor points. In the vicinity of these points, spatial knowledge is detailed and accurate, but the fidelity decreases with the distance of a location from the nearest anchor. The spatial relations between the anchor points, however, are less well known and may indeed be subject to substantial error. The idea is similar to the local chart approach to spatial working memory (see section 7.2.4 and Meilinger 2008), this time, however, applied to spatial long-term memory. Golledge (1999) gives a graph account of the anchor point hypothesis, which is basically a tree representing the sequence of knowledge acquisition. High-level nodes representing "home," "work," and "shopping" are built and connected early in spatial learning and later augmented by adding adjacent nodes in each area as spatial knowledge is growing.

Experimental evidence for the anchor point hypothesis was presented in a study by Couclelis et al. (1987) who asked subjects about the spatial relations of a number of well-known places within an area of some 15 by 6 km in their hometown of Goleta, California. A cluster analysis of the distance estimates and their deviations from the true distances shows that errors are roughly the same for places within the vicinity of one anchor point but differ more strongly between places belonging to different anchors. Couclelis et al. (1987)

describe this as "plate tectonics," where each anchor point defines one plate moving independently of the others in cognitive space. Of course, each of these "plates" is a region in our sense and measures but a kilometer or so in diameter.

Hirtle and Jonides (1985) used the recall of a number of landmark locations in Ann Arbor, Michigan, to define the hierarchical clusterings of these landmarks by individual subjects. Subjects were asked to verbally recall the set of landmarks starting from different "cue" locations presented by the experimenter. From a number of sequences thus produced, a tree graph was calculated for each subject, reflecting the nearness of the landmarks in the sequence. This graph defines landmark clusters as perceived by the subjects. A comparison with distance estimates over the same set of landmarks showed that distances within these clusters were underestimated while landmarks grouped in different clusters were estimated to be further apart than they actually were.

On a larger scale, distance estimation within and across regions was studied by Carbon and Leder (2005), using Germany with its former (1949–1990) division into East and West Germany as an example. Subjects living in Berlin were asked to rate distances between pairs of German cities and consistently overestimated distances between cities in former East and West Germany as compared to pairs within either part. This is surprising since by the time of this study, the German unification had already been established for some fifteen years. The authors also asked subjects about their attitude regarding unification and found a strong interaction: subjects with a negative political attitude to unification showed particularly strong distance overestimation, while the effect was not significant in subjects who were more content with the political situation. This shows that distance perception may also be affected by nonspatial factors, in particular if a geographical rather than a walking scale is concerned.

Perceived direction and distance in laboratory space Studies with experimentally designed regions cues have been carried out in indoor walking arenas. Newcombe and Liben (1982) marked ten locations in a square of about 5 by 5 m divided into four quadrants by movable walls. A central area was left free from barriers so that walking between regions was easily possible. Subjects learned the locations until they met a criterion. Next, they were asked to estimate the distances between pairs of locations in a visual adjustment task. Results show a significant overestimation of distances across boundaries as compared to distances within the four quadrants.

Using a similar setup, McNamara (1986) recorded both distance estimates and reaction times within and between regions. The results reproduce the overestimation of between-region judgments. In addition, response times were also found to be larger for between-region judgments as compared to judgments within regions, but also for shorter as compared to longer distances in between-region cases. This indicates that regions are represented, however, not only in the sense of a graph tree (figure. 8.6a) but also with place-to-place relations across region boundaries (figure 8.6c). In the terminology of McNamara (1986), this would be called a "partial" hierarchy.

Route choice in maps The main purpose of representing hierarchies is to simplify search and path planning. Indeed, the planning algorithms discussed in section 8.2.3 will take longer to find a solution if the number of vertices in the graph increases and will converge faster if long-distance links are available. Explicit representation of hierarchies in graphs, as shown in figure 8.6a–c, provides such long-distance paths by going to higher-level nodes and back from there. As a result, routes can be planned on the level of regions, which reduces the required planning depth. One might therefore expect that regionalizations of spatial representations affect the selection of routes.

Bailenson, Shum, and Uttal (2000) presented subjects with maps showing two alternative routes between a start and a goal location that had equal length but differed in the number of turns or straight segments they were composed of. In this situation, subjects have a clear preference for routes with less turns. This result might have been expected from the measurements of perceived walking distance, which was found by Sadalla and Magel (1980) to be increased along curved paths; see section 5.3.2. In a second experiment, Bailenson, Shum, and Uttal (2000) used two routes that were point-mirrored versions of each other, one with a straight initial segment and a curved section toward the end, and the alternative starting with the curved part and ending with the straight one. In this case, subjects prefer the route with the straight initial segment, presumably because this leads them faster away from their starting point.

This "initial segment strategy" was already mentioned above as a possible consequence of vector-based navigation and is not in itself an indication of representational hierarchy. A further experiment therefore used maps in which regions were marked as shaded rectangles. If subjects were now offered a choice between two routes with equally long initial straight segments, they showed a small but statistically significant preference for the route whose initial segment was completely within the start region. This indicates that regions do influence route planning.

Navigational performances depending on regions In the experiments on spatial hierarchies discussed so far, the dependent measures were nonnavigational tasks such as the marking of a direction on a sheet of paper, the judgment of distances, the naming of places, or the selection of a route in a map. Obviously, the structuring of an environment should also be helpful in wayfinding and the acquisition of spatial information. One interesting approach to this problem that also includes a measure for the granularity of the environment has been presented by Juliani et al. (2016). In a virtual environment, a landscape was modeled as a two-dimensional elevation function with a band-limited power spectrum with the $1/f$-characteristic of fractal functions. By varying the width of the frequency band, the roughness of the surface can be controlled, generating smooth hillocks via low spatial frequencies and craggy cliffs via high spatial frequencies. The $1/f$ characteristic introduces a self-symmetry or fractal structure, and the upper cut-off of the frequency band is therefore described as a fractal dimension. Subjects were asked to navigate inside this environment and performed best for intermediate values of fractal dimension: that is, in environments that lent themselves for useful regionalizations.

Fractal dimension allows an alternative approach to spatial hierarchy by providing a continuous measure of the degree to which regionalization is possible. In this view, regionalization is not all or nothing but may be gradual, as is likely the case with natural environments in general.

An advantage for regular environments allowing meaningful regionalization was also shown in a place-learning task by Wiener, Schnee, and Mallot (2004). Subjects were asked to learn the position of sixteen objects arranged in a 4×4 array. Objects were taken from four semantic categories (cars, animals, buildings, flowers) and arranged either regularly in the four quadrants of the array or in random fashion throughout the array. When asked to navigate to a particular object, learning was faster in the structured condition. Also, once learning is achieved, the routes traveled to the required goal are shorter in the structured condition. Thus, the regionalization inherent in the regular environment is discovered by the subjects and used for better wayfinding performance.

Region-dependent route choice Figure 8.8 illustrates a virtual environment used in a wayfinding experiment by Wiener and Mallot (2003). The environment consisted of twelve places arranged as a hexagonal ring (six places) and another six places at the ends of six radial extensions emanating from the central ring (figure 8.8). Each place was marked by a landmark object, and the objects were taken from three semantic classes: cars, animals, and pieces of art. Hierarchy was defined implicitly via the semantic object categories of landmarks: that is, one region was marked by cars, another one by animals, and a third one by pieces of art. Each region included two adjacent places on the central ring together with their connected dead ends.

Subjects were trained for finding goals in the environment until they met a criterion. The acquired knowledge was then tested in search tasks that involved navigation from one dead-end place to the opposing dead end on the other side of the ring (i.e., the most distant place in the environment). Due to the symmetry of the layout, two routes existed for each task, which did not differ in length, but in the sense of the initial turn (left or right) and the number of regions touched along the path (two or three). Since all dead ends were used as starting points, there was no overall correlation between the initial turn direction and the number of region boundaries crossed.

Results show a clear preference on the order of 65 percent for routes touching only two regions. This effect of "region-dependent route choice" indicates that regional knowledge

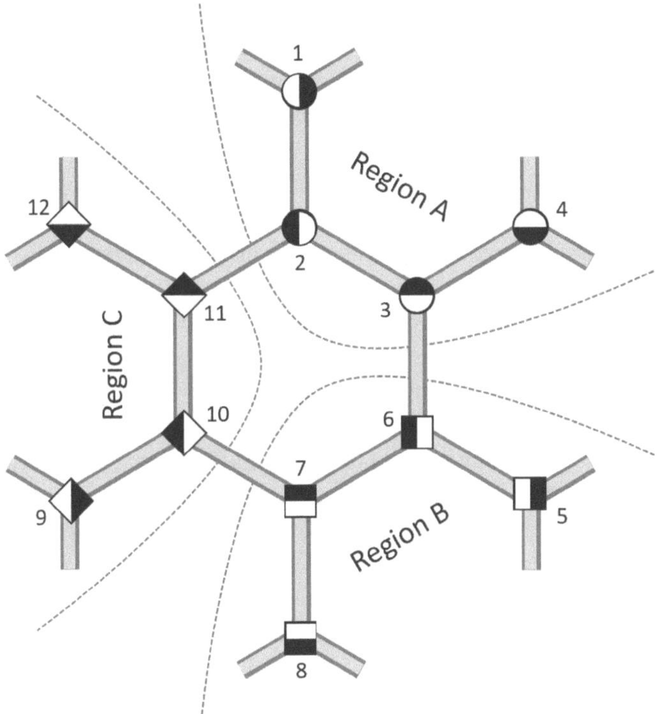

Figure 8.8
Map of the virtual environment used for experiments on region-dependent route choice by Wiener and Mallot (2003). Six places in the central hexagon and six places in the dead ends are marked with landmarks from three semantic categories (symbolized here by circles, squares, and diamonds). The places marked by landmarks of each type can be grouped into regions, as indicated by the dashed lines. In the experiment, all street connections look the same and region boundaries are not marked. When asked to navigate from place 8 to place 1, say, subjects preferably choose the route via places 6 and 3: that is, the one reaching the goal region A as soon as possible. Note that this route starts with a right turn and continues counter-clockwise in the central hexagon, whereas the preferred route from place 9 to place 4 is the one via places 11 and 2, starting with a left turn and progressing clockwise in the central hexagon.

has been acquired during the learning phase, based on the semantic landmark categories, and that this knowledge plays a role in route planning.

Besides this general conclusion, several mechanisms may explain the result. For example, crossing region boundaries might result in some extra cost such that routes with less crossing would be preferred. Wiener and Mallot (2003) therefore performed a second experiment with a rectangular array of three by four places connected by a regular grid of streets. Two regions were defined consistently by landmark category and a central river separating two districts of three by two places each. The river was crossed by three bridges connecting the places on opposite sides. When asked to navigate from one region to a sideways displaced location in the other region, various equidistant routes are possible, one of which leads from the start straight to the river and crosses it right away, while another route starts parallel to the river and takes a 90-degree turn at the height of the goal location. Since both routes cross the region boundary (the river) once, possible preferences cannot be attributed to costs of boundary crossing. Indeed, a region-dependent route choice effect was found: that is, subjects tended to cross the river using the nearest bridge such that they would reach the goal region as soon as possible. This strategy of reaching the goal region as soon as possible also explains the results found in the hexagon and may therefore be the general strategy used by the subjects.

More specifically, Wiener and Mallot (2003) suggest a "fine-to-coarse" planning strategy to explain these results: assume that the environment is represented as a hierarchical graph of the type shown in figure 8.6c. At the start point, a subgraph is considered containing the place nodes of the current region together with all region nodes but not the place nodes of distant regions. A path is planned to the goal region and used to move one step where the procedure is repeated. As soon as a region boundary is reached, the places of the previous region are collapsed to their region node while the newly entered region is expanded to show the places. As a result, the subgraph used for planning will always have a fine resolution at the agent's current position but is coarse for distant regions, initially including the goal. Details of path decisions still down the line will thus be worked out only as they become relevant. For a formal implementation of the fine-to-coarse planning algorithm, see Reineking, Kohlhagen, and Zetzsche (2008).

Further evidence for region-dependent route planning in virtual reality studies will be given in the context of brain imaging below. Here we mention a study from the urban planning literature using agent-based models of pedestrian movements in London and Paris (Filomena, Manley, and Verstegen 2020). If barriers, regions, or both are included in the path-planning algorithm of the agent-based model, the agreement of simulated paths with empirical data is improved. This concerns, for example, the deviation of the length of pedestrian walks from the shortest possible ones or the usage of side streets.

Neural substrates of region representation and region-based planning Recently, a number of fMRI studies have provided evidence as to how spatial information at different levels of hierarchy is represented in the brain. These studies make use of an innovative approach to subtle questions of neural representation called "representational similarity analysis." In this approach, task-dependent activation patterns of the voxels of a brain scan (the "multivoxel activations") are analyzed by methods of multivariate regression to reveal pattern components that encode stimulus or task information.

Peer and Epstein (2021) trained subjects to learn the position of sixteen objects placed in a virtual courtyard that was divided centrally by a river. The buildup of a hierarchical representation of the courtyard and its two halves was verified with a battery of behavioral tasks. For example, subjects overestimated distances between regions as compared to within-region distances. When asked to rank two object pairs for distance ("which two are further apart?"), answering took longer if the two pairs were taken from different sides of the river. Also, in free recall of the objects from memory, subjects tended to report objects region by region. In addition to behavioral testing, subjects underwent fMRI scanning while imagining certain objects or viewing isolated images of them. The voxel-based activity patterns of various brain areas were recorded and the overall similarity (correlation) of these voxel patterns was calculated for each pair of objects. In the retrosplenial cortex (RSC), these "representational similarities" reflected the metric distances of the represented objects in the courtyard. Object positions were also compared within their containing regions, such that two objects in the northeast corner of their region, although metrically separated by about half the diameter of the courtyard, would be considered as being at the same position. Interestingly, the representational similarity in the occipital place area (OPA) and left anterior hippocampus correlated with this region-relative distance measure. This indicates the presence of both regional and global spatial information of the same environment in different parts of the brain.

The neural substrates of hierarchical route planning were investigated in a study by Balaguer et al. (2016). Subjects learned a subway network with four intersecting lines and thirty-five stations, five of which were exchange stations. After training, subjects traveled the network with button clicks and were told the name of their current station and the goal. Routes were rated for various types of cost, including the number of stations traveled, the number of subway lines used, the number of exchange stations visited, and the special cost of a U-turn necessary in some tasks. fMRI activity correlating with the number of lines used but not with the number of stations visited was found in the dorsomedial prefrontal cortex. Also in this area, representational similarity (i.e., the correlation of multivoxel activity) was higher for stations on the same line than for equidistant stations on different subway lines. This may mean that lines are represented as high-level structures and used to simplify route planning. Note that the idea of treating subway lines as high-level entities is similar to the skeleton approach to cognitive mapping depicted in figure 8.2d.

8.4.3 What Makes a Region?

Based on the experimental evidence summarized above, it seems fair to conclude that cognitive space is hierarchically structured in one way or another. This raises the question of how regions are learned or discovered during spatial exploration. One possibility is that regionalization is explicitly "taught" to the explorer, as might be the case with urban divisions identified by names. In this case, called "designation" by Schick et al. (2019), regional structure would not be discovered by navigation but imposed by spatial language;

it would then reflect historical or political distinctions rather than street connectivity or similarities within a neighborhood. This mechanism is clearly involved with politically defined regions such as in the example of Nevada and California, shown in figure 8.7. Region designation is one of the many fascinating interactions between language and cognitive space, in particular the assignment of names to spatial entities. For an overview of place naming, see, for example, Steward (1975).

The discovery of regions solely from spatial exploration can be based on two main types of information: (i) similarities and dissimilarities of places within and between regions, including similarities of function, and (ii) connectivity patterns such as street hubs, barriers, or navigational bottlenecks. We will briefly discuss these two approaches.

Similarity A type of regionalization that cannot be based on connectivity is the perception of sections of a route: in a chain of places, the only available cue to regionalization is the similarity or dissimilarity of the waypoints along the route. Allen (1988) explored this idea with experiments using a series of photographs that were taken at regular distances (20 m) along an outdoor route of 1 km length. Subjects watched the series and were asked to remember the route from the slides. Afterward, they were asked to name regions and judge distances either within or across their reported regions. The route comprised sections in a park, on a university campus, and in different residential areas, and these regions were indeed named by the subjects.

One way to think about regions in exploratory walking is that the transition from a park into a city square, say, may be perceived as a boundary between events. Zacks and Tversky (2001) define events as temporal episodes in a subject's lifeline within which perceptual change is predictable. Thus, when we walk in the park, the scenery will gradually change, but these changes are predictable in the sense that we will always see trees, lawns, and so on. As soon as we leave the park, however, we will see cars and houses: that is, we will encounter a major change in perception, which would be encoded in episodic memory as an event boundary. The definition of regions in space would then be derived from the perception of events in time. Interestingly, Sargent et al. (2018) show that in a large group of subjects, the ability for learning a space from videos of routes walked in that space correlates with the subjects' event and episodic memory abilities.

In the Wiener and Mallot (2003) study (figure 8.8), the cue to regionalization provided to the subjects was the semantic categorization of the landmark objects. Subjects picked up on the object categories and grouped the respective places into regions. This may have happened as a perception of event boundaries, but other explanations are possible, such as the formation of regions in a process of memory consolidation. Subjects might realize only later that the environment features three exhibition areas for cars, animals, and art pieces. Indeed, Noack et al. (2017) show that sleep enhances the understanding of the regionalized structure of the explored environment.

The importance of semantic similarities among the landmarks constituting a region was confirmed in a study by LeVinh and Mallot (2022) who repeated the Wiener and Mallot (2003) study with landmarks whose semantic categorization was ambiguous. For example, a landmark "seagull" could be put in a category "bird" together with a quail and a chicken, or in a category "ocean," together with a steamboat and a ferry. The entire set of landmarks was carefully selected to allow for two alternative categorizations such that the corresponding regions in the hexagonal maze of figure 8.8 would be offset by 60 degrees. Subjects were primed into one or the other categorization scheme in an initial landmark sorting task. During navigation, they then showed region-dependent route choice according to the primed regionalization scheme. Regions were thus induced by the perceived categorization of the landmark set.

Besides similarities generated by semantic categorization of landmarks, other forms of similarity are possible: for example, the functional relatedness of places named "library," "lecture hall," or "lab" that imply a region "campus." Schick et al. (2019) studied the role of these and other similarities in a version of the Wiener and Mallot (2003) experiment using textual place names instead of landmark objects. Results were mixed and leave open the possibility that different types of place similarities affect region formation in different ways.

Connectivity An archipelago in the middle of an ocean clearly qualifies as a region since the islands within the archipelago are connected with each other by more and shorter passages than with the mainland or with other islands far away. In this example, regions are defined by spatial clustering (i.e., derived from metric information) but also by the density of connections. Another example for regions defined by connectivity patterns is given by the floors of a building in which multiple connections exist between the rooms in each floor while the floors themselves are connected only by one central staircase. For an example of a graph with regions defined by connectivity, see also figure 8.6d.

Connectivity-based hierarchies have been studied in the context of reinforcement learning. Wayfinding is treated as an instance of general problem solving: that is, as a search problem in a state–action graph. Once the goal is reached (the problem is solved), a reward or "payoff" is generated. Unlike wayfinding, reinforcement problems are usually defined by just one goal that is to be found many times from different starting points and learning results in a "policy" specifying the appropriateness of each action link for achieving this particular goal. Like a stereotyped route, this policy has to be learned anew if another goal is to be found.

Solway et al. (2014) address the problem of behavioral hierarchy in reinforcement learning by the idea of "bottleneck" states through which any solution path must pass. Hierarchical policies are defined as policies that always pass through one such bottleneck node. Each problem is thereby decomposed into two subproblems: that is, reaching the bottleneck state as a stepping stone, and finding the goal from there. In a regionalized problem space with dense within-region connections and sparse between-region connections, such bottlenecks would be the links leading from one region to another or the states from which such gateways go off: for example, the overseas port at the main island of our archipelago. Solway et al. (2014) show in a simulation study that reinforcement learning yields better results if a hierarchical policy is employed that uses an interregion gateway as a steppingstone. However, hierarchical policies can also lead to reduced performance if inappropriate steppingstones far away from the interregion gates are used.

In a behavioral experiment designed along these lines, Solway et al. (2014) had subjects learn a two-region environment. After learning, subjects reported which places they would visit when navigating to a goal in the other region. Subjects were instructed to click icons of the visited places in the sequence in which the places came to their mind. Interestingly, the "steppingstone places" that allowed the most efficient wayfinding in the simulation study were the places that the subjects reported first. This indicates a hierarchical planning strategy, as would also be predicted by reinforcement learning.

An experiment on connectivity-based hierarchies that involves routes to varying goals was performed by Tomov et al. (2020). Subjects learned a set of ten named places connected in a loop. Training routes were selected such that certain groups of places were always traveled in a row and could thus be learned as chunks between fixed start and end points. In the test phase, subjects were asked to travel to a goal five steps away such that the routes left or right along the loop were equally long. However, they would differ in the number of place chunks learned during training. In this situation, subjects prefer the route involving a lower number of place chunks. Thus, the subjects seem to have learned long-range connections and choose the route requiring a lower number of long-range planning steps.

In section 7.5, we have defined the cognitive map as the declarative part of spatial long-term memory. However, it is quite clear from the discussions on metric and hierarchical structure in the present chapter that our knowledge of space includes more than just if-then statements and the connectivity of places. On the other hand, Tolman's map metaphor does look a bit time-worn, in particular after the advent of interactive, searchable, and multiscale "map" applications in the computer and mobile phone. Indeed, a route-finding app is in many respects closer to our cognitive representation of space than a printed map, and it is probably also closer to Tolman's original idea of the means-ends field. It may therefore be time to drop the term "cognitive map" and replace it by something like "cognitive space," keeping in mind that the word "space" has not only the connotations of mathematical spaces with their different types of internal structure but also that of geographical and architectural spaces and, above all, that of a space for living and behavior.

Key Points of Chapter 8

- Cognitive space is organized as a state–action graph of places and known actions leading from one place to another. It supports wayfinding to variable goals by means of graph search.
- In addition to plain graph structure, cognitive space contains partial or local metric information such as knowledge of distances between places or the headings or heading changes required to travel certain action links. These pieces of information can be thought of as being attached to the nodes and links as "labels" and are not generally tested for global consistency or metric embedding.
- Space graphs with local metric information can be embedded into metric maps, but the extent to which this is done by human subjects is unclear.
- Cognitive space is also hierarchically structured and allows path planning on multiple scales. This concerns hierarchical representations, such as places as parts of superordinate regions, control hierarchies such as steppingstones or gateways in longer routes, or nested metrical charts (e.g., of indoor rooms within their larger environment).
- In wayfinding and other forms of spatial behavior, all mechanisms described in this and the previous chapters interact.

References

Aggarwal, C. C. 2010. "Graph clustering." In *Encyclopedia of machine learning,* edited by G. I. Sammut C. and Webb, 459–467. Boston: Springer US. doi:10.1007/978-0-387-30164-8_348.

Allen, G. A. 1988. "The acquisition of spatial knowledge under conditions of temporospatial discontinuity." *Psychological Research* 50:183–190.

Appleyard, D. 1969. "Why buildings are known." *Environment & Behavior* 1:131–156.

Arbib, M. A., and I. Lieblich. 1977. "Motivational learning of spatial behavior." In *Systems neuroscience,* edited by J. Metzler, 221–239. New York: Academic Press.

Bailenson, J. N., M. S. Shum, and D. H. Uttal. 2000. "The initial segment strategy: A heuristic for route selection." *Memory & Cognition* 28:306–318.

Balaguer, J., H. Spiers, D. Hassabis, and C. Summerfield. 2016. "Neural mechanisms of hierarchical planning in a virtual subway network." *Neuron* 90:893–903.

Balestro, V., Á. G. Horváth, H. Martini, and R. Teixeira. 2017. "Angles in normed spaces." *Aequationes Mathematicae* 91:201–236.

Bast, H., D. Delling, A. Goldberg, M. Müller-Hannemann, T. Pajor, P. Sanders, D. Wagner, and R. F. Werneck. 2016. "Route planning in transportation networks." *Lecture Notes in Computer Science* 9220:19–80.

Bongiorno, C., Y. Zhou, M. Kryven, D. Theurel, A. Rizzo, P. Santi, J. Tenenbaum, and C. Ratti. 2021. "Vector-based pedestrian navigation in cities." *Nature Computational Science* 1:678–685.

Borg, I., and P. Groenen. 1997. *Modern multidimensional scaling: Theory and applications.* New York: Springer Verlag.

Carbon, C. C., and H. Leder. 2005. "The wall inside the brain: Overestimation of distances crossing the former Iron Curtain." *Psychonomic Bulletin & Review* 12:746–750.

Couclelis, H., R. G. Golledge, N. Gale, and W. Tobler. 1987. "Exploring the anchor-point hypothesis of spatial cognition." *Journal of Experimental Psychology* 7:99–122.

Differt, D., and W. Stürzl. 2021. "A generalized multi-snapshot model for 3D homing and route following." *Adaptive Behavior* 29:541–548.

Dijkstra, E. W. 1959. "A note on two problems in connexion with graphs." *Numerische Mathematik* 1:269–271.

Durrant-Whyte, H., and T. Bailey. 2006. "Simultaneous localization and mapping: Part I." *IEEE Robotic & Automation Magazine* 13:99–108.

Filomena, G., E. Manley, and J. A. Verstegen. 2020. "Perception of urban subdivisions in pedestrian movement simulation." *PLoS ONE* 15 (12): e0244099.

Foo, P., W. H. Warren, A. Duchon, and M. J. Tarr. 2005. "Do humans integrate routes into a cognitive map? Map- versus landmark-based navigation of novel shortcuts." *Journal of Experimental Psychology: Learning, Memory, and Cognition* 31:195–215.

Franz, M. O., B. Schölkopf, H. A. Mallot, and H. H. Bülthoff. 1998. "Learning view graphs for robot navigation." *Autonomous Robots* 5:111–125.

Gärling, T., A. Böök, E. Lindberg, and C. Arce. 1990. "Is elevation encoded in cognitive maps?" *Journal of Environmental Psychology* 10:341–351.

Golfarelli, M., D. Maio, and S. Rizzi. 2001. "Correction of dead-reckoning errors in map building for mobile robots." *IEEE Transactions on Robotics and Automation* 17:37–47.

Golledge, R. G. 1999. "Human wayfinding and cognitive maps." In *Wayfinding behavior: Cognitive mapping and other spatial processes,* edited by R. G. Golledge, 5–45. Baltimore: Johns Hopkins University Press.

Golledge, R. G. 2002. "The nature of geographic knowledge." *Annals of the Association of American Geographers* 92:1–14.

Golledge, R. G., and A. N. Spector. 1978. "Comprehending the urban environment: Theory and practice." *Geographical Analysis* 10:403–426.

Hart, P. E., N. J. Nilsson, and B. Raphael. 1968. "A formal basis for the heuristic determination of minimum cost paths." *IEEE Transactions on Systems Science and Cybernetics* SSC-4:100–107.

Hirtle, S. C., and J. Jonides. 1985. "Evidence of hierarchies in cognitive maps." *Memory & Cognition* 13:208–217.

Holyoak, K. J., and W. A. Mah. 1982. "Cognitive reference points in judgments of symbolic magnitude." *Cognitive Psychology* 14:328–352.

Hübner, W., and H. A. Mallot. 2007. "Metric embedding of view graphs: A vision and odometry-based approach to cognitive mapping." *Autonomous Robots* 23:183–196.

Ishikawa, T., and D. R. Montello. 2006. "Spatial knowledge acquisition from direct experience in the environment: Individual differences in the development of metric knowledge and the integration of separately learned places." *Cognitive Psychology* 52:93–129.

Juliani, A. W., A. J. Bies, C. R. Boydston, R. P. Taylor, and M. E. Sereno. 2016. "Navigation performance in virtual environments varies with fractal dimension of landscape." *Journal of Environmental Psychology* 47:155–165.

Kuipers, B. 1978. "Modeling spatial knowledge." *Cognitive Science* 2:129–153.

Kuipers, B. 2000. "The spatial semantic hierarchy." *Artificial Intelligence* 119:191–233.

Kuipers, B. J., D. G. Teucci, and B. J. Stankiewicz. 2003. "The skeleton in the cognitive map: A computational and empirical exploration." *Environment & Behavior* 35:81–106.

LeVinh, L., and H. A. Mallot. 2022. "When your border and my border differ: Spatial regionalisation and route choice depends on perceived landmark categorization." *bioRxiv*. doi:10.1101/2022.11.10.515979.

Luxburg, U. von. 2007. "A tutorial on spectral clustering." *Statistics and Computing* 17:395–416.

Lynch, K. 1960. *The image of the city*. Cambridge, MA: MIT Press.

Madl, T., K. Chen, D. Montaldi, and R. Trappl. 2015. "Computational cognitive models of spatial memory in navigation space: A review." *Neural Networks* 65:18–43.

Mallot, H. A., and K. Basten. 2009. "Embodied spatial cognition: Biological and artificial systems." *Image and Vision Computing* 27:1658–1670.

Mallot, H. A., G. A. Ecke, and T. Baumann. 2020. "Dual population coding for path planning in graphs with overlapping place representations." *Lecture Notes in Artificial Intelligence* 12162:3–17.

Manley, E., G. Filomena, and P. Mavros. 2021. "A spatial model of cognitive distance in cities." *International Journal of Geographical Information Science* 35:2316–2338.

Mardia, K. V., J. T. Kent, and J. M. Bibby. 1979. *Multivariate analysis*. London: Academic Press.

McNamara, T. P. 1986. "Mental representations of spatial relations." *Cognitive Psychology* 18:87–121.

McNamara, T. P., and V. A. Diwadkar. 1997. "Symmetry and asymmetriy of human spatial memory." *Cognitive Psychology* 34:160–190.

Meilinger, T. 2008. "The network of reference frames theory: A synthesis of graphs and cognitive maps." *Lecture Notes in Artificial Intelligence* 5248:344–360.

Meilinger, T., M. Strickrodt, and H. H. Bülthoff. 2016. "Qualitative differences in memory for vista and environmental spaces are caused by opaque borders, not movement or successive presentation." *Cognition* 155:77–95.

Moar, I., and G. H. Bower. 1983. "Inconsistency in spatial knowledge." *Memory & Cognition* 11:107–113.

Montello, D. R. 1993. "Scale and multiple psychologies of space." *Lecture Notes in Computer Science* 716:312–321.

Montello, D. R., ed. 2018. *Handbook of behavioral and cognitive geography*. Cheltenham, UK: Edward Elgar Publishing.

Muller, R. U., M. Stead, and J. Pach. 1996. "The hippocampus as a cognitive graph." *Journal of General Physiology* 107:663–694.

Newcombe, N., and L. S. Liben. 1982. "Barrier effects in the cognitive maps of children and adults." *Journal of Experimental Child Psychology* 34:46–58.

Noack, H., W. Schick, H. A. Mallot, and J. Born. 2017. "Sleep enhances knowledge of routes and regions in spatial environments." *Learning and Memory* 24:140–144.

Normand, E., and C. Boesch. 2009. "Sophisticated Euclidean maps in forest chimpanzees." *Animal Behavior* 77:1195–1201.

Peer, M., and R. A. Epstein. 2021. "The human brain uses spatial schemas to represent segmented environments." *Current Biology* 31:1–12.

Ponulak, F., and J. J. Hopfield. 2013. "Rapid, parallel path planning by propagating wavefronts of spiking neural activity." *Frontiers in Computational Neuroscience* 7:98.

Reineking, T., C. Kohlhagen, and C. Zetzsche. 2008. "Efficient wayfinding in hierarchically regionalized spatial enviropnments." *Lecture Notes in Artificial Intelligence* 5248:56–70.

Restat, J., S. D. Steck, H. F. Mochnatzki, and H. A. Mallot. 2004. "Geographical slant facilitates navigation and orientation in virtual environments." *Perception* 33:667–687.

Ruddle, R. A., A. Howes, S. J. Payne, and D. M. Jones. 2000. "The effects of hyperlinks on navigation in virtual environments." *International Journal of Human-Computer Studies* 53:551–581.

Sadalla, E. K., and S. G. Magel. 1980. "The perception of traversed distance." *Environment & Behavior* 12:65–79.

Sargent, J. Q., J. M. Zacks, D. Z. Hambrick, and N. Lin. 2018. "Event memory uniquely predicts memory for large-scale space." *Memory & Cognition* 47:212–228.

Schick, W., M. Halfmann, G. Hardiess, F. Hamm, and H. A. Mallot. 2019. "Language cues in the formation of hierarchical representations of space." *Spatial Cognition and Computation* 19:252–281.

Schmajuk, N. A., and A. D. Thieme. 1992. "Purposive behaviour and cognitive mapping: A neural network model." *Biological Cybernetics* 67:165–174.

Schölkopf, B., and H. A. Mallot. 1995. "View-based cognitive mapping and path planning." *Adaptive Behavior* 3:311–348.

Solway, A., C. Duik, N. Córdova, D. Yee, A. G. Barto, Y. Niv, and M. M. Botvinick. 2014. "Optimal behavioral hierarchy." *PLoS Computational Biology* 10:e1003779.

Stevens, A., and P. Coupe. 1978. "Distortions in judged spatial relations." *Cognitive Psychology* 10:422–437.

Steward, George R. 1975. *Names on the globe.* New York: Oxford University Press.

Tomov, M. S., S. Yagati, A. Kumar, W. Yang, and S. J. Gershman. 2020. "Discovery of hierarchical representations for efficient planning." *PLoS Computational Biology* 16 (4): e1007594.

Trullier, O., S. I. Wiener, A. Berthoz, and J.-A. Meyer. 1997. "Biologically based artificial navigation systems: Review and prospects." *Progress in Neurobiology* 51:483–544.

Tversky, B. 1981. "Distortions in memory for maps." *Cognitive Psychology* 13:407–433.

Voicu, H., and N. Schmajuk. 2002. "Latent learning, shortcuts and detours: A computational model." *Behavioural Processes* 59:67–86.

Waller, D., and D. B. M. Haun. 2003. "Scaling techniques for modeling directional knowledge." *Behavior Research Methods, Instruments, & Computers* 35:285–293.

Warren, W. H. 2019. "Non-Euclidean navigation." *Journal of Experimental Biology* 222:jeb187971.

Warren, W. H., D. B. Rothman, B. H. Schnapp, and J. D. Ericson. 2017. "Wormholes in virtual space: From cognitive maps to cognitive graphs." *Cognition* 166:152–163.

Wender, K. F., M. Wagener-Wender, and R. Rothkegel. 1997. "Measures of spatial memory and routes of learning." *Psychological Research* 59:269–278.

Wiener, J. M., and H. A. Mallot. 2003. "'Fine-to-coarse' route planning and navigation in regionalized environments." *Spatial Cognition and Computation* 3:331–358.

Wiener, J. M., A. Schnee, and H. A. Mallot. 2004. "Use and interaction of navigation strategies in regionalized environments." *Journal of Environmental Psychology* 24:475–493.

Wyeth, G., and M. Milford. 2009. "Spatial cognition for robots." *IEEE Robotics & Automation Magazine* 16:24–32.

Zacks, J. M., and B. Tversky. 2001. "Event structure in perception and conception." *Psychological Review* 127:3–21.

9 Epilogue: Reason Evolves

After having laid out the many behavioral competences and neural mechanisms of spatial cognition, one remaining question is that of its natural history: How has spatial cognition evolved? What were the driving forces and adaptive values? What are traits in spatial cognition, and how are they distributed throughout the animal kingdom? What neural mechanisms have been invented to carry out the necessary computations, and how can their phylogeny be traced? And finally, how did spatial cognition shape the advent of other cognitive abilities? Many of these questions cannot be conclusively answered at this day, but some evolutionary lines are beginning to emerge. Taken together, they support and flesh out the idea that spatial cognition is a root of reason and cognition at large.

Traits

In evolutionary theory, a trait is a heritable adaptation found in a species or a group of species to which the notion of phylogenetic homology can be applied. Typical examples are the five-fingered limb of the four-legged vertebrates or the opposed thumb of the primates. While these examples are taken from anatomy, the concept of a trait can be extended to physiology and behavior. Here we discuss some important traits occurring in spatial cognition.

Representation The need for dealing with space arises as soon as an organism is able to perform active movements, be it as free migration of an entire animal or as relative movement of body parts in sessile organisms. At the level of taxis and tropism (chapter 4), this requires a reaction to orienting stimuli such that *information* about the stimulus is picked up and processed in the signals and mechanisms underlying each movement. One might be tempted to say that this information is in some sense "represented" in the system, but at least in simple organisms, this would be little more than the empty phrase that an apple needs to "know" the direction toward the earth in order to fall down. The information conveyed by a signal will become a true representation only if further properties obtain. For example, it would be necessary that the information is maintained for some period of time and that it can be used for interaction with other mechanisms, which then also depend on the represented piece of information. Using a formulation from David Marr (1982), it can be said that in order to become a representation, the information has to be made *explicit* in the sense that other mechanisms can also understand and use it.

An early example of an explicit representation in this sense is *heading*: that is, the currently pursued forward direction of the agent. A representation of heading is different from a mere reaction to an orienting stimulus in that it is kept alive also in the absence of the stimulus and updated upon body turns. It would be useful for animals, which preferably move in specific directions relative to their own body. In evolution, the earliest multicellular animals meeting this condition: that is, in which a forward direction and a "head" can at all be defined, are the bilaterians. Unlike sponges or coelenterates (jellyfish, corals, etc.), bilaterians have a left–right symmetry about a body axis extending from a "head" to a "tail." It is interesting to note that the ring attractor network for the representation of heading is found both in insects and in mammals: that is, in representatives of the two major groups into which the early bilaterians split up, protostomia and deuterostomia. This may indicate that it is indeed inherited from a common bilaterian ancestor. If this is true, the ring attractor for the representation of heading might well be the phylogenetically oldest example of a true representation. Attractor dynamics in recurrent neural networks also seems to be a general mechanism in *working memory*, of which the representation of heading may thus be an ancestor.

Path integration This is another trait of spatial cognition that heavily relies on the sense of heading but also needs additional information concerning the distance traveled. In mammals, this is provided by the grid cell system whose interaction with the head direction system is now well understood. In contrast, the distance component of path integration in insects seems to be computed in the same network together with heading (Stone et al. 2017). It is therefore not homologous to the grid cell system in mammals. This interpretation is in line with the mentioned branching of the phylogenetic tree after the basic bilaterians.

Homing and place recognition A further system of spatial cognition that may have evolved in parallel with path integration is the machinery for *homing and place recognition*: that is, the "taxon" system of O'Keefe and Nadel (1978). Together, path integration and place recognition are prerequisites of an ecological niche in which a nest or burrow is maintained for shelter, breeding, and storage and can be reliably found and identified when returning from a foraging excursion ("central place foraging"). As suggested by the term "taxon system", an evolutionary line can possibly be drawn from taxis, in particular menotaxis, to stimulus–response behavior and the usage of stored if-then declarations (see also section 6.4.3). Menotaxis is an innate reflex in which an animal achieves to move in a straight line by keeping a fixed bearing to a salient stimulus such as the sun. In stimulus–response behavior, the sun is replaced by a learned landmark memory, which is now used as a guidance. The mechanism is still "procedural," causing the agent to move in the same way whenever the stimulus occurs. Only if another piece of memory is added, which specifies the expected outcome or goal, can the procedure be turned into an if-then statement that would contribute to a maplike representation of space.

Long-term memory of objects and places The recognition of places requires a basic form of long-term memory, such as the snapshot memories discussed in section 6.2. An evolutionary question in this context is the relation between the *recognition systems for objects and places*. Both require a decision mechanism in which a pattern of cues is compared to a stored description of a class (i.e., classified to belong to this class or not). One might therefore think that places are just another "natural kind" of object such as "predator," "food," or "mating partner." Alternatively, systems for the recognition of places and for objects may have evolved independently, or object recognition might even be derived from place recognition. Neural substrates of place and object recognition are mostly found in medial and inferior temporal cortices, respectively, but are not completely separated. At this point, the relation between the cognitive representations of objects, landmarks, and places remains largely elusive.

Planning and declarative memory A final trait of spatial cognition is spatial planning: that is, *the formation of sequences of actions* required to reach a goal. It rests on declarative memory as a special form of long-term memory and a planning stage in working memory. At each planning step, the agent imagines the situation, probes into possible actions, and arrives at a decision based on declarative contents from long-term memory. The chosen action activates the next remembered state and so on until the plan is complete. This ability relates to *general problem solving* and *imagery* both inside and outside the domain of spatial cognition. Since planning always means to think about things to do in the future, it is also an example of mental time travel as discussed as a component of higher-level cognition by Suddendorf and Corballis (2007). Even the representation of "self," an element of *self-consciousness*, may play a role in spatial planning: for example, if the agent needs to choose between two escape routes, one of which requires a long jump to cross a narrow gorge. The agent will then need to decide whether it will be able to perform this jump, given its current bodily condition.

Computation

How do neurons process information? Early work in the field focused on two main ideas: the emulation of a Turing machine by neural circuits, as suggested by McCulloch and Pitts (1943), and the idea of sensory filtering pioneered by Mach (1865) and Hartline and Ratliff (1958). While the former approach has hence been neglected (but see Penn, Holyoak, and Povinelli 2008), the filtering idea turned out to be more fruitful. Complex stimuli can be detected by matched filters, a mechanism that was suggested as a model of retinal processing by Lettvin et al. (1959). Together with the idea of supervised learning in the perceptron, this led to the modern theory of artificial neural networks and deep learning. However, supervised learning with explicit teacher signals is biologically rather implausible, and deep learning theory is therefore only of limited value in modeling the function of the brain. In computational neuroscience, a number of alternative mechanisms have been identified, many of them in the context of spatial behavior.

Control Natural computation starts with the feedback loop of action and perception as formulated in Uexküll's "functional cycle" reproduced in figure 4.1b. It is first and foremost a matter of control, not recognition or even symbol manipulation, although these latter computations may be part of overall control loops. Control may be realized as a simple reflex arc including just one or a small number of neurons, as is normally assumed in the case of taxis. In these cases, internal processing is mostly a feed-forward flow of signals from the sensors to the effectors. Single neurons in this stream are both *detectors* and *command neurons* but these functions may be gradually separated in multi-synaptic reflex arcs. Still, each neuron will keep a certain computational load in the sense that its firing signals a delimitable part of an event or a meaningful piece of information, as described by Horace Barlow (1972) in his "neuron doctrine of perceptual psychology." A neural network theory for learning relevant subevents by single neurons is known as *sparse coding* (Olshausen and Field 1996).

If internal feedback connections are added to the simple reflex arc, *reverberating neuronal activity* can arise, which allows for further computational operations. Reverberating activity plays a role in coordinated motion such as in the *central pattern generators* observed in the both vertebrates (Grillner 1975) and arthropods (Cruse 1990). In simple motor control, neuronal dynamics and motor action are coupled in real time, but the dynamics of neuronal activity may also exist as a stand-alone process, for example, in motor imagery and mental training. In spatial cognition, the *attractor dynamic* of neuronal activity is the substrate of path integration and heading control.

Population coding Another principle of neural computation, *population coding*, has first been described for the control of arm movements in grasp space (Georgopoulos et al. 1982) but is also found, for example, in the encoding of spatial position in the place cell system (Wilson and McNaughton 1993). At first glance, the fact that location is encoded in the activity of a whole ensemble of cells rather than by just one spatial "grandmother neuron" may appear to conflict with the sparse coding idea discussed above. However, even in sparse coding, the specificity of a neuron cannot be too high since otherwise, it would almost never be used and encoding would become inefficient. Sparse coding and population coding are therefore two sides of the same coin and mark the intermediate level of specificity optimal for neural processing. In addition, population coding has the well-known benefits studied in the vision and motor systems: that is, increased (sub-pixel) resolution, smooth interpolation, and the adjustment of the working range of the representation by adaptation.

State–action graphs State–action graphs and their application to way-finding have been linked to neural networks in the hippocampus by Schmajuk and Thieme (1992), Schölkopf and Mallot (1995), and Muller, Stead, and Pach (1996). Each neuron represents a state of the agent (place or view), and the computation carried out is a *prediction* or forward model of the states reachable from the current one. This is achieved by propagating activity according to the known place adjacencies as stored in the synaptic strengths of the according connections. These predictions may also take into account the possible actions taken by the agent, such as "walk left" or "walk uphill," for example, by gating the state-to-state synapses (Schölkopf and Mallot 1995). In any case, the predictions about reachable states have to be turned into an actual plan by an additional mechanism. One suggestion for planning assumes a wave of activity initiated at the goal state and propagating through the entire network. *Anti-spike-timing-dependent plasticity* (anti-STDP) might then facilitate connections leading in the direction of the goal, as described in section 7.3.2 (Ponulak and Hopfield 2013). Note that in population coded systems, states will not correspond to single neurons but to neuron populations. For an attempt to reconcile state–action networks with population coding, see Mallot, Ecke, and Baumann (2020).

Cognition

Space is a challenge for any agent capable of free movement. Even before path integration and homing, behavioral strategies such as optimally searching a foraging range or moving away from a hazard as fast as possible (i.e., in a straight line) have exerted adaptive pressure on the evolving nervous systems. The evolution of other domains of cognition, such as the understanding of objects, causality, events, or the interaction with other individuals and their cognitive systems in social cognition, took place while a basic apparatus for spatial cognition was already up and running. It is therefore likely that this existing apparatus has influenced the development in the other cognitive domains. If spatial cognition is the evolutionary oldest form of cognition, can we find examples of cognitive traits in other domains that can be traced back to spatial origins?

Spatial representation of nonspatial contents In the study of perception, continuous parameters such as color or pitch are often visualized as spaces such as the chromaticity diagram spanning the color "space" between the primary colors red, green, and blue, or musical scales organizing the pitch "space" or gamut from low to high. Similarly, the set of natural numbers seems to be conceived of as a line from left to right or from right to left, depending on the direction of writing in the subjects' native language (*Spatial-numerical association of response codes*, or SNARC-effect; see Dehaene, Bossini, and Giraux 1993). A deep relation between such spatially organized representations and spatial cognition has been suggested by Constantinescu, O'Reilly, and Behrens (2016), who trained subjects with sets of visual stimuli varying continuously along two independent dimensions; for example, they used silhouettes of birds with various lengths of the neck and legs. Bird images generated by such variation can be arranged in a two-dimensional *metaphorical space* with the leg and neck lengths represented on the axes. Subjects then imagined birds with increasing neck length, decreasing leg length, or combinations of such variations. In the metaphorical space, these transformations mark movements along straight lines with various angles relative to the x-axis. Scanning data from a cortical network including the entorhinal cortex showed a periodic activity modulation with the angle of progression in metaphorical space, at a periodicity of 120 degrees. This is taken as evidence for a representation of position by the hexagonally organized grid cell system (see also section 5.4). The intuition of metaphorical spaces may thus be grounded in the entorhinal apparatus of spatial cognition. If this is true, the grid cell system originally evolved for path integration would have become a "preadaptation" for the representation of objects with continuously varying feature dimensions.

Territories and social cognition Space also affects social cognition. One example for this interaction is the so-called *dear enemy* effect (Fisher 1954). Many animals are known to establish and defend territories, either small ones around their nesting site or larger ones for foraging. This requires a form of spatial knowledge of the territory itself. Beyond that, the owner of a territory generally reacts aggressively to intruders, but many animals show a reduced aggression to territorial neighbors. This makes a lot of sense since neighbors may occasionally trespass by chance, while strangers might threaten to expel the current owner. In songbirds and other vocalizing species, it has been shown that the distinction of neighbors and strangers is based on individual features of familiar songs. In this case, neighbor recognition may fail if the familiar song is played from an unusual position outside the neighbor's territory. This raises the possibility that *individual recognition*, a central steppingstone in the evolution of social cognition, is initially a component of the place code of the neighboring territory and might thus be considered a special type of landmark knowledge.

Territories can also be owned by groups of socially organized individuals such as a pack of wolves or a community of chimpanzees. *Group territories* require a spatial knowledge of the territory itself but also a social knowledge of the group members. In animals, group membership is often marked by pheromones, but in human evolution, this role seems to have been taken over by artifacts. These act as communicative signs that need to be understood by all members of a group and will be distinguished from the markings used by other groups. Such markings might therefore constitute an understanding of group identity and might even become signs for the group itself, thus constituting a relation between the group and its territory, or "homeland." Indeed, Wunn and Grojnowski (2016) argue that in human evolution, group territoriality may have been the origin not only of group identities and group symbols but even of early forms of religion.

Theory of mind Another interaction of spatial and social cognition occurs in the context of perspective taking, where observers may assume an imaginary viewpoint outside their own body. This ability may be useful in path planning, where it allows one to predict views from distant waypoints and to foresee the upcoming navigational decisions. It may however also be helpful in judging the visual information accessible by another person: that is, in "theorizing" about the other person's mind. Taking the perspective from a viewpoint occupied by another person may thus be a starting point for the evolution of a "theory of mind". In any case, the experiences of self-location and imagined or real own viewpoint and their possible dissociations are important elements of self consciousness (Blanke 2012).

Language The relation between wayfinding and *general problem solving* has been discussed at length in section 8.2 and will not be repeated here. It leads the way to the ultimate question of cognitive evolution: that is, the evolution of language. Possible contributions to language evolution might be found in a number of the aforementioned traits of spatial cognition. For example, the understanding of places and landmarks constitutes a conceptual system that allows the inclusion of place references in communication. Bickerton and Szathmáry (2011) suggest that this element of "displacement" (i.e., the ability to communicate about events or planned actions in remote places) is a crucial element of the "cognitive niche" created and developed by language-endowed primates. Displacement is largely absent in animal communication systems (except maybe in the honey bee dance "language") but enabled early hominids to plan cooperative actions at distant places. These are needed in a lifestyle called confrontational scavenging, which is thought to be part of the cognitive niche.

Another spatial concept relating to language is that of the route. Routes are sequential productions generated from internal rules and stored knowledge, much like sentences in language. The underlying system of rules and knowledge is the state–action graph discussed in chapters 7 and 8. It is tempting to consider this graph as a precursor or even an alternative to a Chomskyan generative grammar, another graph structure thought to control the production of verbal utterances. Hauser, Chomsky, and Fitch (2002) discuss these and related types of prelingual cognitive traits as the "faculty of language in a broad sense," which is present in many animals and forms a basis from which language evolution started.

Decision-making and volition As a final and domain-general element of cognition, consider *decision-making*. In reflex behavior, "decisions" are just triggers: that is, the decision is only about when to initiate a behavior, not about which behavior to choose. Advanced decision-making (i.e., purposive behavior in the sense of Tolman 1932) is based on the existence of action alternatives and on the knowledge of their consequences. For many animals, such behavioral alternatives are mostly of spatial nature, including, for example, the choice of a foraging range or a site for nesting. In this case, the knowledge about action consequences is represented in the cognitive map as a simple form of *declarative memory*. Simulations of possible actions and their expected outcomes are carried out in spatial working memory and provide a basis for the final decision. Interestingly, the evolution of decision-making, from automatic stimulus–response behavior to choices based on goals and expected outcomes, has been described as the evolution of "freedom" in a philosophical sense by Dennett (2003). The evolution of reason, as we see it in the invention of the various traits of spatial cognition and particularly in planning and state–action graphs, is an essential part of this process.

References

Barlow, H. B. 1972. "Single units and sensation: A neuron doctrine for perceptual psychology?" *Perception* 1:371–394.

Bickerton, D., and E. Szathmáry. 2011. "Confrontational scavenging as a possible source for language and cooperation." *BMC Evolutionary Biology* 11:261.

Blanke, O. 2012. "Multisensory brain mechanisms of bodily self-consciousness." *Nature Reviews Neuroscience* 13:556–571.

Constantinescu, A. O., J. X. O'Reilly, and T. E. J. Behrens. 2016. "Organizing conceptual knowledge in humans with a gridlike code." *Science* 352:1464–1468.

Cruse, H. 1990. "What mechanisms coordinate leg movement in walking arthopods." *Trends in Neural Sciences* 13:15–21.

Dehaene, S., S. Bossini, and P. Giraux. 1993. "The mental representation of parity and number magnitude." *Experimental Psychology: General* 122:371–396.

Dennett, D. C. 2003. *Freedom evolves.* London: Penguin Books.

Fisher, J. 1954. "Evolution and bird sociality." In *Evolution as a process,* edited by J. Huxley, A. C. Hardy, and E. B. Ford, 71–83. London: Allen & Unwin.

Georgopoulos, A. P., J. F. Kalasak, R. Caminiti, and J. T. Massey. 1982. "On the relation of the direction of two-dimensional arm movements and cell discharge in primate motor cortex." *The Journal of Neuroscience* 2:1527–1537.

Grillner, S. 1975. "Locomotion in vertebrates: Central mechanisms and reflex interaction." *Physiological Reviews* 55:247–304.

Hartline, H. K., and F. Ratliff. 1958. "Spatial summation of inhibitory influences in the eye of Limulus, and the mutual interaction of receptor units." *Journal of General Physiology* 41:1049–1066.

Hauser, M. D., N. Chomsky, and W. T. Fitch. 2002. "The faculty of language: What is it, who has it, and how did it evolve?" *Science* 298:1569–1579.

Lettvin, J. Y., H. R. Maturana, W. S. McCulloch, and W. H. Pitts. 1959. "What the frog's eye tells the frog's brain." *Proceedings of the Institute of Radio Engineers* 47:1950–1961.

Mach, E. 1865. "Über die Wirkung der räumlichen Vertheilung des Lichtreizes auf die Netzhaut." *Sitzungsberichte der mathematisch-naturwissenschaftlichen Classe der kaiserlichen Akademie der Wissenschaften Wien* 52 (2): 303–322.

Mallot, H. A., G. A. Ecke, and T. Baumann. 2020. "Dual population coding for path planning in graphs with overlapping place representations." *Lecture Notes in Artificial Intelligence* 12162:3–17.

Marr, D. 1982. *Vision.* San Francisco: W. H. Freeman.

McCulloch, W. S., and W. Pitts. 1943. "A logical calculus of the ideas immanent in nervous activity." *Bulletin of Mathematical Biophysics* 5:115–133.

Muller, R. U., M. Stead, and J. Pach. 1996. "The hippocampus as a cognitive graph." *Journal of General Physiology* 107:663–694.

O'Keefe, J., and L. Nadel. 1978. *The hippocampus as a cognitive map.* Oxford: Clarendon.

Olshausen, B. A., and D. J. Field. 1996. "Emergence of simple-cell receptive field properties by learning a sparse code for natural images." *Nature* 381:607–609.

Penn, D. C., K. J. Holyoak, and D. J. Povinelli. 2008. "Darwin's mistake: Explaining the discontinuity between human and nonhuman minds." *Behavioral and Brain Sciences* 31:109–130.

Ponulak, F., and J. J. Hopfield. 2013. "Rapid, parallel path planning by propagating wavefronts of spiking neural activity." *Frontiers in Computational Neuroscience* 7:98.

Schmajuk, N. A., and A. D. Thieme. 1992. "Purposive behaviour and cognitive mapping: A neural network model." *Biological Cybernetics* 67:165–174.

Schölkopf, B., and H. A. Mallot. 1995. "View-based cognitive mapping and path planning." *Adaptive Behavior* 3:311–348.

Stone, T., B. Webb, A. Adden, N. B. Weddig, A. Honkanen, R. Templin, W. Wcislo, L. Scimeca, E. Warrant, and S. Heinze. 2017. "An anatomically constrained model for path integration in the bee brain." *Current Biology* 27:3069–3085.

Suddendorf, T., and M. C. Corballis. 2007. "The evolution of foresight: What is mental time travel, and is it unique to humans?" *Behavioral and Brain Sciences* 30:299–313.

Tolman, E. C. 1932. *Purposive behavior in animals and men.* New York: The Century Company.

Wilson, M. A., and B. L. McNaughton. 1993. "Dynamics of the hippocampal ensemble code for space." *Science* 261:1055–1058.

Wunn, I., and D. Grojnowski. 2016. *Ancestors, territoriality, and gods: A natural history of religion.* Berlin: Springer.

Index

A^*-algorithm, 264
action voting, 237
action–perception cycle, 10, 39, 89–113, 209, 304
  augmented, 110–113
adjacency matrix, 259
aftereffect, 84
aging, 135, 137, 180
aliasing, 53, 169, 185, 186
alley experiment, 70
allo-oriented chart, 225
allocentric, 16, 79, 219, 225, 227, 242, *see also* reference frame
allothetic, 22, 36, 37, 127
Alzheimer's disease, 137
anchor point, 285
angle, 267
Ann Arbor, Michigan, 286
ant, 94, 120, 128, 166, 169, 235
ant mill, 96
anti-spike-timing-dependent plasticity, 231, 264
association, 190
attractor, 20, 144, 207, 227, 231
  periodic, 148
  ring shaped, 146
automata theory, 6, 258
average landmark vector, 168
avoidance reaction, 91

Barlow, Horace B., 139, 304
barycentric coordinates, 14
basal ganglia, 232
bat, 20, 140, 144, 196
Bayes estimation, 57, 181
beacon, 37, 189
beewolf, 162
behavior, 21, 89, 258
behaviorism, 4, 91
Betelgeuze (star), 128
bilaterian animals, 92, 302
biological motion, 34
biomorphic robotics, 100
bird migration, 128
bistable perception, 7, 35
blocking, 187
Boston, Massachusetts, 280
boundaries as place cues, 183
boundary vector cell, 20, 174, 192, 227
Bourbaki, Nicolas, 265
Braitenberg, Valentino, 92
braking, 51, 99
Bufo viridis, 106

Calliphora sp., 45
Cambridge, UK, 276
canonical view, 221
Cartwright–Collett model, 165
Cataglyphis sp., 120, 169
catchment area, 161, 168, 171, 175, 228
category, 158
Catharus sp., 128
caudate nucleus, 248
center-of-gravity estimator, 139
centering behavior, 97
central executive, 208, 210
central place forager, 120

child development, 178, 188, 232
chimpanzee, 7, 279
city block norm, 267
clique, 261
closure of visual space, 72
cognitive agent, 9, 209
cognitive evolution, 113, 301–308
cognitive map, 5, 25, 241, 294
  in insects, 242
cognized present, 23
collective building, 96
compass, 22, 37, 127, 128, 130, 163, 271, 279
  polarization, 122
conditioning, 4, 208, 243
confusion area, 161, 168
coordinate system, 13–14, 241, 266
core knowledge, 9
corpus striatum, 184
Corsi block-tapping task, 213
cortex
  entorhinal, 18, 306
  inferotemporal, 78
  intraparietal, 47
  medial parietal, 197
  mediotemporal, 227
  motor, 77
  parietal, 16, 78, 81, 227
  perirhinal, 248
  posterior parietal, 241
  prefrontal, 136, 207, 209, 291
  retrosplenial, 136, 175, 197, 218, 225, 291
  visual, 46, 136
cue conflict, 161
cue integration, 23, 46, 52–58, 69, 180, 238
cultural differences, 137, 214, 223
curvature
  Gaussian, 74
  of a trajectory, 123
curve, differential geometry, 123
cyclovection, 35, 55

Darwin, Charles R., 162
dead reckoning, 120
dead-locking, 106
dear enemy effect, 306
decision point, 188
decision-making, 112, 245, 307
declarative memory, 22, 208, 249, 303, 307
degree of freedom of movement, 32, 42, 54
delayed response task, 207
depth of processing, 175
depth ordering, 69
depth perception, 64
detour behavior, 244
Devonian toolkit, 99
Dijkstra algorithm, 263
dimensionality, 259, 266
direct perception, 38
direction (landmark usage), 190, 234, 279
disparity (binocular), 64
dissociation, 70, 112, 248
distance estimate, asymmetric, 274
distance reproduction task, 134
domains of cognition, 7–9
dorsal stream, 46, 70, 78, 185
driving behavior, 51, 99, 100, 107
Drosophila sp., 148
dual task, 135, 209, 213, 224

Ebbinghaus illusion, 70, 276
ecological psychology, 91
efference copy, 37, 48, 68
ego-acceleration, 36
egocentric, 16, 125, 219, 225, 227
egomotion, 20, 31–58, 75, 82, 119, 123, 146
elevation, 278
embodied cognition, 94
Emmert's law, 70
encoding error model, 133
ensemble code, *see* population code
entorhinal cortex, 142
epipolar plane constraint, 43
episodic memory, 25, 158, 210, 292
Eristalis sp., 95
Euclidean geometry, 11
Euler, Leonhard, 11
event, 157, 210, 292
evolutionary epistemology, 4
exocentric pointing, 72
expansion rate, 50
eye movements, 14, 47, 187, 188
  in driving, 100

farthest plane, 72

firing field, 139, 142
fish, 35, 108
focus of expansion, 40, 48
food depletion, 211
foot placement, 102
foraging, 107, 120
Formica rufa, 166
Fourier transform, 143
fractal dimension, 288
Frenet–Serret theory, 122–124
fusiform gyrus, 197

geocentric, 125, 140
geodesic, 3, 70
geometric module, 176, 183
Germany, East and West, 286
Gibson, James J., 38, 48, 91
global landmark, 238
goal, 5, 105, 107, 164, 232
goal-directed behavior, 241
Goleta, California, 285
granularity, 258, 287
graph, 6, 11, 234, 259–264
  bipartite, 263
  directed, 258
  hierarchical, 281
  labeled, 228, 261, 267
  undirected, 259
graph distance, 258
grasp space, 213
grasping movement, 70
grid cell, 18, 141, 149, 192, 302
  in humans, 144, 306
grid cell remapping, 195
guidance, 96, 189, 234, 302

head direction cell, 18, 129, 138, 140, 146, 193
  in humans, 141
heading, 34, 43, 47, 106, 123, 147, 236, 302
  imagined, 214
  vertical, 140
Helmholtz, Hermann von, 2, 64, 70, 73, 83
here, 23, 63, 157
Hering coordinates of visual space, 13
hexagonal grid, 142, 148
hierarchy, 110, 196, 216, 263, 281
Hillebrand hyperbola, 72
hippocampus, 18, 26, 184, 190, 211, 232
  evolution, 137
  human, 136, 225, 248, 291
  in primates, 196
home vector, 120–131
homeostasis, 91
homing, 120, 163
honey bee, 97, 128, 162, 163, 239
horopter, 67, 70
human MT+ complex, 136
hyperbolic compression, 73
hyperlink, 280

iconic memory, 206
idiothetic, 22, 36, 123
if-then clause, 246
image manifold, 170
imagery, 24, 76, 303
imagined rotation, 80
implicit scaling theory, 274
incentive learning, 245
indigo bunting, 128
indoor navigation, 104, 133, 175, 230, 286
inertial sensor, 36
inheritance of spatial relation, 283
initial segment strategy, 280, 287
inner state, 5–7, 234
instinct, 111, 112
intercepting paths, 107
intrinsic reference axis, 215, 216, 219, 220, 227, 229, 230, 241
intuitive physics, 8
invariance, 17, 173, 185, 261
isovist, 104, 175, 183, 187

James, William, 23
judgment, 64
  of location, 173
  of relative direction, 79, 214–219

Königsberg bridge problem, 11
Kant, Immanuel, 2, 3, 158, 222
kinetic depth effect, 68
kite-shaped room, 180, 181

Labidus praedator, 96
landmark, 140, 184–190, 236

distant, 127
global, 188
street junction as, 185
vs. object, 184, 197
landmark recall, 240
landmark replacement, 161, 180
landmark selection, 186
language, 190, 222, 223, 307
Lasius fuliginosus, 95
lateral inhibition, 145
learning, 208
associative, 4, 187, 208
incidental, 161, 183, 184, 188
latent, 242, 243
reinforcement, 4, 184, 208, 243, 293
left–right balancing, 92, 98
legibility of the environment, 103, 263
Lévy flight, 108
Lima, Peru, 285
limbic system, 140
limited lifetime random dots, 181
local chart, 219
local image variation, 171
local metric, 277
local position information, 20, 168, 170
locale system, 26, 96, 193
London, UK, 224, 248, 290
long-term memory, 24, 162, 208, 258, 274
looming, 51
Lorenz, Konrad, 4, 112, 249
Lynch, Kevin, 103, 263

Mach, Ernst, 35, 303
machine learning, 140
magnetic sense, 128
map reading, 79
Marr, David, 64, 112, 175, 302
matched filter, 45
mathematical angle convention, 271
maximum likelihood estimation, 56, 139, 181, 212
maze
elevated, 244
figure-of-eight, 211
iterated T-, 243
iterated Y-, 237, 238
plus-, 159
radial arm, 24, 211
rectangular grid, 197
T-, 245
Melophorus bagoti, 169, 235
memory consolidation, 292
menotaxis, 92, 189, 302
mental rotation, 75
mental scanning, 223
mental state, *see* inner state
mental transformation, 75
metric maps, 264–281
metric, partial, 267
Miami, Florida, 285
Milano, Italy, 219
minimal variance estimator, 56
modularity, 112
monkey, 207
moon, perceived size, 73
morphing algorithm of visual homing, 169
motion parallax, 68
motion platform, 46
Müller, Johannes, 138
multidimensional scaling (MDS), 268
multigraph, 259

navigation, 119, 233
in three dimensions, 124, 135, 141, 196, 278
Neumann, John von, 23, 208
neuronal specificity, 138, 207
for optic flow, 45
norm, 266

object, 7, 69, 74
object vector cell, 183, 193, 227, 241
occipital place area, 197, 291
odometry, 120
ontology of space, 103, 261
optic flow, 38–52, 97, 107, 133, 166
optic flow discontinuity, 43
optokinetic response, 35
overshadowing, 184

Panama Canal, 283
parahippocampal place area, 175, 197
parallax, 42
Paramecium sp., 91, 245
Paris, France, 290

path finding, 104
path integration, 119–149, 192, 302
as optimization, 130
computational neuroscience of, 138
error, 135
idiothetic, 123, 272
in the blind, 131
patient study, 219
percept, 7, 34, 84
perception, 64
perceptual constancy, 16, 69, 78, 158
perspective, 39
perspective taking, 9, 78, 84, 214, 306
pheromone trail, 94
Philanthus triangulum, 162
piloting, 189
pitch rotation, 33
place, 157–162
place cell, 18, 190, 212
in three dimensions, 196
out-of-place firing, 212, 231
place cell remapping, 195
place code, 158, 181
place graph, 261
place learning, 159
in regionalized environment, 288
place naming, 292
place recognition, 302
neurophysiology of, 190
planning, 212, 224, 231, 244, 263, 280, 287, 303
fine-to-coarse, 290
plasticity, 208
pointing, 131
Polaris (star), 128
polyhierarchy, 283
Popper, Karl, 4
population code, 15, 18, 45, 77, 138, 190, 212, 258, 262, 304
in grid cells, 141
position-dependent recall, 220, 224
postural sway, 35
potential landscape, 105
preadaptation, 2, 306
preattentive vision, 175, 186
precuneus, 83
prediction, 305
predictive coding, 75
presubiculum, 140
priming, 240
principle of least commitment, 112
prism adaptation, 83
probability summation, 55
problem solving, 257, 307
projective geometry, 74
proprioception, 31, 37, 54, 57
psychological distance, 1

qualitative reasoning, 257

random walk, 108, 130
rat, 20, 185, 192, 194, 211, 227, 243, 244
realignment of represented and actual space, 216
recalibration, 57, 83
recognition-triggered response, 25, 190, 234
reference frame, 14–16, 31, 82, 126, 194
reflex, 92
region, 281
as route section, 292
induced by barriers, 286
neural representation, 290
region-dependent route choice, 288
relative nearness, 40
Reno, Nevada, 284
representation, 12, 34, 53, 64
amodal, 17, 31, 222, 223
analogic vs. propositional, 77
coordinate free, 14
vs. information, 301
representational neglect, 219
representational similarity analysis, 81, 140, 218, 290
response learning, 159
retrospective verbal report, 224
rigid body motion, 32
roll rotation, 33
rotation, 40
route, 6, 190, 224, 232–241
route choice, 287
Russell, Bertrand, 158

San Diego, California, 284
San Francisco, California, 280
Santa Barbara, California, 278

scale–velocity ambiguity, 42, 48, 51
schema, 234
Scottish Highlands, 101
search, 107, 121
Seattle, Washington, 284
sensorimotor alignment, 216
sensory modality, 22, 31, 222
sequence learning, 240
sequentiality, 239
sex differences, 78, 137
short-term memory, 206
shortcut, 247
size constancy, 69, 72
sketch map, 220
skyline cues in visual homing, 169
SLAM, 53, 263, 273
sleep, 212
slope, 37, 185, 278
snapshot, 175, 179, 261
snapshot homing, 162–174, 189, 232
  in humans, 173
SNARC effect, 305
social cognition, 8, 80, 306
solar azimuth, 122, 128
space
  absolute, 2, 74
  environmental, 277
  Euclidean, 11, 266
  geographical, 13
  metaphorical, 306
  metric, 265
  non-Euclidean, 3, 70, 73, 74, 276, 281
  normed, 266
  peripersonal, 12, 63–85, 158, 175, 213, 225
  topological, 12, 265
  vista, 13, 277
spatial image, 75, 225
spatial imagery, 214
spatial layout, 175, 183, 187, 197
spatial presence, 84
spatial updating, 73, 81, 134, 213, 222
splay angle, 101
state–action association, 234, 235
state–action graph, 258, 293, 307
stereopsis, 64
stigmergy, 97
stimulus–response behavior, 94, 99
street crossing, 107
strip map, 235
structure from motion, 68
subiculum, 20
subsumption, 110
supplementary motor area, 83
survey knowledge, 232, 242

taxis, 4, 91–96
taxon system, 26, 96, 193, 302
teleportation, 161
template matching, 44
termite, 97
territory, 306
texture gradient, 69
thalamus, 18, 140, 225
theory of mind, 8, 80, 307
thigmotaxis, 99
time perception, 135
time to collision, 35, 48, 107
Tinbergen, Nikolaas, 3, 112, 161, 162, 170
Tolman, Edward Chace, 6, 159, 234, 241, 245
topological navigation, 257, 265
Toronto, Canada, 284
torus, 149
Tower of Hanoi, 257
trail following, 94
trait, 3, 301–303
transfer of spatial knowledge, 247
translation, 33, 40
traveling salesperson problem, 109
tree (graph theory), 281, 286
trial-and-error behavior, 91, 242
triangle completion, 129, 131
  from long-term memory, 276
triangle, sum of angles, 276
triangular inequality, 265, 266, 268
triangulation, 271
tropotaxis, 92, 94
Tübingen, Germany, 220, 224
tuning curve, 138

Uexküll, Jakob von, 10, 72, 90, 92, 304
Umeå, Sweden, 278
unconscious inference, 64
Urbana-Champaign, Illinois, 216

vanishing point, 39
vection, 35
vector navigation, 279
vector space, 266
ventral stream, 70, 185
vergence, 14, 67, 68
vervet monkey, 109
Vespula vulgaris, 163
vestibular cue, 55
vestibulo-ocular reflex, 35
view cell, 196, 227
view graph, 227, 238, 261
virtual environment, 73, 81, 131, 132, 160, 185, 187, 188, 196, 224, 236, 279, 287, 288
  impossible, 280

walking straight, 130
walking trajectory, 102
walking without vision, 57, 73, 83, 132
water maze, 159, 180, 187
waterstrider, 35
wayfinding, 6, 225, 246
working memory, 10, 23, 64, 75, 146, 205–231, 258
  capacity, 209
  computer analogy, 208
wormhole, 281

yaw rotation, 33